Zweiter Internationaler Kongreß für Histo- und Cytochemie

Second International Congress of Histo- and Cytochemistry

Deuxième Congrès International d'Histochimie et de Cytochimie

Frankfurt/Main, 16.-21. August 1964

Herausgegeben von

Editorial board

Publié par

T. H. Schiebler, Würzburg

A. G. E. Pearse, London

H. H. Wolff, Würzburg

Mit 32 Textabbildungen

SPRINGER-VERLAG
BERLIN HEIDELBERG GMBH 1964

ISBN 978-3-662-22687-2   ISBN 978-3-662-24616-0 (eBook)
DOI 10.1007/978-3-662-24616-0

Ursprünglich erschienen bei Springer-Verlag 1964
Softcover reprint of the hardcover 1st edition 1964

Library of Congress Catalog Card Number 64-25705

Titel Nr. 1253

Vorwort

Der II. Internationale Kongreß für Histo- und Cytochemie greift aus der Vielzahl der interessanten und bedeutungsvollen Einzelthemen der Histo- und Cytochemie diejenigen für Hauptreferate, Panel Diskussionen und Symposien heraus, die den gegenwärtigen Stand dieser Spezialwissenschaft in besonders eindringlicher Weise widerspiegeln und zukünftige Entwicklungen andeuten. Es soll versucht werden, bereits Erarbeitetes kritisch zu werten und auf neue Gebiete aufmerksam zu machen. Besonders liegt dem Kongreß daran, die der Histo- und Cytochemie eng benachbarten Arbeitsgebiete wie Biophysik und Mikrochemie ausreichend zu Wort kommen zu lassen. Neben mehr methodisch orientierten Themen werden einige angewandt-histochemischen Inhalts behandelt.

In Ergänzung zu den von der Kongreßleitung vorbereiteten Veranstaltungen finden 24 Parallelsitzungen statt, in denen die Teilnehmer des Kongresses ihre neuesten Ergebnisse in Form von Kurzvorträgen mitteilen. Hier zeigt sich die Entwicklung der Histochemie besonders deutlich. Viele der Freien Vorträge ergänzen die Hauptthemen des Kongresses. Insgesamt hoffen wir, den Kongreß als eine geschlossene Einheit vorstellen zu können.

Der vorliegende Band wurde darauf ausgerichtet, es den Kongreßteilnehmern während der Tagung und denjenigen, die nicht zum Kongreß kommen können, nachträglich zu ermöglichen, den Gang des Kongresses zu verfolgen. Alle Hauptreferate werden im vollen Wortlaut gedruckt. Von den Panel Diskussionen und den Symposien liegen Zusammenfassungen der einzelnen Beiträge vor. Die Kurzvorträge, von denen ebenfalls nur die Zusammenfassungen veröffentlicht werden, sind so gut als möglich geordnet. Auch hier soll der rote Faden des Kongresses in Erscheinung treten. Die Reihenfolge der Zusammenfassungen entspricht der Vortragsfolge während des Kongresses. Für den Inhalt der Vorträge und ihre Zusammenfassungen sind allein die Autoren verantwortlich.

Die Vollendung dieses Werkes zu Kongreßbeginn war nur durch die bereitwillige Mitarbeit aller Referenten, Moderatoren, Symposienleiter sowie aller Vortragenden möglich. Hierfür danken die Herausgeber auch an dieser Stelle. Dennoch wäre ohne die bis zum Äußersten entgegenkommende und sorgfältige Arbeit des Springer-Verlages dieses Buch nie termingerecht fertig geworden. Die Herausgeber möchten in besonderer Weise das große Verständnis von Herrn Dr. H. Goetze, Mitinhaber des Springer-Verlages, die sorgfältige und äußerst prompte Arbeit von Fräulein H. Schröder vom Springer-Verlag bei der Herstellung, die Tätigkeit aller Helfer in der Druckerei und die Übersetzung der deutschen und französischen Texte ins Englische durch Fräulein R. Liske, Cambridge/England, hervorheben.

Würzburg und London
August 1964

T. H. Schiebler
A. G. E. Pearse
H. H. Wolff

Preface

In selecting the lectures, panel discussions and symposia of the Second International Congress of Histo- and Cytochemistry it has only been possible to select a few of the large number of interesting and important topics available. Those which have been chosen reflect most markedly the present situation of histochemistry and its future development. The aim of the Congress is to discuss known facts critically as well as to call attention to new topics. Other concerns of the Congress are subjects closely related to histo- and cytochemistry, such as biophysics and microchemistry. Together with the discussion of methodology attention will be given to applied histochemistry.

In addition to the sessions prepared by the Congress Committee, 24 parallel sessions will be held. During these sessions the members of the Congress will have the opportunity to present their latest results in a series of short communications. This part of the Congress will probably reflect most clearly the development of histochemistry. Many of the short communications supplement the main topics of the Congress. Thus we hope to reach certain uniform conclusions.

The aim of this volume is to help members of the Congress — and those who are unable to attend — to follow the course of the meetings. All principal lectures are published in full, whereas the contributions to the panel discussions and symposia are published in the form of summaries. The short communications — resumés of which are published — are classified as clearly as possible. The order of the summaries corresponds to the main thread of the lectures during the Congress. Responsibility for the content of each lecture and its summary lies with its authors.

The Editors are greatly indebted to all speakers, moderators, and leaders of symposia, whose ready collaboration has made the completion of this volume possible. However, the publication in time for the Congress would not have been possible without the immensely helpful and detailed assistance of the „Springer-Verlag". The particular thanks of the Editors must be offered to: Dr. H. GÖTZE, joint proprietor of the „Springer-Verlag", for his professional understanding and help, to Miss H. SCHRÖDER ("Springer-Verlag") for her accurate and ready assistance, to all the employees of the printing house, and to Miss R. LISKE, Cambridge/England, for the translation of the German and French originals into English.

<table>
<tr><td>Würzburg and London
August 1964</td><td>T. H. SCHIEBLER
A. G. E. PEARSE
H. H. WOLFF</td></tr>
</table>

Avant-Propos

Le deuxième congrès d'histochimie et de cytochimie a choisi parmi les nombreux sujets intéressants et importants de ces deux disciplines, ceux qui, se prêtant à des exposés, à des discussions de la part des membres du jury et à des colloques, reflètent le mieux l'état actuel de cette branche de la science et en indiquent les développements ultérieurs. On essayera de mesurer à leur juste valeur les résultats acquis et d'attirer l'attention sur des domaines nouveaux. L'une des idées directrices de ce congrès est de laisser s'exprimer les représentants de disciplines très voisines de l'histochimie et de la cytochimie, comme la biophysique et la microchimie. Outre les sujets de pure méthode, seront traités quelques points d'histochimie appliquée.

En complément du programme organisé par la direction du congrès, auront lieu 24 séances parallèles, au cours desquelles les participants donneront connaissance, sous forme de courts exposés, de leurs derniers travaux, le but recherché étant de mettre clairement en évidence le développement de l'histochimie. Beaucoup de ces exposés complèteront ceux faits au cours des séances principales, et nous espérons que, dans l'ensemble, le congrès donnera une impression d'unité.

Le présent volume a donc pour objet de permettre de suivre le déroulement du congrès, pendant les séances, à ceux qui y participeront, a postériori, à ceux qui n'auront pu s'y joindre. Tous les exposés faits au cours des séances principales seront donnés in extenso, les compte-rendus des discussions du jury et des colloques, sous forme de résumés, ainsi que les courts exposés, que l'on classera aussi logiquement que possible. Cette publication doit aussi mettre en évidence l'idée directrice du congrès. L'ordre dans lequel seront présentés les résumés sera celui observé au cours du congrès, les auteurs étant seuls responsables du fond et de la forme.

Cet ouvrage n'a pu être prêt à la date prévue pour le début du congrès que grâce à la collaboration de tous les rapporteurs, présidents de séance, directeurs de colloques et de tous les conférenciers. Qu'ils en soient remerciés ici! Toutefois, ce livre n'aurait pas été terminé à temps sans le travail soigneux et minutieux à l'extrême de la Springer-Verlag. Nous voudrions tout particulièrement témoigner notre reconnaissance à Monsieur H. Götze, co-propriétaire de la maison d'édition, de la compréhension dont il a fait preuve à notre égard, à Mademoiselle H. Schröder, de la même maison, de son travail soigneux et rapide, à tous ceux qui nous ont aider à imprimer ce livre et à Mademoiselle R. Liske, de Cambridge, de ses traductions en anglais des textes allemands et français.

Würzburg et London   T. H. Schiebler
Août 1964   A. G. E. Pearse
                                                                    H. H. Wolff

Inhaltsverzeichnis · Contents · Table des Matières

Festansprache · Opening Speech · Discours d'inauguration

Referate · Reviews · Résumés

Podiumsgespräche · Panel Discussions · Discussions du Jury

Application of Histochemistry to Electron Microscopy

Moderator: R. J. BARRNETT

Physical-Optical Methods in Histochemistry: Instruments

Moderator: T. CASPERSSON

A. *Principle and Application of Different Methods and/or Instruments*

B. *Methods in Interference Microscopy*

Physical-Optical Methods in Histochemistry: Comparison of Results with Various Independent Methods

Moderator: W. SANDRITTER

A. Prologue

B. DNA

C. RNA

D. Proteins

Comparison and Criticism of Autoradiographic Methods

Moderator: S. R. PELC

Normal and Pathologic Interphase Growth

Moderator: J. Fautrez

A. Unicellular Organisms

B. Tissue Cultures

C. Oocytes

D. Tissues

E. Pathological Growth

Immunohistochemistry

Moderator: A. H. Coons

A. Identification of Cell Products

B. Identification of Structural Components

Symposien · Symposia · Symposia

Comparison of Different Fixation Methods and their Significance for Histochemistry

Chairman: R. E. STOWELL

Histochemistry of Esterases and Peptidases

Chairman: E. P. BENDITT

Histochemistry of the Skin, Especially of Normal and Pathological Epidermal Differentiation

Chairman: O. BRAUN-FALCO

A. Histotopochemistry in Relation to Epidermal Proliferation

Interference Microscopy and Histology

Chairman: R. BARER

Neurohistochemistry

Chairman: A. POPE

Histochemistry of the Mitochondrial System

Chairman: T. BARKA

Histochemistry of Carbohydrates in the Intercellular Space

Chairman: J. A. SZIRMAI

Freie Vorträge · Short Communications · Conférences libres

Montag, Monday, Lundi, 17. August 1964

Dienstag, Tuesday, Mardi, 18. August 1964

Haut · Skin · Peau

Zelle und Zellbestandteile · The Cell and its Components · La cellule et ses éléments

Enzyme · Enzymes · Enzymes

Fixierung · Fixation · Fixation
Verschiedenes · Divers · Divers

Metachromasie · Metachromasia · Métachromasie
Mucopolysaccharide · Mucopolysaccharides · Mucopolysaccharides
Bindegewebe · Connective Tissue · Tissu conjonctif

Tumoren · Tumors · Tumeurs   Seite

Donnerstag, Thursday, Jeudi, 20. August 1964

Haut und Epithel · Skin and Epithelium · Peau et tissu épithélial
Verschiedenes · Divers · Divers

Physikalische und quantitative Methoden · Physical and quantitative Methods
Méthodes physiques et quantitatives
Autoradiographie · Autoradiography · Autoradiographie

Magen-Darm · Gastro-intestinal Tract · Tube digestif

Freitag, Friday, Vendredi, 21. August 1964

Urogenitalsystem · Urogenital System · Système urogénital

Immunologie · Immunology · Immunologie
Amyloid · Amyloid · Amyloïde
Verschiedenes · Divers · Divers

Seite

Zelle und Zellbestandteile · The Cell and its Components · La cellule et ses éléments

Muskulatur · Musculature · Musculature
Verschiedenes · Divers · Divers

Read by Title

Institute for Medical Cell Research, Karolinska Institutet Stockholm

Histochemistry in Modern Biological and Medical Research

By

Torbjörn Caspersson

One of the most striking developments within science during the recent decade has been the progress made in fundamental biology. The rate of progress in this field is still increasing, with a corresponding increase in the flow of valuable results for many fields, above all for medicine.

It is particularly striking how in recent years the boundaries between what used to be well-defined, stable areas of research have been broken down, at the same time as entirely new fields have grown up along these boundaries. The interaction of experience, methods and ways of thinking from several of these traditional fields has often resulted in the dynamic growth of a new line of research. Histochemistry is a classical example — a meeting-ground between the vast material collected over the years by microscopic morphology on the one hand and the more recent findings in the fields of experimental genetics, cytophysiology and above all biochemistry on the other.

Classical histology, with the microscope as its primary aid, developed the current conception of the cell over a hundred years ago. I am referring to the observations that the organs and tissues of all higher organisms are constituted by cells which are in themselves independent units and possess the essential prerequisites for independent life. They can multiply, grow in size, carry out chemical processes which supply the energy necessary for life, and also — in response to the demands of their surroundings — bring about a wide variety of structure building processes of a chemical nature. The concept of the cell is basic to the whole of modern biology and to all its medical ramifications.

Once the individual cell had been shown to be the fundamental unit in the organism, the unit in which the processes essential to life occur, it was only natural that efforts should be made to penetrate the internal mechanisms of the cell itself. That the cell has a complex structure had been demonstrated in the 19th Century, first the body of the cell and then the nucleus inside the body; later on it was demonstrated that the nucleus carries the factors that transmit heredity from generation to generation, and also that it governs the development and function of the individual cell. An enormous amount of observations were gradually compiled on cell structures, the collective term for all the small elements constituting the individual cell. Certain structures, such as the cell nucleus, were shown to be universal, while others occurred only in certain types of cells.

Many of the results of these labours were of fundamental importance to medicine. The cellular pathology, for instance, was developed largely here in Germany. Pathological changes in the organism were referred back to disturbances in individual processes in the cell; these processes could then be identified and investigated more fully through microscopy of tissues and cells. This approach is still the primary weapon in pathologico-anatomical diagnostics.

It became clear at an early stage in this work that the cell must be a complex chemical machine, and that the various cell structures revealed by the microscope must represent different parts of this machine. The structures found in all cells, such as the nucleus with its chromosomes must in all reason be involved in basic processes common to all cells, such as growth, multiplication and the transmission of hereditary factors from generation to generation. Others, on the other hand, are to be regarded as special structures, enabling a cell to fulfil a particular function — say, in a gland or a support — in the composite organism.

The desire to know more about this chemical machinery was naturally intense, not least because of its implications for medicine. A chemical interpretation of the changes in the cell that accompany pathological processes would clearly be an enormous advance.

To do this it would be necessary to determine the chemical composition of the individual cell structures, as well as of the changes which occur in these during the course of the various vital processes. This is what is directly implied by the words Histochemistry and Cytochemistry, which are to be the theme of this Congress. Cytochemistry implies chemical analysis within the individual cell. Histochemistry is a wider term — strictly speaking, it means the chemistry of tissues, but as a rule it is held to include the chemistry of the individual cells. The two terms are often interchanged, however.

The microscopic techniques of classical histology supply only morphological data. Throughout this century, however, many attempts have been made to combine microscopic techniques with chemical methods of determination, in order to provide histochemical information. Unfortunately, the technical difficulties were very great and progress was generally slow. Decades passed and histochemistry remained a relatively undeveloped field.

Meanwhile, however, development continued in other fields of cell research and in some of these, such as experimental genetics and general cytophysiology, there arose an acute need to penetrate the chemical functions of the cell structures. The explosive development of histochemistry in recent years, with the result that it is now an independent discipline with large international congresses such as this one, has been conditioned by the demand from these fields combined with another factor — the development of modern biophysics as an efficient science in the service of biology.

Experimental genetics supplied a more concrete foundation for the concept of hereditary factors or genes, which had previously had a somewhat metaphysical air. These genes are the elements which must be present in every cell to guarantee continuity from generation to generation, not only in respect of the individual cell but also of the adult organisms. It was clear that these factors must be material, and localized to definite cell structures — the chromosomes of the cell nucleus which at mitosis are divided exactly equally between the daughter cells. The next major advance concerned the discovery of how these hereditary factors for genes regulate the development of the cells — it was shown that they induce the formation of special enzymes, which in combination conduct the chemical processes during cell growth, to bring about not only the development of the adult cell but of the whole organism as well. In addition, they induce and regulate the special processes of chemical synthesis that I have already mentioned, and which enable for instance a gland cell in a composite organism to carry out its specialized function.

Here, then, we have a large complex of observations which directly show that certain structural elements in the cell must have chemical functions of decisive importance for such basic processes as the transmission of heredity, mitosis, cell development and function. In consequence, there gradually grew up a demand for data on the details of the chemical processes in the cell structure — neither the conventional methods of experimental genetics nor the histological procedures I have already mentioned were able to give much direct information: experimental genetics works in the main with indirect and statistical procedures.

General cellphysiology has flourished in the last few decades and from many of its fields have come demands for further penetration, using direct analysis of the cell structures. One example among many here is the work done on groups of cell structures which have been shown to perform definite tasks in cell metabolism. It would take too long to deal more deeply with the large field of cytophysiology. Let me therefore say a few words about one sector only — developmental physiology. As I mentioned earlier, the genes of the cell nucleus impart the continuity from one generation to the next. As the fertilized egg cell develops into the adult organism, the genes regulate not only the chemical processes in every individual cell and thereby the way in which it develops, but also the way in which cells with different specialized functions together build up the new organism largely as a replica of its parents. It is well known that all this occurs through a series of chemical processes entirely controlled by the system of genes. Extremely complicated processes must clearly be involved — a wealth of interactions between different types of genes and groups of genes. If we are ever to get to the bottom of these relationships, we must — parallel with other cytophysiological methods — develop and apply procedures that will enable us to follow the basic individual chemical processes within the complex cell structures. In other words here we have yet another field in which the demand for direct cytochemical analysis has grown stronger and stronger in recent years. Now and again, it is said that some types of experimental work in developmental physiology having to do with exotic zoological or botanic material, have little relation to the general line of approach and hence to practical applications — however theoretical one is, there must always be something practical on the horizon. Nothing could be further from the truth. In medicine, a primary role is played in particular by disturbances in the general development of cells — both in individual organs and in the organism as a whole. What is needed at present is increased knowledge on the course of development in normal cells in order to reveal the background to these disturbances and hence be able to deal with them. One such problem, of particular medical relevance, is the background to the development of tumours. It is also interesting in the present context because the experience from this speciality during the last two years has particularly stimulated the development of cytochemical approaches.

Thus we have already two large fields of research, crying out for means with which to attack chemical problems in the structure of the individual cell. And, indeed, during the last two decades the field has gradually widened — though very slowly until the last few years. It may seem surprising that the early development did not go faster but the explanation lies mainly in the exceptional technical difficulties presented by this field. A number of isolated problems may well be relatively accessible but the more general attack on the many different groups of substances involved in cytometabolism demands such a very large amount of work, not to mention the development of new methods, that overwhelming arguments were required to incite the laborious work which led to real progress on the field.

As we all know, such progress has been made in recent years and histochemistry has become an independent and major area of research, the significance of which is demonstrated by this great international congress.

But apart from the demands from the two fields of research I have already described, two further factors have contributed to this rapid advance by histochemistry, one of them biochemical and the other biophysical.

Biochemistry has come very far in recent decades and our knowledge of the processes of metabolism in the living substance has increased enormously. Naturally, too, this has generally reinforced the interest in histochemistry, but the development of one field in particular has contributed to an unusual extent, namely the chemistry of nucleic acids and protein. Thus we now have some idea of what happens in the two most fundamental chemical processes in living matter, on the one hand the synthesis of protein in the cell, which is the ultimate vehicle for the vital processes, and on the other the mechanism for the transport of genetic information between the generations. The decisive part is played by the nucleic acids in both processes. The genetic information is carried from one generation of cells to the next by long chains of nucleic acids in the chromosomes of the cell nucleus, and we are aquainted in principle with the chemical composition of these chains. The synthesis of all the different types of protein that compose the main substance of the cell and which induce and mediate the main part of the chemical metabolism within the cell, occurs in similar nucleic acid chains according to principles which now seem to be clear, at least in outline. There is little doubt that this is in fact one of the greatest advances in biological research. We can now for the first time get right to the bottom of some of the chemical events that lie behind such fundamental vital processes as cell multiplication and cell growth. Meanwhile, however, we must not forget that these observations relate only to the principles for some of the central functions of the cell. Their outstanding value almost certainly lies in that they indicate the direction for further, more extensive research into the extremely complex chemical processes occurring in the cell during its development, differentiation and function.

Isolated results within the chemistry of the nucleic acids and protein have already directly stimulated several lines of histochemical research. Even so, I believe that the greatest importance of the development of histochemistry lies at present in the psychological factor that the concept of the gene, which previously had such strong metaphysical overtones, has become a chemically well-defined, material reality. It seems possible and reasonable to attack the entire field of research surrounding genetic functioning, cell development and the development of the organism. One circumstance worthy of note in this context is that the primary observations of the fundamental role of the nucleic acids in protein synthesis were made with purely histochemical methods, which have given rise to the later work of a general chemical and physico-chemical nature which lies behind the major part of our present knowledge in this field.

And now let me turn to the last of the fields which have given rise to the recent rapid development of modern histochemistry — I refer to *biophysics* in its modern sense.

First, however, I must try to present the general nature of histochemical work, primarily why it is often technically so difficult to perform. I am by no means going to go into individual procedures or principles — these are to constitute the major part of the congress itself.

A general histochemistry must essentially be a rather comprehensive and complex subject. We know from biochemistry that the cells contain a very large number of different substances, and consequently a number of different procedures are required for indentifying or determining these. Some of them will be found in relatively large quantities or at least concentrations, but there are others, such as the enzymes, where the quantities concerned are so infinitesimal that, as in ordinary macrochemistry, they must usually be determined indirectly. It is further desirable in certain cases that methods should be developed for determining the speed of individual metabolic processes.

There are two circumstances that have given the methods of histochemistry their peculiar status in the field of general chemistry. One of them is the smallness of the object to be investigated, the other is the difficulty of manipulating these objects. The ordinary biochemical procedures for identification and determination, work with quantities down to the milligram range, while what is known as microchemistry deals in hundredths or possibly thousandths of a milligram. When we come to investigating single mammalian cells, however, we find ourselves faced with millionths of a milligram and a cell structure in the 1 micron range weighs only a thousand millionth of a milligram. Where it is possible to isolate the tissue or cell part to be investigated so that — without suffering from the manipulation — it is free from extraneous material, such as the surrounding cells, it is sometimes possible to perform substance indentification or even determination using sensitive ordinary chemical procedures; and some such approaches were crowned with success early in the development of histochemistry. On the cellular level this can, however, only be done in material with exceptionally large cells. As interest gradually concentrated on what one might call ordinary cell material, such as mammalian tissues, where the cells are small, work has gradually had to be performed by other procedures.

When it is not possible to isolate the cell part to be analysed, indentification or determination must be performed within the structural array of the tissue or cell, and it is this circumstance above all which has given modern histochemical work its special profile. When it is simply a question of indentifying a substance — of finding out whether it is present in a part of a cell or not — the technical problems are often not very great. Even in the early years of staining histology, towards the end of the 19th century, qualitative such work was attempted with stains that selectively attach themselves to individual cell substances. This is, then, the simplest combination of a chemical procedure — the selective, staining reaction — and a purely physical one — microscopy. Although a great number of such simple qualitative procedures have been developed, they are generally of only limited value. What one wants in the great majority of cases is to be able to make quantitative determinations or at least approximate estimations of the amounts of various substances — what is termed semi-quantitative work. But this requires more refined methods than simple conventional staining combined with visual microscopy. The object must be pre-treated in some suitable way, and controls and models must be built up in order to follow how the method functions in each particular case. Finally, if real measurements are to be made, often special instruments have to be constructed, and these have shown themselves to tend to be complicated and often difficult to operate.

This is where biophysics comes into the picture — a multitude of methods for observation and measurement taken from experimental physics and specially adapted to biological problems. It is from this "marriage" between cytobiology and biophysics that the most vital branches of modern cytochemistry have sprung.

Methodology is the predominant theme of this Congress and within this subject come the procedures of a biophysical nature. I shall not bore you with a long list of these but should prefer instead to go a little deeper into the interaction between cytochemistry and biophysics.

In recent years modern biophysics has stimulated the development of cytochemistry mainly in two related respects. The first of these is the way in which the range of problems that can be tackled has been extended through the application of various physical methods for observation and measurement. While the microscopic methods still predominate, they are no longer limited to the visible region but include ultra-violet and even infrared. The shortwave X-ray region is also utilized in a corresponding way. Electron microscopy, which makes it possible to penetrate even smaller morphological elements, is being introduced to histochemical work, as are interferometric methods, which in recent years have helped to extend the possibilities for quantitative work. Fluorometric measurements, which have long been used to a limited extent in histology, now enjoy a more general application and permit further penetration thanks to the improved quantitative methods of recent years. Finally, there is the major field of the radioactive isotopes — even on the histochemical level they supply a longfelt need to study the kinetics of chemical processes. It is thus clear that the work of histochemistry is performed with tools borrowed and modified from a large number of biophysical fields.

The second contribution from modern biophysics to chemico-physiological research into tissues and cells, is the quantitative work. Time and again in the history of science we find that the development of an empirical field is greatly stimulated by the introduction of truely quantitative methods. Such has been the aim for a long time in histochemistry, but the technical difficulties have been great and even though methods for certain purposes have long been developed, their complexity has prevented them from spreading more rapidly. The current rapid expansion of work in this field is reflected in the programme of this Congress. This is not only thanks to the development of methodology — many of these methods have been available for some time. The primary cause, I believe, is that biologists, particularly in recent years, have come to realize the necessity of working on a wide front, often with quite complex biophysical apparatus, and also that in recent years people have grown less afraid of what appear at first sight to be complicated physical working procedures.

What I have just said gives a picture of cytochemistry as a sector in general cell biology where many fields of biology and medicine meet, and where, with the application of findings from biochemistry and — above all — the more recent biophysics, an independent and important area of research is growing up fast. This is the background against which the complex programme of this congress has to be seen.

Which, then, are the fields where cytochemistry and histochemistry can be expected to give particularly valuable results in the near future ?

I have earlier referred to cell multiplication, cell growth, cell development and cell function as the most important of the processes generated and controlled by the chemical machinery within the individual cell; this is in effect the answer to this question. We can look forward to a greatly extended knowledge of these basic processes, assuming that development continues as it has begun.

It goes without saying that increased knowledge in this area must be of value for many different fields of the life sciences.

To take just one field, medicine, which of obvious reasons is of central interest, one can say that pretty well every new observation from cytophysiology has a certain importance. Increased knowledge of the function of even individual types of cells, for instance gland cells or nerve cells, is clearly of interest both for the possibilities of developing methods for opposing disturbances in their work and also for indentifying disturbances in tissues and in cells. Histochemical procedures, particularly certain simple staining procedures, already play a part in pathologico-anatomical diagnosis.

Of still greater and more general importance are the possibilities of penetrating disturbances in the processes of cell growth and also in the processes of differentiation in the context of tumour pathology. Quantitative cytochemical procedures have, for instance, already been used to demonstrate biochemical disturbances in cell populations that build up tumours, and in all probability these will prove of value in diagnostic work in the future. These changes occur namely early in the development of tumour growth, and can be used for identification of the early stages of tumour development which is an important medical problem. Cyto-chemical work of this nature has also proved valuable in the development of tumour chemotherapy. In general one can say that one of the primary causes of the flourishing of cytochemistry in recent years is just this interest for the potential it offers in tumour research.

Among other fields of particular medical interest can be cited virology, where cytochemistry makes it possible to study the interaction between virus and host cell. Growing importance will no doubt be attached also to the work on the immunity mechanisms in the body, a complex of problems that is attracting increasing interest for a variety of medical reasons.

A glance at the programme of this Congress will serve to show other fields in biology and medicine that are benefitting from histochemistry and cytochemistry; thus it is not necessary for me to take up your time with further examples.

Finally, a few words about the present position of histochemistry at our institutions for teaching and for research. In many places in recent years histochemistry has already been adopted as a special subject in conjunction with instruction in biochemistry and histology. Characteristically, this seems to have been particularly the case at seats of learning with a medical leaning. Histochemical research is nowadays being conducted at many different types of biological and medical laboratories and it is to be expected that such work will be extended rapidly; for one thing, the more complex biophysical apparatus needed for more elaborate work is now rapidly becoming increasingly available on the commercial market. Previously such was difficult to acquire and as a rule an institution had to build up its own, which naturally limited the spread of these methods.

The fact that histochemical work is tending to be more and more dependent on apparatus has had several repercussions, including the question of finance. Quantitative histochemistry in particular demands a relatively extensive instrumentation, and a more general programme on such fields demands quite expensive equipment. I personally believe that it would be a great advantage for the work in this field if well-equipped laboratories could be stablished which were made easily available to visiting researchers who need access to special histochemical apparatus and experience for particular problems. These laboratories would also serve as training centres for research workers. We have in fact a parallel here to the development in experimental physics with its international laboratories. Histochemical equipment is perhaps not yet so expensive as to motivate the creation of entirely international institutions, but an efficient international utilization of well-equipped national laboratories, in respect of both instruments and personnel, would be a great step forward. If this Congress, in addition to stimulating the field in general through the exchange of ideas and experience, also furthered thoughts in this direction, much will have been gained.

Professor Dr. T. CASPERSSON, Institute for Medical Research,
Karolinska Institutet, Stockholm (Sweden)

From the Departments of Surgery, Sinai Hospital of Baltimore and The Johns Hopkins
University School of Medicine, Baltimore, Maryland

Some Recent Trends and Advances in Enzyme Histochemistry*

By

ARNOLD M. SELIGMAN

With 12 Figures in the Text

In the past four years there has been a distinct tendency to abandon the
earlier methods of enzyme histochemistry that were characterized by a general
and diffuse cytoplasmic localization and to direct more interest and attention
to methods capable of demonstrating mitochondria or other organelles. Dis-
enchantment with former methods has undoubtedly been accelerated by the
perfection of methods for the dehydrogenases, which demonstrate mitochondria
clearly and readily [*38, 59, 60*]. It has been finally recognized, that much of
the diffuseness obtained with hydrolytic enzymes was due to the soluble and
diffusible component of the enzymes (lyoenzyme). By using various methods
of fixation this component could be effectively allowed to diffuse from the tissue
leaving behind for histochemical demonstration the more precisely localized, less
soluble component of the enzymes (desmoenzyme) [*36*], made even more insoluble by appropriate fixation. The use of better fixation, better substrates
and enzyme inhibitors has been instrumental in delineating the droplet form of
some hydrolytic enzymes, and these localizations have been correlated with
DE DUVE's interesting and important biochemical work on the lysosomes [*11,
23, 40*]. Most popular have become methods for various dehydrogenases using
the ditetrazolium hydrogen acceptor, nitro-BT [*57*] and for cytochrome oxidase
using analogues of phenylene diamine and naphtholic derivatives [*7, 33*]. The
methods for the hydrolytic enzymes have been improved by using formalin-
fixed tissue [*51*] and substrates and coupling agents that yield insoluble and
substantive dye pigments, such for example as the demonstration of esterase
with fast blue BB and naphthol-AS acetate, a substrate first proposed by Dr.
GOMORI [*16*]. Much effort has been directed with success toward proper fixation
in order to preserve organelle morphology [*12, 18, 59, 60*] while maintaining significant enzymatic activity and at the same time allowing diffusible enzyme to
leave the tissue block. Similar improvements in localization were also produced
by exclusively inhibiting the diffuse enzymatic components while inhibitor-
resistant components in droplets were sharply localized [*25, 26, 56*]. The evaluation by SABATINI *et al.* [*46*], of a host of new aldehyde fixatives with both the
light and electron microscope, gives promise of further improvement in this
area. It has also become evident that the protective colloids other than the
isotonic effects of sucrose offer little improvement when the right fixative is
used [*60*]. Equally important effort has been expended in the design of new

* This work was supported by a research grant (CA-02478-10) from the National Cancer
Institute, National Institutes of Health, Department of Health, Education and Welfare,
Bethesda, Maryland.

reagents that react rapidly to produce a significantly faster capture reaction [*34, 35*], that are substantive for proteinaceous components of the cell, and that are sufficiently insoluble in lipid to discourage crystallization. Although ultimate limits of ideal properties have not yet been reached, the more precisely the localization of such reaction products can be made to occur within or on intracellular organelles, the more likely are the possibilities of extending this methodology with profit to the electron microscope [*54*].

The ultimate aim of histochemistry today is to perfect methods to the degree that their application to *EM-cytochemistry* will add to our knowledge of the relationship of enzymatic activity and structure. This goal has not been reached with many enzyme methods offered for electron microscopy. Some investigators have been willing to publish results with crystalline end-products, or to focus narrowly on seemingly attractive streaks, granules or droplets which on broader inspection are surrounded by a sea of debris or on comparison with appropriate controls reveal areas of contrast that are of questionable significance. This criticism can be leveled at some of the metal salt methods [*55*] and the tellurium method for dehydrogenase [*4*]. On the other hand, valuable information has been obtained in the case of some lead salt methods [*5, 15, 24, 31, 47*]. I should like to encourage the more frequent inclusion in the literature of parallel light photomicrographs, as well as corresponding electron micrographs. Only in this way is the reader given ample opportunity to evaluate the methods used and to observe the degree of surrounding artefact. Since there are severe limitations on the developmental aspects of metal salt methods, the hope for further success beyond the first obvious trials with metal salts appeared to reside in the organic dye methods. On the other hand, the use of organic end-products so far have produced images that are too vague and too deficient in contrast to satisfy our ultimate aims [*2, 41, 47, 50*]. The problem has not been advanced especially by those of us who append the statement to our histochemical papers that since the final products chelate with metals, the methods will find application in electron microscopy. To hope that someone else will make them work for electron microscopy, while we retain a certain measure of credit for the advance in methodology, may have been justifiable a few years ago when research in this direction was just beginning, but this is no longer original or impressive. Since I have both made such statements and tried subsequently to make chelating methods work for electron microscopy, I am well aware of the pitfalls of chelate EM-cytochemistry, of the tremendous hurdle to be cleared and of the small fraction of the task which has been accomplished by the demonstration that a histochemical reaction product will chelate with a heavy metal. A good deal of difficulty arises from the fact that most metal chelates of organic dyes retain a significant solubility in lipid and some of the best metals for this purpose cannot be introduced until after the enzymatic reaction has been completed.

In order to illustrate some recent trends in this direction, I will discuss in some detail the evolution of methods for four enzymes. I have chosen four methods which have been brought successfully to the level of the electron microscope in our laboratory by the establishment of *a new cytochemical principle*, by the design and preparation of new reagents which contain reactive sulfur and react with osmium tetroxide to yield osmium black, an ideal end-product

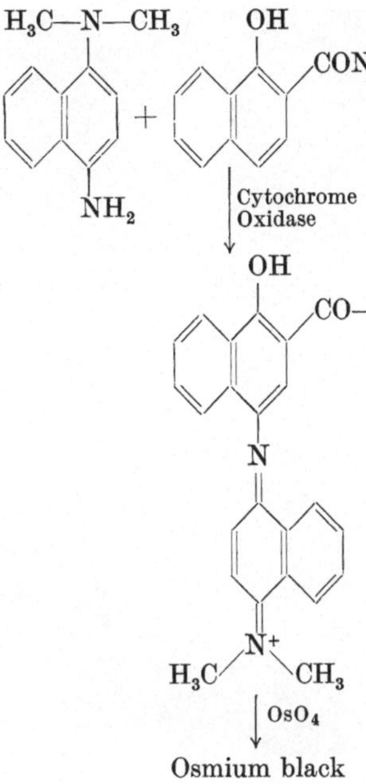

Osmium black
Chart 1

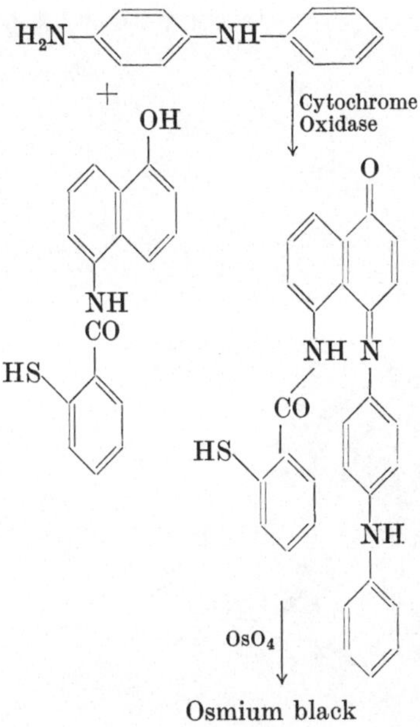

Osmium black
Chart 2

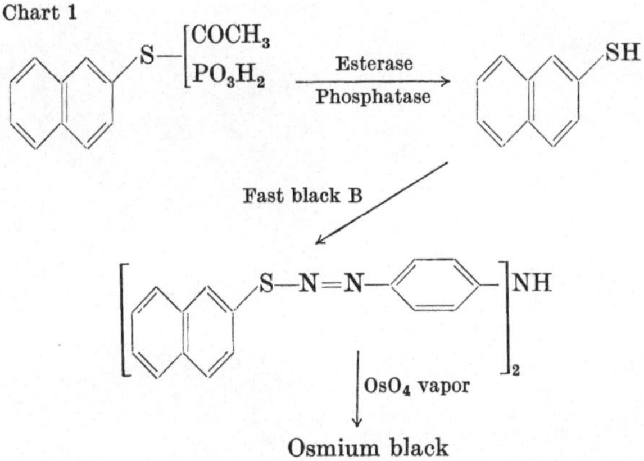

Osmium black
Chart 3

Chart 1. Reagents that may be used for the cytochemical demonstration of cytochrome oxidase for both light and electron microscopy

Chart 2. Reagents that may be used for the cytochemical demonstration of cytochrome oxidase for both light and electron microscopy

Chart 3. Reagents that may be used for both light and electron microscopic demonstration of esterases and phosphatases

for electron microscopy as well as for light microscopy [20—22]. The first is the oxidative enzyme, *cytochrome oxidase*, and the other three are the hydrolytic enzymes, *esterase, acid phosphatase* and *alkaline phosphatase*.

Cytochrome Oxidase

The Nadi reaction for cytochrome oxidase was applied to tissue sections by MOOG in 1943 [32] and improved reagents were developed by NACHLAS et al., in 1958 (4-amino-1-N,N-dimethyl naphthylamine [ADN]) [33] and by BURSTONE

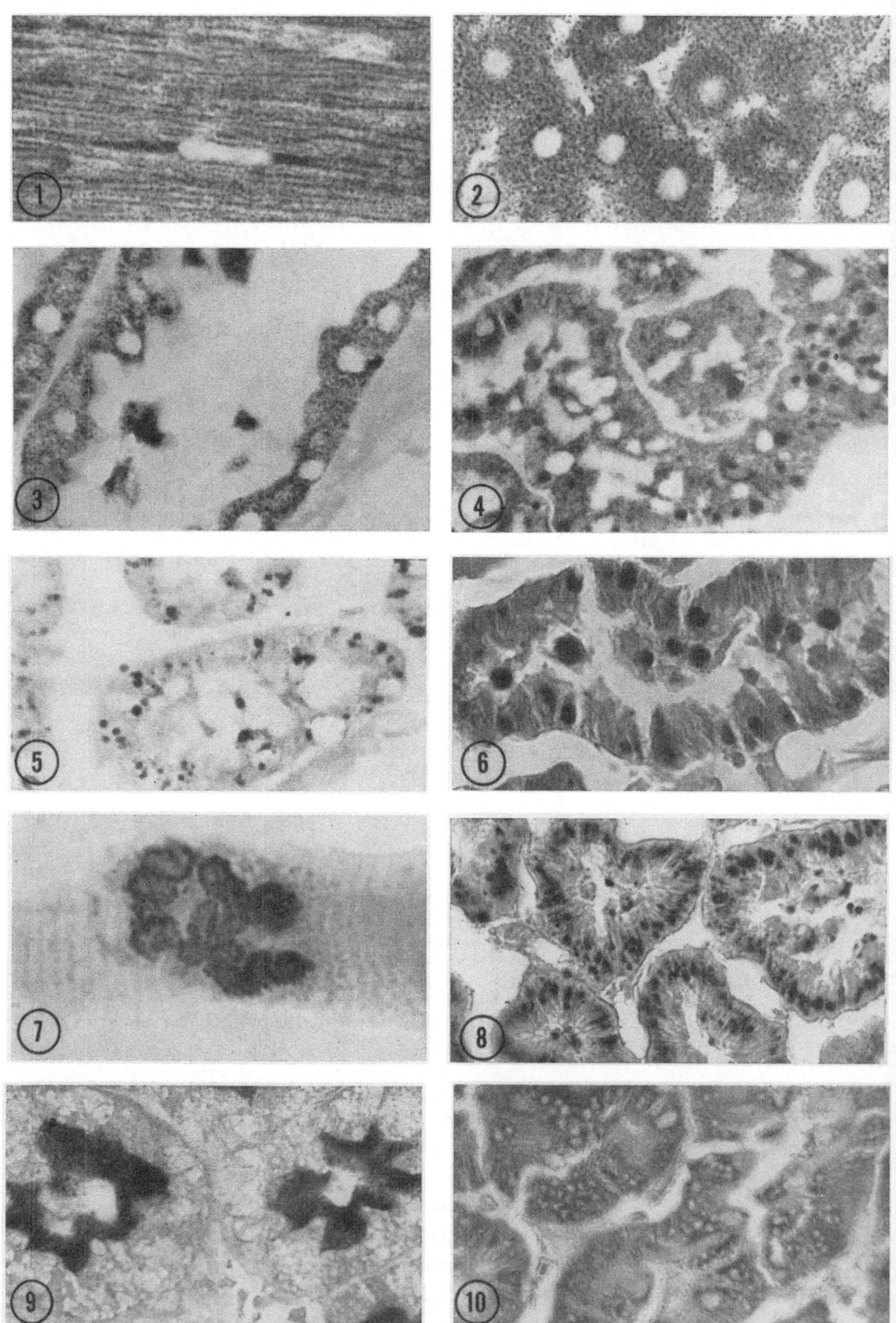

Figs. 1—10 (Legends see p. 13)

in 1959 (p-aminodiphenylamine) [7]. Cytochrome oxidase demonstrated with either reagent gives identical localizations. Although both methods provide pigments of greater stability and with less tendency to crystallize than the original Nadi reaction, they suffer from the drawback of a certain degree of lipid solubility. In cardiac and skeletal muscle and in the retina the enzyme appears to be confined to mitochondria, but in kidney, gut and liver the numerous granules cannot be identified as mitochondria under the light microscope. This discrepancy in localization makes the extension of cytochrome oxidase to EM-cytochemistry especially important. If we could replace the organic pigment of the current methods for cytochrome oxidase with the pigment, osmium black, not only would the histochemical methods be improved, but the possibility would be good of visualizing sites of activity with the electron microscope. A variety of agents with functional groups capable of reacting with OsO_4 were prepared by our organic chemist, JACOB HANKER and tested for this purpose [21]. The sulfhydryl group was found to be particularly useful because it readily reacts with OsO_4 to form osmium black, a possible mixture of osmium mercaptide, lower oxides of osmium and osmium metal. The sulfhydryl group did not interfere with the enzymatic reaction when incorporated into the naphtholic component of the histochemical reagents. Its lack of lipid solubility and fine amorphous character make osmium black a more permanent pigment than the original organic pigment produced by the histochemical reaction and allows for dehydration and embedding in either acrylic or epoxy resins. Either 4-amino-1-N,N-dimethylnaphthylamine (ADN) or 4-aminodiphenylamine may be used with 1-hydroxy-4′-mercapto-2-naphthanilide or 5-[2′-mercaptobenzamido]-1-naphthol. The formulation of the reactions is shown in Charts 1 and 2. Histological sections

Fig. 1. *Rat heart*. Photomicrograph of cytochrome oxidase in a fresh frozen section stained with the reagents shown in Chart 1 or 2 and exposed to OsO_4 vapor. The reaction is confined to mitochondria

Fig. 2. *Rat liver*. Photomicrograph of cytochrome oxidase in a fresh frozen section treated as in Fig. 1. The reaction appears to be more widespread than would be expected if confined to the mitochondria alone

Fig. 3. *Rat kidney*. Esterase activity at 37° C for 20 minutes appears in granules evenly distributed in the cytoplasm of the tubular cells of formalin-fixed tissue cut in frozen sections. After the reaction, osmication was accomplished in OsO_4 vapor for 40 minutes, when the pale yellow pigment was replaced by osmium black. The substrate is phenylthiolacetate and the diazonium salt is fast blue BBN

Fig. 4. *Rat kidney*. Same as Fig. 3 except that the histochemical reaction was carried out at 0° C for 1 hour. Note the inhibition of cytoplasmic esterase at 0° C and the enhancement of droplet esterase by 1 hour's incubation even at this low temperature. The substrate and diazonium salt are the same as Fig. 3

Fig. 5. *Rat kidney*. Same tissue as Figs. 3 and 4 but the substrate is 1-thiolacetoxy-2-benzanilide for 20 minutes. Note the intense droplet reaction and the slight reaction in the cytoplasm

Fig. 6. *Rat kidney*. Same tissue as Figs. 3 and 4 but the substrate is octadecylthiolacetate. Note the intense droplet reaction. The droplets are larger than noted before because of overstaining

Fig. 7. *Mouse intercostal muscle*. Formalin-fixed and frozen section stained for acetylcholinesterase with 2-naphthyl thiolacetate and fast blue BBN followed by OsO_4 vapor. Myoneural junctions were readily stained in 10 minutes

Fig. 8. *Rat kidney*. Formalin-fixed, frozen sections stained for *acid phosphatase* at pH 5.3 with phenylthiolphosphate and fast blue BBN followed by OsO_4 vapor. Note droplet distribution of acid phosphatase activity

Fig. 9. *Rat kidney*. Same tissue at pH 9.0 stained for *alkaline phosphatase* with the same reagents as in Fig. 8. The alkaline phosphatase activity is confined to the brush border

Fig. 10. *Rat kidney*. Same tissue as Fig. 8 stained for acid phosphatase with naphthol-AS phosphate and fast blue BBN. Note negative image of the droplets in azo dye stained cytoplasm in contrast to the droplet stain with phenylthiophosphate shown in Fig. 8

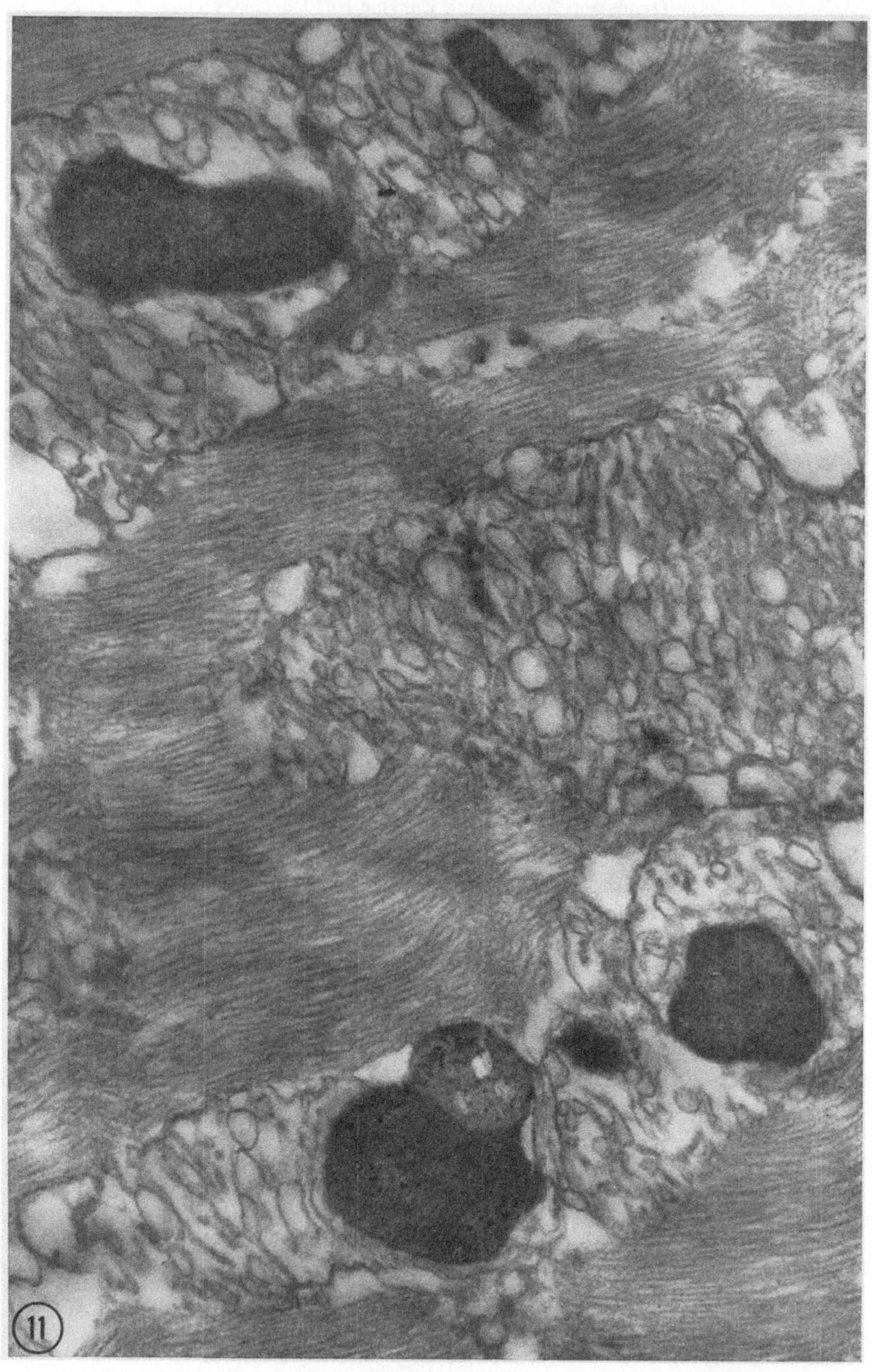

Fig. 11. *Rat heart*. Electron micrograph of a section prepared from a 50 μ fresh frozen section, stained for 20 minutes as for Fig. 1, and then osmicated in OsO_4 vapor for 40 minutes, dehydrated, embedded in the epoxy resin, Maraglas, cut, stained with lead, and viewed with a RCA EMU3 Electron Microscope. Note that the reaction is confined to droplets within mitochondria. Two degrees of intensity may be seen in these droplets

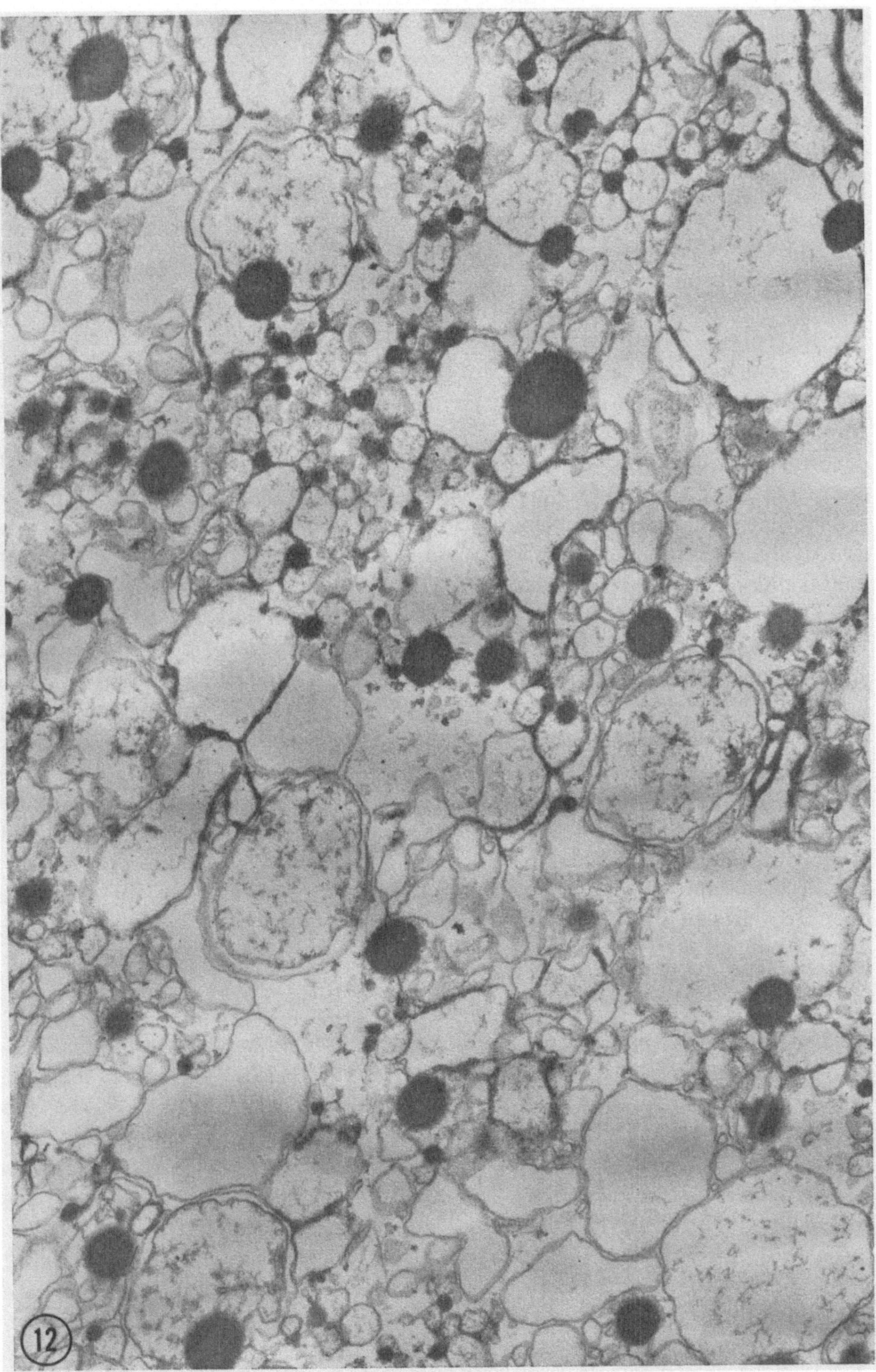

Fig. 12. *Rat liver*. Electron micrograph of a section prepared as in Fig. 11. See Fig. 2 for light micrograph. Lead staining was not used. Note that the small droplets are associated with the membranes of the endoplasmic reticulum as well as at the outer limits of some of the mitochondria. This provides strong presumptive evidence that fractionation techniques for liver have been in error in locating this enzyme exclusively in mitochondria of liver

reveal a distribution of the osmium black identical to that of the organic dye before osmication, except that the osmium black is permanent and withstands dehydration and mounting in Permount. The mitochondrial distribution in a photomicrograph of rat heart is shown in Fig. 1 and the numerous granules in the hepatic cell, which cannot be identified as mitochondria are shown in Fig. 2.

Reagents for EM-Cytochemistry of Cytochrome Oxidase

EM-cytochemistry of cytochrome oxidase poses a particularly difficult problem because the enzyme does not withstand any of the currently used fixatives, which would be helpful in preserving the fine structure of the tissue. SABATINI et al., have recently proposed hydroxyadipaldehyde as a suitable fixative for cytochrome oxidase [46] and they have published results with rat heart using Burstone's method (p-aminodiphenylamine and 3-amino-9-ethylcarbazole) followed by fixation in OsO_4. Although they attributed electron opacity to the final organic product, the degree of density shown in their photograph was undoubtedly produced by chelation of osmium to the carbazole component of the organic dye produced by the enzyme reaction. In our laboratory, Dr. ARLENE SEAMAN has examined both 50μ unfixed frozen sections and tissues fixed in hydroxyadipaldehyde and then frozen and sectioned, stained for cytochrome oxidase with our new reagents, exposed to OsO_4 vapor, dehydrated and embedded in epoxy resin (Maraglas) [13] for sectioning. Lead staining was used to enhance background fine structure when needed. Sucrose (0.44 M) was included in all reaction solutions and washes. The prominent, discrete localization of cytochrome oxidase in the mitochondria of rat myocardial cells is shown in Fig. 11, and the localization of similar deposits on the membranes of the endoplasmic reticulum as well as at the outer limits of mitochondria of a rat hepatic cell is shown in Fig. 12 [48]. This provides strong presumptive evidence that fractionation techniques have disrupted cytochrome oxidase activity in non-mitochondrial sites of liver heretofore or that the reagents used for showing cytochrome oxidase activity are capable of demonstrating some other electron transport system as well.

Esterase

The first azo dye method for esterase required 2-naphthyl acetate and diazotized 1-naphthylamine [37]. When it was shown that better localization could be obtained with 1-naphthyl acetate and tetrazotized diorthoanisidine [17], the improvement was first attributed to the greater insolubility of the pigment [17], later to the greater speed of coupling [39], and finally the best results with these agents were obtained when the soluble components of the esterase in tissue were allowed to diffuse out of the sections during periods of fixation with formalin [36]. Improvement in methodology was enhanced by increasing insolubility of the final pigment with hexazotized pararosaniline [30], by increasing the rate of coupling by an increase in the concentration of tetrazonium salt [34], or by special techniques of fixation [12, 18]. Nevertheless, it is important to realize that when 1-naphthyl acetate is used as a substrate under the best conditions now possible, one rarely gets the same distribution of esterase in rat tubular epithelium or liver, for example, as one obtains with the haloindoxyl acetate method of HOLT [25] or the naphthol-AS acetate method of GOMORI [16]. When first introduced, Holt's method was considered to be closer to the true localization of esterase than heretofore, because discrete droplets were demonstrated with it in tubular epithelium of the adult male rat, and in the peribiliary region of rat liver. The valuable concept of substantivity for protein was presented by HOLT to explain the superior discrete localizations of the dye [25]. It was later shown that the beautiful and important localizations could be produced with both methods and were due in large part to the inhibition of esterase activity in the non-droplet portion of the cytoplasm by a variety of inhibitors including formalin, ferricyanide, atoxyl or fluoride. The esterase in the droplet proved to be inhibitor-resistant [56]. It is rarely possible to duplicate this picture with 1-naphthyl acetate, no matter what coupling agents or inhibitors are used. This is a very interesting and significant observation. The implication of these differences was suggested by GOMORI when he first observed unique staining of Brunner's glands with naphthol-AS acetate [16] which he could not duplicate with other esterase methods. He suggested that there were at least two different esterases, which he called "naphthyl esterase" and "naphthol-AS esterase". Although little attention was paid to his suggestion at first, the evidence obtained by the action of inhibitors with both naphthol-AS acetate and haloindoxyl acetate [56], has made his original suggestion more noteworthy. Further weight has been given to this concept by similar observations made by WACHSTEIN and MEISEL [58] with the method of CREVIER and BÉLANGER [10], which requires thiolacetic acid as substrate. Additional support has been provided in our own unpublished work from a variety of new esterase substrates. We have been able to demonstrate esterase activity in formalin-fixed, renal tubular epithelium as follows: (1) in both droplets and cytoplasm with naphthol-AS acetate (control), (2) only in cytoplasm without a sign of droplets with 1-naphthyl acetate, (3) only positive droplets and very faint cytoplasmic stain with 3-acetoxy-4-formyl-2-naphthanilide, and (4) only cytoplasm and negative holes where droplets should be with 5-(2-acetoxy-3-naphthamide)2,4-dihydroxypyrimidine. These experiments place the results with inhibitors in their proper perspective, and show clearly that there are two types of esterase with definite substrate preferences and different susceptibility to inhibitors when certain substrates are used and not with others. These varieties of esterase should not be confused with isozymes which differ by virtue of their different protein content and consequent differing rate of movement in a strong electric field. The substrate preferences of the latter may be similar and for that matter, the inhibitor-sensitive and inhibitor-resistant esterases may be found to move at similar rates in an electric field, when this experiment is done. The recent work of KOMMA [29] reveals the difficulties ahead in correlating electrophoretic results with histochemistry. Particularly important for future investigators is the important lesson to be gained from the example of esterase. Caution should be exercised by reviewers in recommending one method over another as being more accurate, by which we should mean the true *in vivo* distribution of an enzyme, on the exclusive basis of appealing morphology, such as "crispness" or activity confined to droplets or other organelles.

The naphthol-AS nucleus of Dr. GOMORI has been established as a particularly useful base for the design of new substrates by the early success with naphthol-AS acetate [16], later with naphthol-As phosphate [3] and more recently with naphthol-AS glucuronide [44] and naphthol-AS nonanoate [1]. However, claims for superior localization of esterase based upon the use of even more insoluble anilide derivatives of naphthol-AS [43, 8], available from the dye industry, have not been substantiated in other hands and this particular approach to the advance of esterase methodology appears to me to be on the whole an unprofitable blind alley. When reagents become so insoluble that large amounts of solvent are required to effect their solution, the final pigments are also made more soluble and little advantage is gained other than the precarious benefits of uncontrolled enzyme inhibition.

Reagents for EM-Cytochemistry of Esterase

Our first efforts at designing reagents for EM-cytochemistry involved highly iodinated organic reagents and chelatable organic pigments. We were discouraged from further pursuit of this approach because the final pigments proved to be too soluble in the organic solvents used for dehydration and in the embedding material. These reagents could possibly be useful only in demonstrating functional groups of macromolecules which themselves render the final products sufficiently insoluble [49].

A more profitable approach was developed [21] by the incorporation of the diazothioether group into the final organic pigment which could in a later step react with OsO_4 to form osmium black, an ideal pigment both for light and electron microscopy. For the purpose of testing these ideas we synthesized 2-naphthyl thiolacetate (Chart 3), 3-thiolacetoxy-2-naphthanilide, phenylthiolacetate, 1-thiolacetoxy-2-benzanilide, octadecyl thiolacetate and triphenylmethyl thiolacetate. After hydrolysis by esterase, the thiol group couples very rapidly at either high or low pH with diazonium salts in a simultaneous capture reaction to yield insoluble, yellow diazothioether compounds. The slides were then exposed to OsO_4 vapor for 30 to 40 minutes to decompose the diazothioether compounds, presumably yielding a thiol group that rapidly reacted with OsO_4 to form osmium black at the sites of esterase activity. In preliminary experiments with 2-naphthalenethiol we found that the tetrazonium rather than diazonium salts favored production of non-crystalline insoluble diazothioether compounds, although this difference was less evident with the lower melting benzenethiol. The greatest rate of reaction with OsO_4 was observed when diazonium salts with electron withdrawing rather than electron donating groups were used. For example, tetrazotized diorthoanisidine (Fast blue B) and diazotized 4-benzamido-2,5-diethoxy aniline (Fast blue BN) gave diazothioether compounds that required 30 minutes to react completely with OsO_4 vapor, whereas tetrazotized diamino-diphenylamine (Fast black B) gave a diazothioether compound that reacted in a few minutes and p-nitrobenzendiazonium chloride gave a diazothioether compound that did not react with OsO_4 vapor at all.

A comparison of results with the various thiolsubstrates on rat kidney revealed a similarity of preference for the two main types of esterase in the tubular epithelium to that noted previously with the oxy analogues. Only cytoplasmic esterase without droplets was observed with phenylthiolacetate (Fig. 3) or 2-naphthyl thiolacetate, droplet accentuation was noted with 1-thiolacetoxy-2-

benzanilide (Fig. 5) or 3-thiolacetoxy-2-naphthanilide, and the esterase of the droplets was the only stained component with the two aliphatic substrates, octadecyl thiolacetate (Fig. 6) and triphenylmethyl thiolacetate. It is worth noting that when the reaction of phenyl thiolacetate was conducted at 0^0 instead of room temperature and for 1 hour instead of 20 minutes, the cytoplasmic stain was largely inhibited and the droplet stain was brilliantly shown (Fig. 4). Myoneural junction was particularly well shown (Fig. 7).

In preparing tissues for EM-cytochemistry, they were processed in the same way as for histochemistry, except that 0.44 M sucrose was incorporated in the formalin and in all washes. Frozen sections were cut no thicker than $10\,\mu$ when phenylthiolacetate was used and no thicker than $40\,\mu$ when 1-thiolacetoxy-2-benzanilide was used and the enzymatic reaction was performed and followed immediately by a brief wash. Osmication in OsO_4 vapor for 40 minutes was then performed with the tissue at room temperature and with the floor of the vapor chamber in a bath at 45^0. These sections were dehydrated and embedded in epoxy resin for electron microscopy. Results will be shown at the Congress. These EM-preparations were made in collaboration with Dr. ARLENE SEAMAN. M-band of fresh frozen rat heart did not stain, providing confirmatory evidence to that of KARNOVSKY and HUG [28] that M-band staining is not due to acetyl cholinesterase or any other esterase activity (in collaboration with Dr. LEON WEISS.).

Acid Phosphatase

The first azo dye method for acid phosphatase was proposed in 1949, using calcium 1-naphthyl phosphate as substrate and diazotized 1-amino-anthraquinone as the coupling agent [52]. These agents were selected in order to make possible coupling at the acid pH optimum for this enzyme and to allow long incubation periods required by all tissues except prostate. Although the method was first improved by our synthesis of the more soluble sodium salt [14] and the recommendation of other diazonium salts [6, 19, 42], coupling with 1-naphthol at pH 5.0 was never sufficiently rapid to completely satisfy histochemical requirement. For this reason the post-incubation coupling rationale was introduced in 1950 by the synthesis of 6-benzoyl-2-naphthyl phosphate [53]. The hydrolysis product was sufficiently insoluble and sufficiently substantive for protein to anchor it until coupling with an appropriate diazonium salt could be accomplished later at higher pH [45]. The synthesis of naphthol-AS phosphate has made possible a simultaneous coupling method [3, 9] with the advantages first shown by Dr. GOMORI for naphthol-AS acetate. The high degree of substantivity of naphthol-AS compensates for the slow coupling rate at acid pH. Both azo dyes are too lipid soluble to enable extension to EM-cytochemistry.

Reagents for EM-Cytochemistry of the Phosphatases

We have explored extension of our new principle for EM-cytochemistry by the preparation of phenyl thiolphosphate and 2-naphthyl thiolphosphate, which are hydrolyzed by the phosphatases quite readily [21]. The hydrolysis product, benzenethiol or 2-naphthalenethiol, coupled rapidly at both low and high pH with diazonium salts to form an insoluble diazothioether, which then reacted with OsO_4 vapor to yield osmium black at the sites of enzyme activity in thin sections as described for esterase. The formulation of the reaction for phosphatase is shown in Chart 3, and the droplet distribution of acid phosphatase activity in formalin-fixed renal tubular epithelium is shown in Fig. 8, which can be compared with the brush border localization of alkaline phosphatase when high pH is used (Fig. 9). It is noteworthy that acid phosphatase with phenylthiol-phosphate is revealed mainly in the droplets of renal tubular epithelium, whereas with naphthol-AS phosphate, acid phosphatase is diffusely distributed in the cytoplasm, occasionally revealing negative droplets (Fig. 10). It would appear that there are two varieties of acid phosphatase in the kidney tubular cell, with rather distinct substrate preferences.

This new cytochemical principle has been used for demonstrating diglycols of macromolecules such as glycogen with the reagent thiosemicarbazide after periodic oxidation [22], for alkaline phosphatase [20, 21] (Fig. 9) and peroxidase [20, 21], and is being extended to the dehydrogenases, glucuronidase, sulfatases, monamine oxidase, aldolase, lipase and aminopeptidase by the incorporation of an osmiophilic moiety into tetrazolium salts, diazonium salts and appropriate substrates.

Bibliography

[1] ABE, M., S. P. KRAMER, and A. M. SELIGMAN: J. Histochem. Cytochem. 12, 364—383 (1964).
[2] BARRNETT, R. J.: In: Fourth Internat. Conference on Electron Microscopy, vol. II p. 91—100, 1960.
[3] — R. BRESSLER, and A. M. RUTENBURG: Anat. Rec. 124, 255—256 (1956).
[4] —, and G. E. PALADE: J. biophys. biochem. Cytol. 3, 577—588 (1957).
[5] — — J. biophys. biochem. Cytol. 6, 163—169 (1959).
[6] BURSTONE, M. S.: J. Histochem. Cytochem. 2, 88—94 (1954).
[7] — J. Histochem. Cytochem. 7, 112—122 (1959).
[8] — J. nat. Cancer Inst. 18, 167—172 (1957).
[9] — J. nat. Cancer Inst. 21, 523—539 (1958).
[10] CREVIER, M., and L. F. BÉLANGER: Science 122, 556 (1955).
[11] DUVE, C. DE: In: Subcellular particles (T. MIHYASHI ed.), p. 128—158. New York: Ronald Press 1959.
[12] ERÄNKÖ, O.: Acta anat. (Basel) 16 Suppl. 17, 3—60 (1952).
[13] FREEMAN, J. A., and B. O. SPURLOCK: J. Cell Biol. 13, 437—443 (1962).
[14] FRIEDMAN, O. M., and A. M. SELIGMAN: J. Amer. chem. Soc. 72, 624—625 (1950).
[15] GOLDFISCHER, S., E. ESSNER, and A. B. NOVIKOFF: J. Histochem. Cytochem. 12, 72—95 (1964).
[16] GOMORI, G.: Int. Rev. Cytol. 1, 323—335 (1952).
[17] — J. Lab. clin. Med. 35, 802—809 (1950).
[18] — Microscopic Histochemistry; Principles and Practice. Chicago, Ill.: Chicago Univ. Press 1952.
[19] — Stain Technol. 25, 81—85 (1950).
[20] HANKER, J., A. SEAMAN, L. WEISS, H. UENO, H. DMOCHOWSKI, L. KATZOFF, and A. M. SELIGMAN: J. Histochem. Cytochem. (in press) (1964).
[21] — — — — — — H. STORM, H. WASSERKRUG, L. KATZOFF, and A. M. SELIGMAN: Science (in press) (1964).
[22] — L. WEISS, H. DMOCHOWSKI, L. KATZOFF, and A. M. SELIGMAN: J. Histochem. Cytochem. (in press) (1964).
[23] HOLT, S. J.: Exp. Cell Res., Suppl., 7, 1—27 (1959).
[24] —, and R. M. HICKS: J. biophys. biochem. Cytol. 11, 47—66 (1961).
[25] — J. Histochem. Cytochem. 4, 541—552 (1956).
[26] —, and R. F. J. WITHERS: Nature (Lond.) 170, 1012—1014 (1952).

[27] KARNOVSKY, M. J.: J. biochem. biophys. 11, 729—732 (1961).

[28] —, and K. HUG: J. Cell Biol. 19, 255—260 (1963).

[29] KOMMA, D. J.: J. Histochem. Cytochem. 11, 619—623 (1963).

[30] LEHRER, G. M., and L. ORNSTEIN: J. biophys. biochem. Cytol. 6, 399—406 (1959).

[31] MÖLBERT, E., F. DUSPIVA, and O. H. v. DEIMLING: J. biophys. biochem. Cytol. 7, 387—390 (1960).

[32] MOOG, F.: J. cell. comp. Physiol. 22, 223—231 (1943).

[33] NACHLAS, M. M., D. T. CRAWFORD, T. P. GOLDSTEIN, and A. M. SELIGMAN: J. Histochem. Cytochem. 6, 445—456 (1958).

[34] — T. P. GOLDSTEIN, D. H. ROSENBLATT, M. KIRSCH, and A. M. SELIGMAN: J. Histochem. Cytochem. 7, 50—65 (1959).

[35] — B. MONIS, D. ROSENBLATT, and A. M. SELIGMAN: J. biophys. biochem. Cytol. 7, 261—264 (1960).

[36] — W. PRINN, and A. M. SELIGMAN: J. biophys. biochem. Cytol. 2, 487—502 (1956).

[37] —, and A. M. SELIGMAN: J. nat. Cancer Inst. 9, 415—425 (1949).

[38] — K. C. TSOU, E. DE SOUZA, C. S. CHENG, and A. M. SELIGMAN: J. Histochem. Cytochem. 5, 420—436 (1957).

[39] — A. C. YOUNG, and A. M. SELIGMAN: J. Histochem. Cytochem. 5, 565—583 (1957).

[40] NOVIKOFF, A. B., and E. ESSNER: Amer. J. Med. 29, 102—131 (1960).

[41] OGAWA, K., and Y. SAITO: J. Electron Micr. (Japan) 12, 68—71 (1963).

[42] PEARSE, A. G. E.: Histochemistry, Theoretical and Applied. p. 442—443. Little Brown & Co. 1960.

[43] — Int. Rev. Cytol. 3, 329—358 (1954).

[44] PUGH, D., and P. G. WALKER: J. Histochem. Cytochem. 9, 105—106 (1961).

[45] RUTENBURG, A. M., and A. M. SELIGMAN: J. Histochem. Cytochem. 3, 455—470 (1955).

[46] SABATINI, D. D., K. BENSCH, and R. J. BARRNETT: J. Cell Biol. 17, 19—58 (1963).

[47] — F. MILLER, and R. J. BARRNETT: J. Histochem. Cytochem. 12, 57—71 (1964).

[48] SEAMAN, A., J. HANKER, H. STORM, H. DMOCHOWSKI, L. KATZOFF, and A. M. SELIGMAN: J. Cell Biol. (in press) (1964).

[49] SEAMAN, A. R., J. S. HANKER, and A. M. SELIGMAN: J. Histochem. Cytochem. 9, 596—597 (1961).

[50] SEDAR, A. W., and C. G. ROSA: Anat. Rec. 130, 371 (1958).

[51] SELIGMAN, A. M., H. H. CHAUNCEY, and M. M. NACHLAS: Stain Technol. 26, 19—23 (1951).

[52] —, and L. H. MANHEIMER: J. nat. Cancer Inst. 6, 427—434 (1949).

[53] —, and A. M. RUTENBURG: J. Amer. chem. Soc. 72, 3214—3216 (1950).

[54] — Sinai Hosp. J. (Baltimore) 5, 90—102 (1956).

[55] SHELDON, H., H. ZETTERQUIST, and D. BRANDES: Exp. Cell Res. 9, 592—596 (1955).

[56] SHNITKA, T. K., and A. M. SELIGMAN: J. Histochem. Cytochem. 9, 504—527 (1961).

[57] TSOU, K. C., C. S. CHENG, M. M. NACHLAS, and A. M. SELIGMAN: J. Amer. chem. Soc. 78, 6139—6144 (1956).

[58] WACHSTEIN, M., and E. MEISEL: J. Histochem. Cytochem. 8, 317—318 (1960).

[59] WALKER, D. G., and A. M. SELIGMAN: J. biophys. biochem. Cytol. 9, 415—427 (1961).

[60] — — J. Cell Biol. 16, 455—469 (1963).

Dr. A. M. SELIGMAN,
Sinai Hospital of Baltimore, Belvedere Avn. at Greenspring,
Baltimore 15, Maryland/USA

University of London

The Future of Academic and Applied Enzyme Histochemistry

By

A. G. EVERSON PEARSE

With 3 Figures in the Text

Enzyme Histochemistry is already divided for me, by the title of my address, into two parts. In both cases it is impossible to discuss the future without first considering the past and the present. These contain the structure on which the future has to be built. You will have heard something already, from Dr. ARNOLD SELIGMAN'S stimulating lecture, of recent events in several fields of Enzyme Histochemistry. Moreover, you are about to hear, from Dr. RUSSELL BARRNETT, something more on the vastly important subject of electron enzyme cytochemistry. Sandwiched between two such eminent speakers I have to try somehow to link the two together. It is fortunate for me that they are both very good friends of many years' standing.

I will make no apology for starting my talk with my favourite dictum from the works of RUDOLPH VIRCHOW (1858) — *"If we would serve Science we must extend her limits, not only as far as our own knowledge is concerned, but in the estimation of others"*. The sting of these remarks is in the tail for, as KUBIE (1953) lucidly expressed it, "the scientist . . . stakes his all on professional achievement and *recognition*". According to EIDUSON (1962), moreover, "Scientists are idols-oriented". That New American, GEORGE GOMORI, whom we honour in this session today, was a type of great rarity in scientific circles. We may well take him as our model. All too short as it was, he acquired plentiful recognition and appreciation during his lifetime, and we recognize him as the true founder of the science of Enzyme Histochemistry. Posterity will do likewise.

Let me make a start, before the time comes to finish, by recalling to you then the developmental phases of enzyme histochemistry.

The History of Enzyme Histochemistry

Before 1900 there were only a few methods (Table 1), and up to 1939 still only a handful, which were concerned almost entirely with the localization of oxidases. Enzyme histochemistry can truly be considered to date from the year 1939. In this year GEORGE GOMORI first described his method for alkaline phosphatase. We should not forget, however, that an essentially similar method was independently described in the same year by Professor TAKAMATSU, in Japan.

Table 1. *Available enzyme methods*

Before 1900		Before 1939	
Peroxidases	Guaiac	Peroxidases	Benzidine and zinc-leuco
Indophenol		Phenoloxidases	α-naphthol
(Catochrome)		DOPA-oxidase	Bloch-Laidlaw
oxidase		Myrosulphatase	Peche

It was not so much the introduction of the new method but of a *new principle* which distinguished GOMORI'S contribution to the science of Enzyme Histochemistry. In his hands, and in the hands of others, the principle gave rise to a whole series of what may now be called the *metal precipitation techniques*.

Somewhat later (in 1944), with what I once described as "a stroke of genius", MAUD MENTEN and her associates JUNGE and GREEN, produced a *second principle* with the first of the now very large series of *coupling azo dye methods* for hydrolytic enzymes.

Later still, in 1948 and 1949, a *third principle* came into being with the use of *tetrazolium salts* as hydrogen acceptors in dehydrogenase reactions. It is difficult to detect the precise origins of tetrazolium histochemistry, but it is in the hands of ARNOLD SELIGMAN and his associates that the greatest steps forward have been made.

The situation in 1953, as a result of the efforts to which I have just referred, was that some 26 enzymes could be demonstrated *in situ* in tissue sections or smears (Table 2). Shortly afterwards (in 1955), TAKEUCHI and his collaborators finally produced an acceptable variant of the methods for phosphorylase initiated earlier by YIN and SUN (1947) and COBB (1949). With this a *fourth principle* came into being, which we can perhaps describe briefly as the *synthetic principle*.

A fifth principle, practised fitfully and somewhat crudely until 1957, we now associate particularly with the name of ROGER DAOUST (1957, 1959). This is the *substrate film principle*. Already a respectable number of enzymes can be demonstrated and the degree of accuracy of localization is surprisingly good (Fig. 1). You will hear later in the Congress from Dr. DAOUST about improvements in his technique for ribonuclease.

Table 2. *Available enzyme methods*

Up to 1953	
Alkaline Phosphatases	Cholinesterases
Acid Phosphatases	Phenolsulphatase
5-Nucleotidase	β-Glucuronidase
Aldolase	β-Galactosidase
ATP'ase	β-Glucosidase
Glucose-6-Phosphatase	Amine Oxidase
Phosphorylase	Dehydrogenases (5)
Phosphamidase	Diaphorases (2)
Lipase	Carbonic anhydrase
Esterase (non-specific)	
A.S. Esterase	
Chloroacetylesterase	

The sixth principle, a very interesting one, stems from the original methods devised by TAKAMATSU and WADA (1954) for the demonstration of proteinases. These methods have given rise to the *marked substrate principle* which has already been extended to the demonstration of amylase (TAKAMATSU and YOSHINO, 1954; YOSHINO, 1960).

A number of methods do not fit into any of the 6 principles, such as DOPA oxidase (Pigment-Forming) and cytochrome oxidase (Oxidation Coupling). Moreover, some principles, such as *Intramolecular Rearrangement*, have not yet been entirely successful, while others *(Fluorescent Antibody)*, really belong elsewhere.

In the intervening years these 6 principles (Table 3) were applied, three of them particularly intensively, so that by 1960 when the second edition of my book appeared, there were methods available for about 45 enzymes (Table 4). Today there are certainly over 100 methods for no less than 60 enzymes.

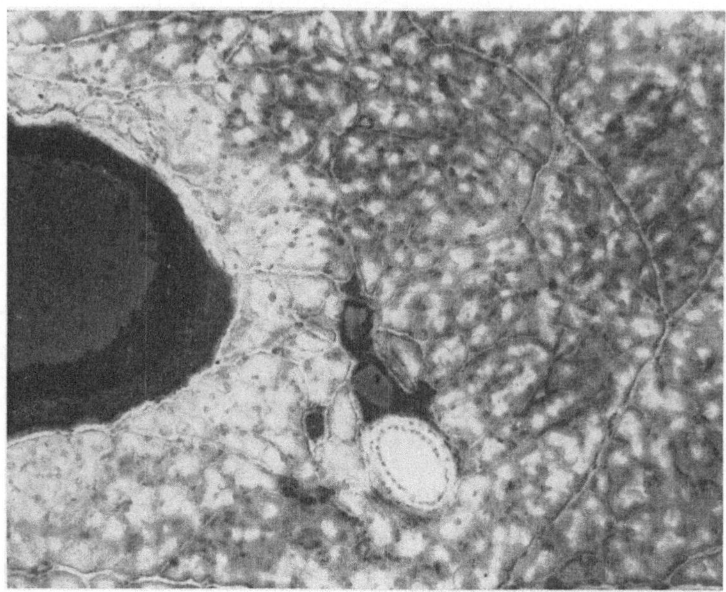

Fig. 1. Rat Pancreas. Fresh frozen cryostat section mounted on starch film and stained after incubation. Amylase activity (white areas) is greatest in the pancreatic duct and in the acini. Stained with iodine. × 220

Table 3. *The 6 principles of enzyme histochemistry*

1) Metal Precipitation	4) Synthetic
2) Coupling Azo Dye	5) Substrate Film
3) Tetrazolium Reduction	6) Marked Substrate

Table 4. *Enzyme methods in 1960*

Alkaline phosphatases	Amine oxidase
Acid phosphatases	Choline oxidase
Desoxyribonuclease	D-Amino-acid oxidase
5-Nucleotidase	DPNH-diaphorase
Adenosine triphosphatase	TPNH-diaphorase
Aldolase	Succinate dehydrogenase
Glucose-6-phosphatase	Glucose-6-phosphate dehydrogenase
Phosphorylase	6-Phosphogluconate dehydrogenase
Amylo-1,4-1,6-transglycosidase	β-Hydroxybutyrate dehydrogenase
Pyrophosphatase	Alcohol dehydrogenase
Phosphamidase	Lactate dehydrogenase
Lipases	Isocitrate dehydrogenase (TPN)
Carboxylic acid esterases	Isocitrate dehydrogenase (DPN)
"AS-type" esterases	Glutamate dehydrogenase
Acetylcholinesterases	α-Glycerophosphate dehydrogenase
Pseudocholinesterases	Malate dehydrogenases
Phenol sulphatases	α-Hydroxysteroid dehydrogenase
β-Glucuronidase	β-Hydroxysteroid dehydrogenase
β-Galactosidase	Cytochrome oxidase
β-Glucosidase	Tyrosinase
Acetyl-β-glucosaminidase	Leucine aminopeptidase
Carbonic anhydrase	Cathepsins
Peroxidase	Catalase

We must temper our pride in the advances of academic Enzyme Histochemistry by recalling that some 700 enzymes have been described and that there are probably at least 350 important ones in mammalian tissues.

The Future of Academic Enzyme Histochemistry

So much for the past. It is obvious that the future depends, in the first place, on the development of new principles and of new techniques within these and existing principles. Coverage must be extended to those enzyme groups not at present represented (Table 5), particularly to the many subgroups of transferring enzymes, to the phosphokinases, and to the so-called ligases or synthetases.

It is essential that properly critical studies should continue to be carried out on existing techniques, some of which have too lightly been accepted as specific, and as accurate in localization. In this context perhaps I may refer to the studies of JOHAN AHLQVIST (1963) on lipid partition of substrates. These have important

Table 5. *Main enzyme groups and subgroups*

Hydrolases	Tansferases	Adding enzymes
Peptidases	*Oxidases*	Synthetases
Deamines	*Reductases*	Enzymes adding groups
Carboxylic esterases	Transaminases	to double bonds:
Phosphatases	Transpeptidases	Fumarase
Sulphatases	Phosphokinases	Aconitase
Glycosidases	Transphosphorylase	*Aldolase*
Pyrophosphatases	Transmethylases	*Carbonic anhydrase*
Decarboxylases	Transacylases	
	Transglycosylases	

implications for us all. I would like to refer also to the work of ZDENEK LOJDA and his associates, on the azo coupling techniques for acid phosphatase, which have underlined and supported the efforts of MARVIN BURSTONE, TIBOR BARKA and others to improve the localization of acid phosphatase.

There are, however, two additional developments in academic Enzyme Histochemistry whose importance transcends all others. These are:

1. Increased employment of biochemical or microchemical controls.

2. Quantitation.

The two are to some extent interrelated. One of the most important control disciplines is the zymogram technique which we usually associate with the starch gel method and the names of HUNTER and MARKERT (1957) although many other media as well as starch gels can be used for separation of enzymes. The second type of control to which I refer is the assay or microassay, carried out with appropriate biochemical characterization of any enzyme whose histochemical localization has been claimed. As an example one may cite the work of ELIZABETH PATTERSON (PATTERSON et al. 1963) and the studies carried out by SYLVÉN and his colleagues (SYLVÉN and BOIS, 1962, 1963; SYLVÉN and SNELLMAN, 1964) on the nature of the peptidase hydrolyzing leucyl-β-naphthyl-amides which is loosely and erroneously called leucine aminopeptidase by histochemists. Homogenization studies with E. M. control, carried out in parallel with histochemical methods, are to be regarded as an essential ancillary technique in advancing our understanding of enzyme activity at the level of intracellular organelles. As an excellent model for such studies I may quote you the work of GUSTAV DALLNER (1963) on the structural and enzymic organization of the microsomes. Perhaps you may think such work on homogenates should be carried out by biochemists — I will tell you that you should *be* biochemists and biochemists should likewise be histochemists. It is no longer possible to tolerate the rift between these two sciences.

There must be at all times the closest possible relationship between histo-enzymology and biochemical knowledge and practice. There is often a quite unwarranted delay in the application of known and described activators and inhibitors to histochemical techniques, for instance.

Quantitation. The two main methods of quantitative histochemistry have both been applied to the enzyme field. It is permissible to hope that the justly famous optical industry of our host country will provide us, perhaps even at this Congress, with the automatic recording compensating cytophotometer which will solve our problems. Until such time, it is necessary to conclude that quantitative enzyme histochemistry must be carried out by combinations of microdissection and microassay techniques. These are associated particularly with the names of LINDERSTRØM-LANG, LOWRY and HYDÉN. We shall hear something more of these techniques on Friday, and I am eagerly looking forward to the session headed by Dr. OLIVER LOWRY.

The Future of Applied Enzyme Histochemistry

I could probably save my breath, and cover the subject quite adequately, by referring you to the contributions recorded in the Abstracts for this Congress. No fewer than 129 papers, or 44% of the total, fall within the province to which I have given the major part of my time during the past two decades.

Applied Enzyme Histochemistry has several broad subdivisions. The first of these, which may be called *Morphological*, embraces the use of methods purely as markers for cells or tissues. As an example one may quote the successful use of the alkaline phosphatase reaction, by Hertig and his associates (PINKERTON et al. 1961) in order to trace the developing sex cells in the primitive gonad. In embryology morphological methods are particularly suitable and it is possible, for instance, to detect the developing pronephros in the mesenchyme of the intermediate cell mass by means of several reactions; the best of these, perhaps, is acid phosphatase (Fig. 2). One can use this same reaction to trace the developing Purkinje cells in the cerebellum. The specific cholinesterase technique of KOELLE (1954) has become, in the hands of Drs. SHUTE and LEWIS (1963) a magnificant morphological tool for the mapping of pathways in mammalian brain. The piling up of the enzyme on the cell body side of a lesion has enabled them to discover the polarity and distribution of cholinesterase-containing tracts in the brain.

A second subdivision, really part of the first, may be called *Structural Assay.* Here enzymes known to be associated with a single cell organelle or component are studied, solely in order to detect alterations in the condition or distribution of such components. I refer to the so-called *mitochondrial assay* using techniques for bound dehydrogenases, to *lysosomal assay*, using the acid phosphatase technique,

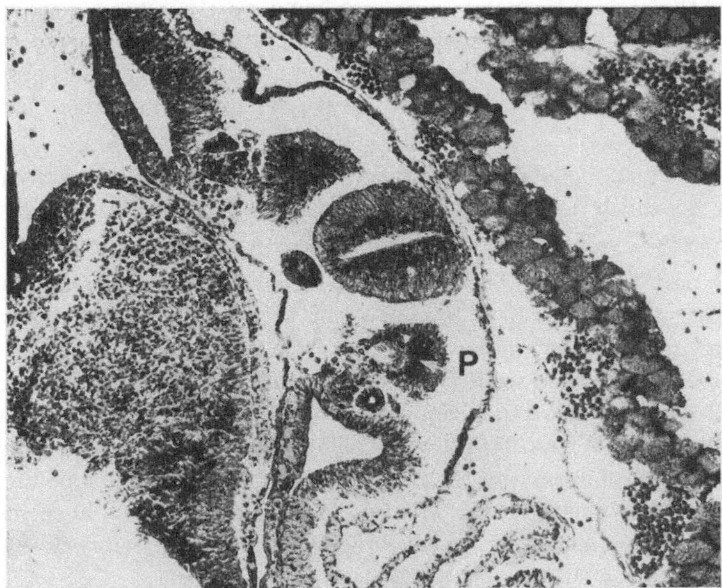

Fig. 2. 48-hour Chick Embryo. Fresh frozen cryostat section. Right, yolk sac; centre, developing embryo. Strong enzyme activity in the pronephric duct and in the developing pronephros (*P*). Acid Phosphatase. ×100

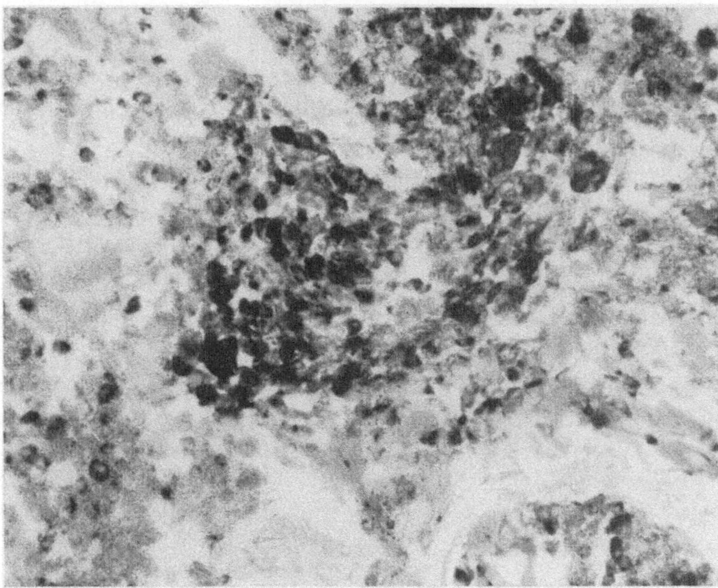

Fig. 3. Transitional cell carcinoma of human bladder. A very strong E. 600 resistant esterase is present in the tumour cells. Indoxyl esterase. ×145

to *microsomal assay*, with glucose-6-phosphatase, and to *membrane assay*, with ATPase or even alkaline phosphatase.

I cannot even begin to ennumerate the applied studies already made in these fields but it is notable that in the past year the electron microscope has begun to appear on the scene. In many cases it is being used purely as a micromorphological control for light microscopic observations. The obvious, and indeed the essential, development here is towards electron cytochemistry. This process I will leave for today, but not otherwise, entirely in the capable hands of Dr. Barrnett.

Before I finish there are two other subdivisions of applied Enzyme Histochemistry to which I should like to refer. These are the *Diagnostic* and the *Topobiochemical*.

In the past 5 years a great number of isolated observations on the enzyme cytochemistry of tumours have been made by pathologists all over the world. No truly coherent pattern would have emerged from these observations if it were not for the more concentrated studies of a few institutions. I refer, particularly, to the work of Roelof Willighagen and his associates in Leiden (Willighagen and Planteydt, 1959; Willighagen et al. 1963). Personally, I have been especially interested in the esterases and the E. 600-resistant esterases of tumours and their diagnostic implications (Fig. 3). The future of tumour enzyme histochemistry is clearly a bright one and I would encourage all, who can possibly do so, to enter this field.

Let me finish with brief reference to the truly functional, or Topobiochemical aspect of applied Enzyme Histochemistry. This is the use of techniques not only to demonstrate the existence, and thus the morphology, of lesions in the tissues, but in order to discover their biochemical nature. I would like to draw your attention to recent work by Niles, Bitensky, Chayen and Cunningham at the Royal College of Surgeons in London (1964), on the analysis of myocardial dysfunction after open heart surgery. It is unlikely that the biochemical lesions shown by their work could possibly have been demonstrated in any other way than by means of enzyme histochemistry.

When we understand more of the true significance of the techniques we are using we shall be able to make more and more valuable contributions to the elucidation of biochemical mechanisms in health and disease.

In 1958 I ended my contributions to the Virchow Centenary Meeting of the British Association of Clinical Pathologists by referring to two other quotations as well as the one with which I started this address. I would like to refer to them once again: — *"Cytology cannot be content to go on indefinitely finding new visible parts of the cell . . . what it needs is to discover methods that will describe the organization of cellular events"* — J. Z. Young (1956). *"New methods must be invented that will be sensitive enough to detect early functional changes"*. — G. R. Cameron and Muzaffar Hasan (1958).

The new methods referred to by J. Z. Young and by Sir Roy Cameron, are with us today. They are the methods of the Science of Functional or Enzyme Histochemistry, begun by George Gomori and carried on by others who are present here at this Congress.

Their application in the future cannot fail to add increasing lustre to the discoveries of the past and to open up new horizons in all the fields to which they are intelligently applied.

References

AHLQVIST, J.: Acta path. microbiol. scand. **59**, 171 (1963).
BARKA, T.: Nature (Lond.) **187**, 248 (1960).
BURSTONE, M. S.: J. Histochem. Cytochem. **9**, 146 (1961).
CAMERON, G. R., and S. MUZAFFAR HASAN: J. Path. Bact. **75**, 333 (1958).
COBB, J.: Anat. Rec. **103**, 531 (1949).
DALLNER, G.: Acta path. microbiol. scand., Suppl. 166 (1963).
DAOUST, R.: Exp. Cell. Res. **12**, 303 (1957).
— Exp. Cell. Res., Suppl. **7**, 40 (1959).
EIDUSON, B. T.: Scientists. Their psychological world. New York: Basic Books 1962.
HUNTER, R., and G. MARKERT: Science **125**, 1294 (1957).
KOELLE, G. B.: J. comp. Neurol. **100**, 211 (1954).
KUBIE, L. S.: Amer. Scientist **41**, 596 (1953).
LOJDA, Z., B. VECEREK, and H. PELICHOVA: Histochemie **3**, 428 (1964).
NILES, N. R., L. BITENSKY, J. CHAYEN, G. J. CUNNINGHAM, and M. W. BRAIMBRIDGE: Lancet **1964** I, 963.
PATTERSON, E. K., S.-H. HSIAO, and A. KEPPEL: J. biol. Chem. **238**, 3611 (1963).
PINKERTON, J. H. M., D. G. McKAY, E. C. ADAMS, and A. T. HERTIG: Obstet. and Gynec. **18**, 152 (1961).
SHUTE, C. C. D., and P. R. LEWIS: Nature (Lond.) **199**, 1160 (1963).
SYLVÉN, B., and I. BOIS: Histochemie **3**, 65 (1962).
— — Histochemie **3**, 341 (1963).
—, and D. SNELLMAN: Histochemie **3**, 484 (1964).
TAKAMATSU, H., and K. WADA: Trans. Soc. path. jap. **43**, 204 (1954).
— and F. YOSHINO: Trans. Soc. path. jap. **43** (1954).
TAKEUCHI, T., and H. KURIAKI: J. Histochem. Cytochem. **3**, 153 (1955).
VIRCHOW, R.: (1858) Cellular pathology. English translation by F. CHANCE. London: Churchill 1860.
WILLIGHAGEN, R. G. J., R. O. VAN DER HEUL, and TH. G. VAN RIJSSEL: J. Path. Bact. **85**, 279 (1963).
—, and H. T. PLANTEYDT: Nature (Lond.) **183**, 263 (1959).
YIN, H. C., and C. N. SUN: Science **105**, 650 (1947).
YOSHINO, F.: Proc. Jap. Hist. Ass. **1**, 94 (1960).
YOUNG, J. Z.: Endeavour **15**, 5 (1956).

Dr. A. G. E. PEARSE,
Postgraduate Medical School of London, Ducane Road Shepherd's Bush,
London, W 12 (England)

Department of Botany, University of California, Berkeley, California

Progress and Results in the Histochemistry of Plants

By

William A. Jensen

With 4 Figures in the Text

Introduction

The histochemistry of plants is not a new area of research. Begun in 1825 by Raspail, the French botanist, hailed as the father of histochemistry, the field has a respectable beginning. But the interest of botanists in histochemistry has not been consistent over the past 150 years. The peak in botanical histochemistry was between 1880 and 1920. Before and during this period there were few papers on plant structure which did not use some histochemical procedures to establish the composition of the parts. Between 1920 and 1950 interest declined in the histochemistry of plants by most botanists. As this corresponds with the period of the awakening of interest on the part of workers using animal tissue, many zoologists have the impression that botanical histochemistry either is non-existent or is a very recent development. Within the past decade there has been renewed interest on the part of botanists in histochemistry, and at present there is an ever increasing use of histochemical methods in botanical research.

There are two reasons for the current revival of interest in the histochemistry of plants. One is the improvement of histochemical methods, particularly in their application to plant tissue. The second is in the types of problems now of interest to botanists. Discussing these two points in order, I shall first consider the development of techniques.

Advances in the Techniques of Botanical Histochemistry

One of the major obstacles in applying many histochemical procedures to plant tissue has been the difficulty of preserving and preparing the tissue. There are two characteristics of plant cells which cause difficulty. One is the wall, while the second is the vacuole.

The problems caused by the presence of rigid cell walls can be demonstrated by the history of the application of freeze-drying to plant tissue. The introduction of modern freeze-drying to histochemistry of animal tissue was the result of work by Gersh [10], published in 1932. This fact is well known, yet it is frequently overlooked that in 1934, only two years later, the first paper on the freeze-drying of plant tissue was published by Goodspeed and Uber [15]. The apparatus they designed was a marked improvement and was widely copied by workers using animal tissue. Yet few, if any, botanists other than Goodspeed and Uber attempted to apply the method to plant tissue. The reason was that plant tissue, while it could be successfully frozen and dried (although excess drying times were required), could not be infiltrated with paraffin. What infiltration did take place was partial or limited to certain tissues. No further attempts were made to freeze-dry plant tissue until 1954 [18]. After twenty years, while the drying time was greatly reduced, the infiltration was little improved.

The apparatus used by GOODSPEED and UBER and by JENSEN used high vacuum drying. In 1954 JENSEN [19] introduced the idea of using a stream of moving gas for the freeze-drying of plant tissue. The tissue not only dried faster but more of it became infiltrated with paraffin. Paraffin infiltration was still far from satisfactory and no explanation was available for the operation of the machine. The major breakthrough came with the work of BRANTON and JACOB-SON [9]. They further refined the moving gas apparatus and introduced the concept of a controlled moisture content for the gas. The idea of deliberately adding moisture to the dehydration chamber appears counter to all the data obtained from animal tissue, but it results not only in decreased drying time, but in an increase of paraffin infiltration.

The explanation for these improvements is in the nature of the plant cell, particularly the wall. The wall is a complex structure composed of cellulose, hemicellulose, pectic substances, and, frequently, lignin as well as several other materials. The cellulose component is organized in the form of long microfibers with the other substances located as a matrix between them. The wall is usually freely permeable to all types of compounds and large molecules readily diffuse through it. BRANTON and JACOBSON found that in tissue dried in a high vacuum of extremely dry gas, the texture of the wall changed and it became imperme-able even to the passage of water molecules. This change resulted in a decrease in water loss during drying and interfered with paraffin infiltration. When low amounts of water were present in the gas (2.6 to 4.0 μg H_2O/liter), the original wall structure was retained. In this condition, water readily diffused from the tissue and paraffin into it.

The Branton-Jacobson freeze-dry apparatus has been used in my laboratory for the past five years on a wide range of tissue. The results have been uniformly excellent. No advantage, however, is gained in using this approach with most animal tissue. The cell walls are the source of the difficulty, and where they are not present other freeze-dry systems are more effective.

The availability of freeze-dried plant tissue means that many problems in plant physiology and morphogenesis can be attacked which have not been open to botanists. These problems include the mechanisms of translocation, the mechanism of floral induction, the initiation of secondary roots, embryogenesis, and many others.

The second feature of the plant cell which causes difficulty in the preparation of the tissue is the vacuole. Many plant cells contain a large central vacuole. The main chemical component of this vacuole is water, but high concentrations of sugar, minerals, amino acids and proteins may be present. The vacuole is consequently extremely difficult to freeze rapidly. This causes trouble in freeze-drying and even more if the tissue is cut on a freezing microtome. The use of the freezing microtome has been of little value to the botanist. The advent of the cold microtome, where the tissue is frozen as for freeze-drying and then sectioned at a temperature well below freezing, should be of great importance to the histochemist working with plant tissue. While the procedure has been used in only a few cases with plant tissue, much wider application can be expected.

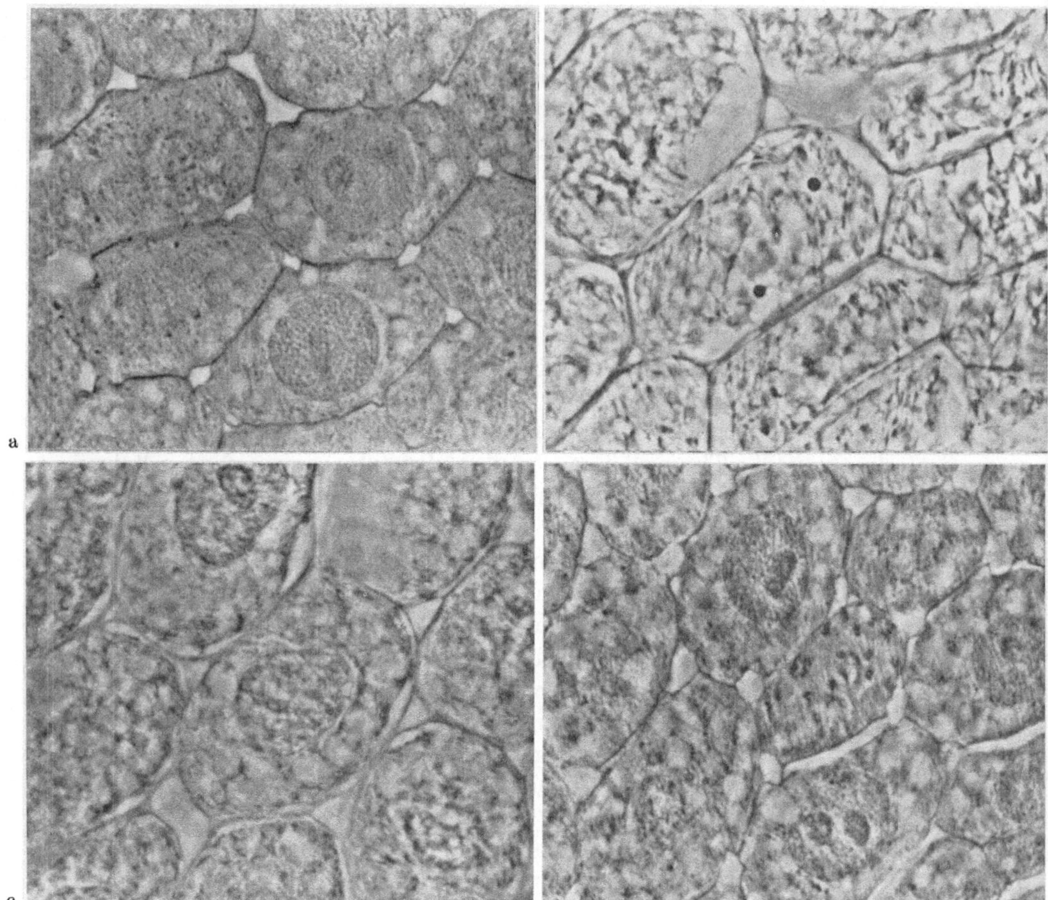

Fig. 1a—d. Freeze-dried onion root tip sections spread on slides by four different methods: a Pressed on the slide dry; b Floated on with water; c Floated on with 95% alcohol; and d Floated on with 4% formalin. In each case, the paraffin was removed by a minimum treatment with xylene and cover slips mounted with a phase contrast medium. Photographed in phase contrast

How wide this application will be, however, remains to be seen. The tremendous vacuoles present in some cells or organs, the central vacuole in the embryo sac, for example, may cause problems in adapting the cold microtome procedures to plant tissue. Clearly no one method of handling plant tissue will solve all of the various requirements for histochemical procedures.

A problem with the use of both freeze-dried and freeze-sectioned material is the lack of information on how best to handle the tissue when it is obtained. Thus, how freeze-dried tissue is placed on the slide has a great effect on tissue preservation and, presumably, on enzyme activity. In Fig. 1, the same freeze-dried tissue (onion root tips) were placed on slides, the paraffin removed and examined with phase contrast. The sections which were pressed directly on the slides (Fig. 1a) showed the best preservation. Water (Fig. 1b) caused a striking and deleterious change in the tissue. Alcohol (Fig. 1c) and formalin (Fig. 1d) caused less change but still had marked effects. Similar problems may be expected

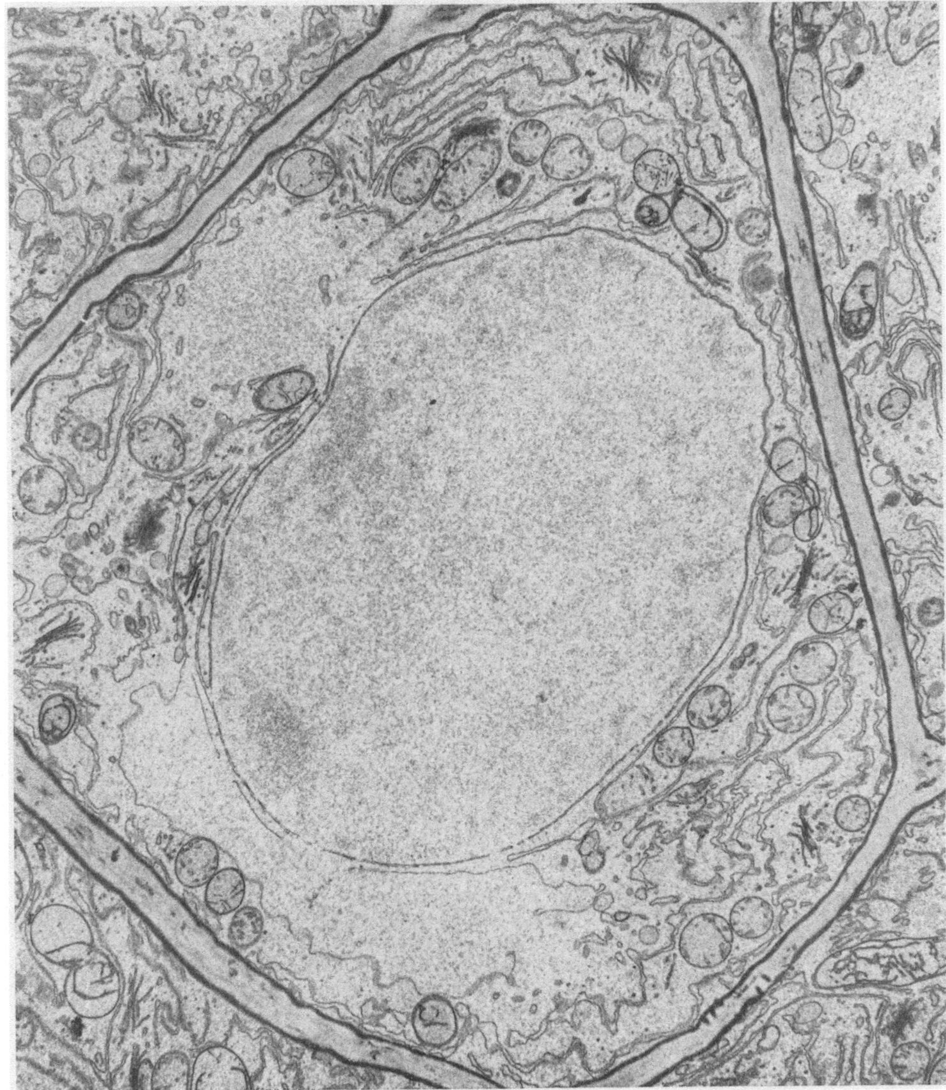

Fig. 2. A nucellar cell from cotton fixed 12 hours at 4° C with 2% KMnO₄. Tissue stained with uranyl nitrate and embedded in epon

to occur in the use of frozen sections. The danger is in assuming that the methods used with success with animal tissue can be directly applied to plants. In this area of the preparation of tissue, botanical histochemistry lags seriously and careful studies are urgently needed if the histochemistry of plants is to progress.

Before leaving the question of the preparation of plant tissue, the problem of the fixation of plant material for electron microscopy should be discussed. Again, plant and animal tissue respond differently to the same treatment. Thus, while osmium tetroxide has become the standard fixative for animal tissue, it has not been widely used on plants, although some successful preparations have

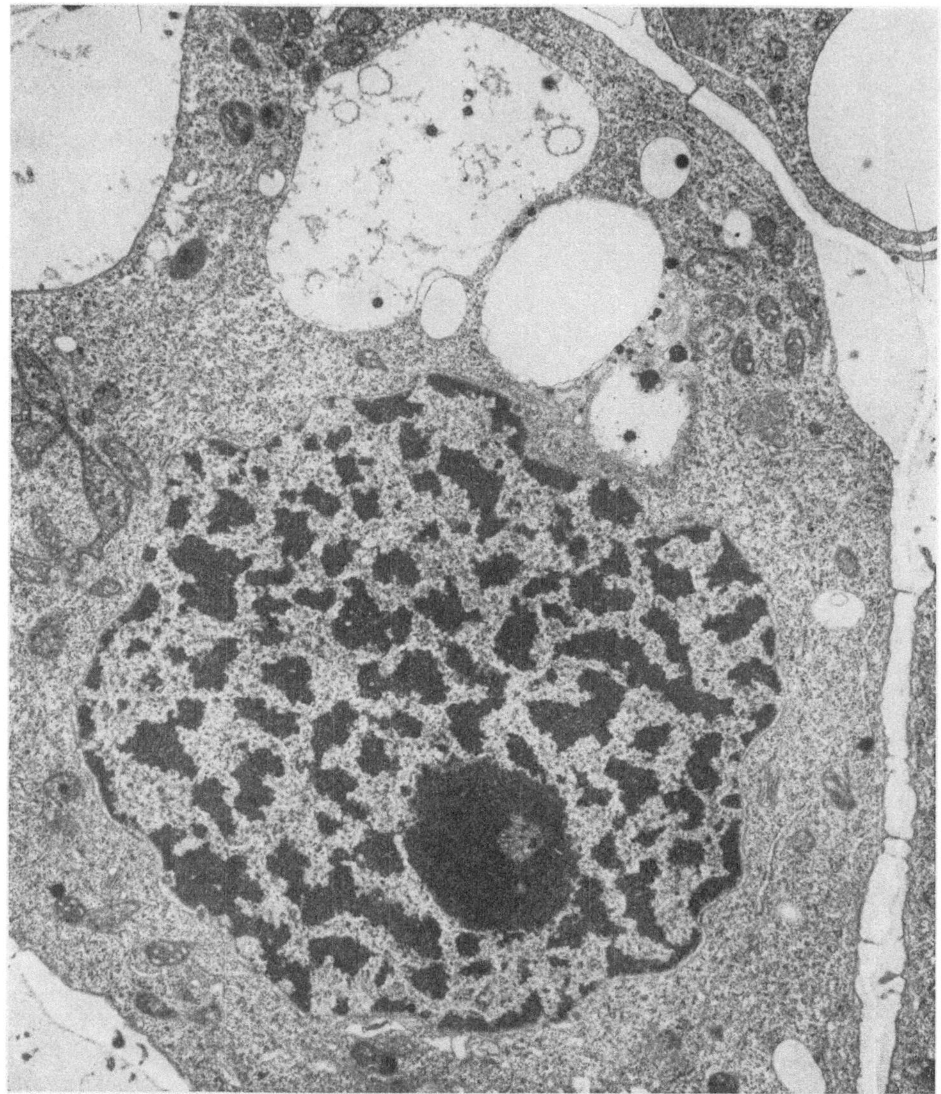

Fig. 3. A nucellar cell from cotton fixed 12 hours at 4° C with 6% gluteraldehyde (pH 6.8), washed, and subsequently fixed in 2% OsO₄ at 4° C for 12 hours. Tissue stained with uranyl nitrate and embedded in epon

been made [27]. The most widely used fixative for plant tissue is potassium permanganate (KMnO₄) [25]. With this fixative the membrane systems of the cell are revealed in dramatic contrast, but the ribosomes are not present, having been removed by the fixative (Fig. 2). Recently, the introduction of glutaraldehyde as a fixative has yielded extremely promising results with plant tissue [23]. Tissue treated first with gluteraldehyde and then with osmium has been well preserved (Fig. 3).

The advantage of this procedure is that after the gluteraldehyde fixation the tissue can be treated with ribonuclease or a substrate for an enzyme location, and then fixed with osmium. Limited use has been made of this procedure to date, but there is every reason to believe it will be widely employed in the future.

The histochemical methods available to the botanist have been largely developed by zoologists and have seen their greatest use with animal tissue. This is particularly true for the protein and enzyme localization procedures. In recent decades few original procedures have been developed using plant tissue. One exception to this has been autoradiograph, in which much of the preliminary research was carried out on plant tissue.

The fact that a procedure works on animal tissue is no assurance that it will be applicable to plant tissue. An excellent example of this is the original Gomori method for alkaline phosphatase. The substrate, glycerophosphate, provides excellent localization in the case of animal tissue but is completely negative in the case of plant tissue. Other enzyme procedures show similar difficulties. Many methods, however, developed for animal tissues can be applied directly to plant tissue with considerable effect.

Two procedures are the periodic acid-Schiff (PAS) reaction for carbohydrate and the Azure B reaction for nucleic acids. The PAS reaction has been particularly useful in studies on the cell wall as it is positive for all wall components. Lignin, because of the large numbers of aldehydes naturally present, gives a reaction directly with the Schiff reagent. The other wall components react after treatment with periodic acid. The PAS reaction can be combined with cell wall extraction procedures giving a basis for studies on wall composition [20]. The Azure B reaction gives excellent results for DNA and RNA in plant cells and has several interesting side benefits. One is that all components of the wall give a negative reaction with the exception of lignin, suberin, and callose. The first two give an intense blue-green color, as does DNA, while the last colors a cherry red. Thus, with RNA giving a purple-blue reaction, an Azure B preparation yields a large amount of information.

Applications of Histochemical Methods to Plants

Histochemical methods have been applied to many botanical problems and a wide of plant tissues. Much of this research has been reviewed recently [21, 33, 34] and will not be covered here. Within the past few years, however, there has been an increased interest in histochemical methods by a wide variety of botanists. This is the result of a mounting interest in understanding problems in terms of the morphology and biochemistry of the cells involved. Many of these problems are concerned with development and involve numbers of cells too small to be analyzed by conventional biochemical methods. Thus, the investigator turns to the methods of histochemistry. An example of this is the current research on the problem of floral induction.

Under the proper light conditions many plants can be induced to flower. When induction occurs, the shoot tip, which is a highly meristematic region regularly producing leaves, is converted to a floral meristem which produces the flower. This conversion is directed by a hormone which is formed in the leaves and moved to the shoot tip through the stem. For many years the problem has been studied on two separate levels. One was the study in purely physiological terms of the nature and production of the flower hormone. The second was the study in purely morphological terms of the development of the flower.

Recently, the question of the nature of the biochemical and physiological changes which occur in the cells of the stem apex during and after induction has become important. This necessitates examining the shoot tip in combined morphological and biochemical terms. However, the relatively small number of cells that are involved in the process are surrounded by large numbers of cells not active in the formation of the flower. These restrictions mean that histochemical methods become the methods of choice.

An example of the type of research beginning in this area is that of GIFFORD and co-workers [11—14]. They have studied the development of the vegetative tip and the early stages of flowering using microscopic histochemical methods for the nucleic acids and protein. Through the alkaline fast green procedure they were able to demonstrate [12, 14] a loss in stain capacity associated with the early stages of flowering. They have interpreted their results in terms of possible DNA binding of histones and the relation of such binding, or lack of it, to the process of cell differentiation.

There are many indications that such work is only the beginning of attempts to further explore the problem in histochemical terms. The correlation of such data to the ultrastructure of the cells involved appears the next logical step.

Plant embryology is another area which is receiving increased attention because of the advances in histochemical methodology. The field of plant embryology is rich in some areas of study and poor in others. Thus, while there is a wealth of data on the morphology of embryo formation and development, there is little known of the biochemistry of embryo growth in plants. There is no botanical equivalent of the biochemical embryology in animals.

A major reason for this lack of development is the difficulty of working with plant embryos. In all plants with true embryos, the embryo is invariably surrounded by parental tissue. The amount of such tissue is highly variable but the embryo is never free. Consequently, to study the biochemistry of the plant embryo, it must be isolated by dissection. This has proven too laborious a procedure to interest most workers using conventional methods. Through the use of histochemical methods, both quantitative and microscopic, the embryos can be analyzed either by isolating small numbers or by using fixed material. Recently, a third alternative has become possible. This is through the mass culture of isolated embryos [16, 31]. The method is still too recent to have been used widely, but it offers a rare opportunity to botanists to study plant embryos in bulk. In the meantime, histochemical methods are receiving constantly wider application in plant embryology and many problems will yield only to histochemical analysis.

The investigation by BELL and MÜHLETHALER [6] of the changes associated with fertilization in a fern *(Pteridium aquilinum)* is an example of some of the work being done. They examined nucleic acid (DNA) distribution during the formation of the egg and zygote using stain procedures, fluorescence methods, and autoradiography [3—8]. These data BELL and MÜHLETHALER correlated with ultrastructural changes in the cells. Finally, they used autoradiographs on the electron microscope level to study DNA distribution directly in terms of the fine structure of the system. They interpret their data as indicating the break-down of all cytoplasmic organelles in the egg before fertilization. New organelles, they believe, are formed by the nuclear membrane of the egg with DNA being transferred from the nucleus to the forming cell parts.

Studies of embryo development in flowering plants are also increasing in number [22, 28, 32]. Again a wide range of techniques have been employed to study the changes in composition and enzyme activity in the developing embryo. In some cases, the histochemical data has been correlated to the ultrastructure of the developing cells. I would like to illustrate this approach with work from my own laboratory on the cotton embryo.

One problem we have studied is the role of the synergids in the metabolism of the embryo sac and their possible relation to the growth of the pollen tube. The synergids are two large haploid cells closely associated with the egg. The egg plus the two synergids are frequently termed the egg apparatus and occupy one end of the embryo sac. When the pollen tube grows into the embryo sac, it enters one of the two synergids. This synergid disintegrates and the pollen tube enters the central portion of the embryo sac where the two male nuclei are discharged, one fusing with the egg nucleus to form the zygote, the second fusing with the two polar nuclei to form the primary endosperm nucleus.

The synergids have long been thought to be associated with the growth of the pollen tube [22] from morphological evidence, but little was known of their composition. When examined with the light microscope they have three prom-inent features. The first is the filiform apparatus. This structure occupies the third of the cell at the end adjacent to the parental tissue and is shaped like a many-fingered hand. The second feature is a large nucleus located near the middle of the cell, while the third feature is a series of vacuoles which fills the third of the synergid which projects into the embryo sac.

This picture of the synergid has been elaborated on by using histochemical methods (Fig. 4) and correlating the data collected with the fine structure of the cell as revealed by the electron microscope. The filiform apparatus has been found to be an elaborate proliferation of the wall of the synergid and its general composition has been established. The fact that the filiform apparatus is a wall means that the surface area of the plasma membrane is increased manyfold. This and data from the histochemical analyses suggest that the synergids are functioning as absorption centers for material entering the embryo sac.

The vacuoles also prove interesting, as they can be shown, through micro-incineration studies (Fig. 4 d), to contain substances high in ash content. While the ash cannot be identified, it is of interest as calcium is known to be a chemotropic agent in the growth of some pollen tubes [24]. The presence of high concentra-tions of calcium or other chemicals in the vacuoles may be the mechanism through which the synergids contribute to the directional growth of the pollen tube.

In other research we have been examining the oxidative pathways present in the developing embryo and the distribution of the enzymes within the cells of the embryo [22]. Using a Cartesian diver, FORMAN measured the rate of oxygen uptake in single isolated embryos. Then, working with inhibitors and added substrates, he was able to determine the major oxidative pathways present in the developing embryo. Through the use of Nitro-BT it was possible to map

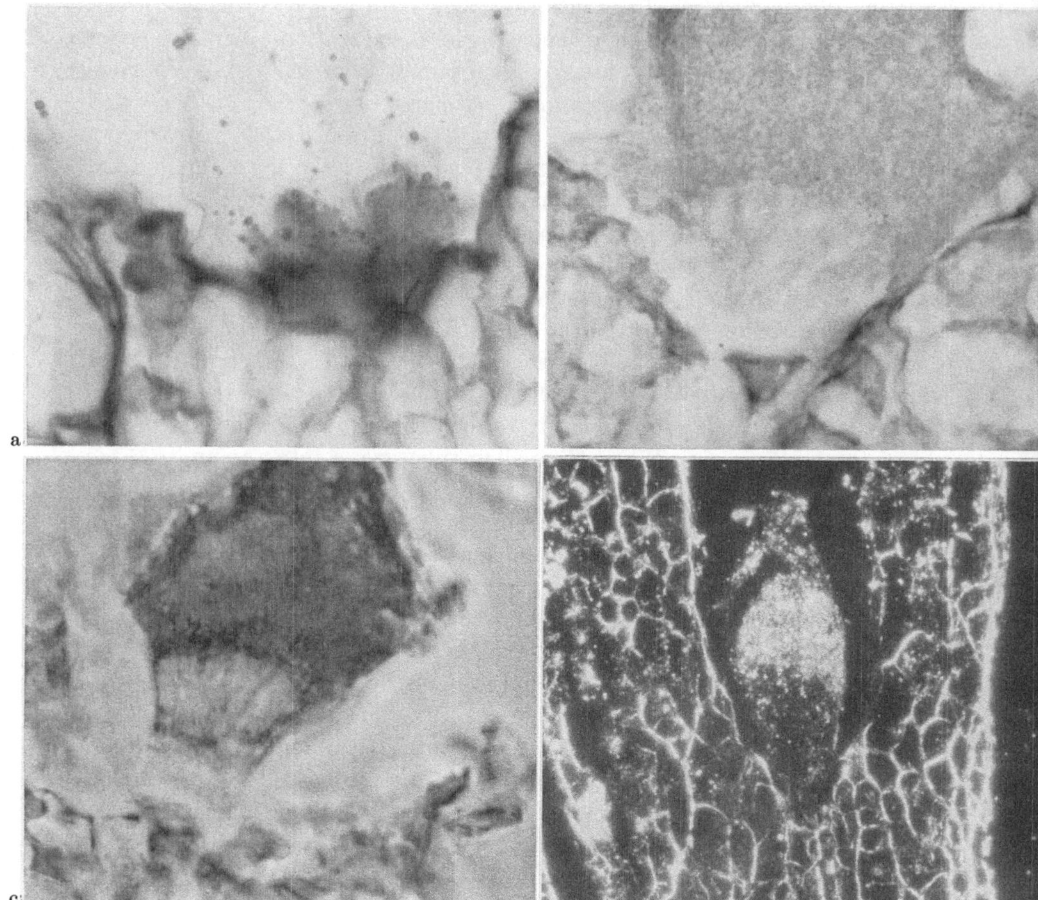

Fig. 4a—d. The synergid of cotton after the following histochemical procedures: a periodic acid-Schiff, b Azure B, c Ninhydrin-Schiff, d microincineration

the distribution of the cells with the highest enzyme activities and to analyze their position with regard to the morphogenetic development of the embryo. Finally, using the same method it was possible to study the localization of the enzyme in terms of the fine structure of the cells.

The pollen tube has been the object of another group of workers [30] using combined histochemical and electron microscope procedures. ROSEN and co-workers [30] studied the distribution of lipid, protein, carbohydrates, and nucleic acids in cultured pollen tubes. They were able to compare patterns of distribution of cell parts as seen in the electron microscope to specific zones high in various chemical components. Their data sheds considerable light on the mechanism of growth of the pollen tube tip.

In most of the studies discussed thus far the histochemical investigations were correlated with electron microscope data, but the localization procedures were not performed at the EM level. Little true ultracytochemistry has been performed with plant tissue. Two examples of this approach using plant tissue are the work of JACOBSON, SWIFT, and BOGORAD [*17*] on plastids and ALBERS-HEIM [*1, 2*] on pectin localization.

The possible occurence and distribution of RNA in plastids was the problem which concerned JACOBSON and co-workers. Using autoradiographs of tritiated cytidine and Azure B on the microscopic level, they found evidence that RNA was present in the plastids. Controls treated with RN'ase confirmed this conclusion. Then, in electron microscope preparations they observed a particulate component of the plastid. To demonstrate that these particles contained RNA, they treated the tissue with RN'ase. This procedure removed the particles and they concluded that the particles were composed of RNA. This procedure, which is relatively simple and which gives good results, should become a standard method for identifying the composition of particles seen in the electron microscope.

Cell wall structure and composition is of great interest to botanists. ALBERS-HEIM's work [*1, 2*] deals with the distribution of pectic substances in the cell wall. The procedure which he adapted to the electron microscope was one developed by REEVE [*29*] for the light microscope. The method involves treating the tissue with hydroxylamine and then with $FeCl_3$. The red color which results is visible in the light microscope and the iron is electron dense in the electron microscope. The procedure yields discrete localization and has been used in the study of cell wall growth and synthesis.

Summary

The field of botanical histochemistry is still relatively underdeveloped. Basic questions, for example, the best methods to prepare and handle plant tissue for histochemical procedures, remain only partially answered. On the other hand, interest in histochemical methods by botanists and plant physiologists is increasing and greater use of these procedures can be expected in the future. In particular, methods combining histochemistry and electron microscopy should be used in much greater numbers in the near future on plant tissue. The challenges are present, the rewards will be great, the future of botanical histochemistry is bright.

Bibliography

[*1*] ALBERSHEIM, P., and URSULA KILLIAS: Histochemical localization at the electron microscope level. Amer. J. Bot. **50**, 732—745 (1963).

[*2*] —, D. MÜHLETHALER, and A. FREY-WYSSLING: Stained pectin as seen in the electron microscope. J. biophys. biochem. Cytol. 8, 501—516 (1960).

[*3*] BELL, P. R.: Failure of nucleotides to diffuse freely into the embryo of *Pteridium aquilinum*. Nature (Lond.) **191**, 91 (1961).

[*4*] — Interaction of nucleus and cytoplasm during oogenesis in *Pteridium aquilinum* (L.) KUHN. Proc. roy. Soc. B **153**, 421 (1961).

[*5*] — The cytochemical and ultrastructural peculiarities of the fern egg. J. Linnean Soc. Bot. **58**, 353 (1963).

[*6*] —, and K. MÜHLETHALER: A membrane peculiar to the egg in the gametophyte of *Pteridium aquilinum*. Nature (Lond.) **195**, 198 (1962).

[*7*] — The fine structure of the cells taking part in oogenesis in *Pteridium aquilinum* (L.) KUHN. J. Ultrastruct. Res. **7**, 452 (1962).

[*8*] — The degeneration and reappearance of mitochondria in the egg cells of a plant. J. Cell Biol. **20**, 235—248 (1964).

[*9*] BRANTON, D., and L. JACOBSON: Freeze-drying of plant material. Exp. Cell Res. **22**, 559—568 (1961).

[10] GERSH, I.: The Altmann technique for fixation by drying while freezing. Anat. Rec. **53**, 309 (1932).

[11] GIFFORD jr., E. M.: Incorporation of H³-thymidine into shoot and root apices of *Ceratopteris*. Amer. J. Bot. **47**, 834—838 (1960).

[12] — Developmental studies of vegetative and floral meristems. Brookhaven Symposia in Biol. **16**, 126—136 (1963).

[13] —, and H. B. TEPPEN: Ontogeny of the inflorescence in *Chenopodium album*. Amer. J. Bot. **48**, 657—666 (1961).

[14] Histochemical and autoradiographic studies of floral induction in *Chenopodium*. Amer. J. Bot. **49**, 706—714 (1962).

[15] GOODSPEED, T. H., and F. M. UBER: Application of the Altmann freezing-drying techniques to plant cytology. Proc. nat. Acad. Sci. (Wash.) **20**, 495—501 (1934).

[16] HALPERIN, W., and D. F. WETHERELL: Adentive embryony in the tissue cultures of the wild carrot, *Daucus corota*. Amer. J. Bot. **51**, 274—283 (1964).

[17] JACOBSON, ANN B., H. SWIFT, and L. BOGORAD: Cytochemical studies concerning the occurrence and distribution of RNA in plastids of *Zea mays*. J. Cell Biol. **17**, 557—570 (1963).

[18] JENSEN, W. A.: A new approach to freeze-drying of tissue. Exp. Cell Res. **7**, 572—574 (1954).

[19] — The application of freeze-dry methods to plant material. Stain Technol. **29**, 143—150 (1954).

[20] — The composition of the developing primary wall in onion root tip cells. II. Cytochemical localization. Amer. J. Bot. **47**, 287—295 (1960).

[21] — Botanical histochemistry. San Francisco: W. H. Freeman & Co. 1962.

[22] — Cell development during plant embryogenesis. Brookhaven Symposia in Biol. **16**, 179—202 (1963).

[23] LEDBETTER, M. C., and K. R. PORTER: A "microtubule" in plant cell fine structure. J. Cell Biol. **19**, 239—250 (1963).

[24] MASCARENHAS, J. P., and L. MACHLIS: Chemotropic response of the pollen of *Antirrhinum majus* to calcium. Plant Physiol. **39**, 70—77 (1964).

[25] MOLLENHAUER, H.: Permanganate fixation of plant cells. J. biophys. biochem. Cytol. **2**, 215—218 (1959).

[26] MÜHLETHALER, K., u. P. R. BELL: Untersuchungen über die Kontinuität von Plastiden und Mitochondrien in der Eizelle von *Pteridium aquilinum* (L.) KUHN. Naturwissenschaften **49**, 63 (1962).

[27] PORTER, K. R., and R. D. MACHADO: Studies on the endoplasmic reticulum. IV. Its form and distribution during mitosis in cells of onion root tip. J. biophys. biochem. Cytol. **7**, 167—180 (1953).

[28] PRITCHARD, H. N.: A cytochemical study of embryo sac development in ʹStellaria media. Amer. J. Bot. (In press).

[29] REEVE, R. M.: A specific hydroxylamine-ferric chloride reaction for histochemical localization of pectin. Stain Technol. **34**, 209—211 (1959).

[30] ROSEN, W. G., S. R. GAWLIK, W. V. DASHEK, and K. A. SIEGESMUND: Fine structure and cytochemistry of Lilium pollen tubes. Amer. J. Bot. **51**, 61—71 (1964).

[31] STEWARD, F. C.: Growth and organization in free cell cultures. Brookhaven Symposia in Biol. **16**, 73—88 (1963).

[32] TAKAO, A.: Histochemical studies on the formation of some leguminous seeds. Jap. J. Bot. **18**, 55—72 (1962).

[33] VAN FLEET, D. S.: Histochemical localization of enzymes in vascular plants. Bot. Rev. **18**, 354—398 (1952).

[34] — Histochemistry of enzymes in plant tissues. In: Handbuch der Histochemie, Bd. VII/2, S. 1—38. 1962.

Professor Dr. WILLIAM A. JENSEN,
Department of Botany, University of California,
Berkeley/California (USA)

Neuere Methoden und Ergebnisse der Autoradiographie

Von

W. Maurer

Dieser Text gibt nicht den Wortlaut des Referates wieder. Er bringt lediglich einen Überblick über den Inhalt des Referates. Auf die Wiedergabe von Abbildungen und Tabellen wurde hier verzichtet.

Bereits 1924 wiesen Laccassagne und Latès die Verteilung des natürlich radioaktiven α-Strahlers Polonium-210 in der Niere mit Hilfe einer photographischen Platte nach. Bis zur Entdeckung der künstlichen Radioaktivität im Jahre 1934 war diese Methode auf die Untersuchung des Stoffwechselverhaltens von wenigen natürlich radioaktiven Elementen am Ende des Periodischen Systems beschränkt. Heute stehen bis auf wenige Ausnahmen von allen physiologisch interessanten Elementen Radioisotope zur Verfügung, und die Autoradiographie ist so zu einer Methode geworden, welche wie keine andere Einblicke in den Stoffwechsel der Zelle und ihrer Strukturen gestattet.

Die *Methode der Autoradiographie* besteht darin, daß biologische Objekte nach Markierung mit radioaktiven Substanzen in Kontakt mit einer photographischen Emulsion gebracht werden. Die radioaktive Teilchenstrahlung erzeugt dann in den AgBr-Kristallen der Emulsion sog. latente Bilder, woraus bei der photographischen Entwicklung einzelne Silberkörner entstehen. Die Zahl der Silberkörner ist unter bestimmten Bedingungen ein Maß für die im Schnitt inkorporierte Radioaktivität. An die Stelle des Geiger-Müller-Zählers oder des Szintillations-Zählers bei biochemischen Versuchen tritt in der Autoradiographie die photographische Emulsion. Der Impulsrate des Zählers entspricht die Silberkornzahl der Autoradiogramme.

Für *autoradiographische* Zwecke stehen heute eine Reihe von *Spezial-Emulsionen* zur Verfügung. Alle Emulsionen haben einen hohen Gehalt an Silberbromid und eine kleine Korngröße. Auf die verschiedenen Emulsionen und die Technik der Herstellung von Autoradiogrammen soll hier nicht eingegangen werden. In diesem Zusammenhang sei auf das Referat von Leblond auf dem I. Internationalen Kongreß für Histo- und Cytochemie, Paris, über die Verwendung von flüssigen Emulsionen verwiesen.

Das *Auflösungsvermögen der autoradiographischen Methode* wird dadurch bestimmt, daß eine radioaktive Quelle im histologischen Schnitt in direktem Kontakt mit einer photographischen Schicht steht. Es kommt dann keine eigentliche Abbildung der Quelle im optischen Sinne zustande. Vielmehr liefert eine punktförmige Quelle von Aktivität im Schnitt eine Silberkornverteilung, welche eine gewisse Breite hat. Die Halbwertsbreite bestimmt das autoradiographische Auflösungsvermögen. Letzteres ist besonders groß, wenn Schnitt und Emulsion dünn sind und wenn beide unmittelbar aufeinander liegen. Weiterhin ist das Auflösungsvermögen um so größer, je kleiner die Reichweite der radioaktiven Teilchenstrahlung ist. Das Auflösungsvermögen beträgt mit P-32 etwa $5\,\mu$ und mit C-14 sowie S-35 2—$3\,\mu$.

Eine bevorzugte Stellung nimmt das Wasserstoff-Isotop H-3 (HZ = 12,6 Jahre) ein. Sein wesentlicher Vorzug besteht in der extrem geringen Reichweite seiner β-Strahlung. Die maximale Reichweite der H-3-β-Strahlung beträgt ca. 5 μ (nasses Gewebe). Der größte Teil der β-Teilchen hat aber nur eine Reichweite von 1—2 μ. Mit H-3 ist die autoradiographische Auflösung besser als 1 μ und damit vergleichbar mit der optischen. Auch die Inkorporation in so kleinen Strukturen wie dem Nukleolus ist damit der Messung zugänglich.

Fitzgerald war 1951 der erste, der H-3-markierte Substanzen in die Autoradiographie einführte. Mit der zunehmenden Verfügbarkeit von H-3-markierten Verbindungen der verschiedensten Art hat die H-3-Autoradiographie in den letzten Jahren immer mehr an Bedeutung gewonnen.

Die *Entwicklung* geht dahin, daß sich diese hochauflösende H-3-Autoradiographie auch *zu einer quantitativen Autoradiographie* entwickeln wird. Die geringe Reichweite der H-3-β-Strahlung hat auf der anderen Seite den Nachteil, daß bereits im Material des Schnitts eine je nach Schnittdicke mehr oder weniger große „*β-Selbstabsorption*" stattfindet. Dadurch erreicht nur ein Teil der β-Teilchen die Emulsion. Die Bestimmung dieser β-Selbstabsorption ist auch deshalb von besonderer Bedeutung, weil der Gehalt an Trockenmasse bei Nukleolus, Kernchromatin und Cytoplasma sehr verschieden ist, was beim Nukleolus zu einer besonders großen β-Selbstabsorption und damit zu einer „viel zu kleinen" Silberkornzahl führt. Das letzte Ziel einer quantitativen H-3-Autoradiographie kann also nur über eine genaue Bestimmung der lokalen β-Selbstabsorption erreicht werden. Dazu ist die Kenntnis der lokalen Massenverteilung (trocken) innerhalb der untersuchten Zellen erforderlich. Die autoradiographischen Untersuchungen müssen also durch *Massenbestimmungen* etwa mit dem Interferenz-Mikroskop oder nach der Methode der Radiographie (Absorption von sehr weichen Röntgenstrahlen) ergänzt werden. Insgesamt kann erst dann aus der Silberkornzahl auf die relative Aktivität von Strukturen oder auch auf die absolute Aktivität geschlossen werden. Erst dann leistet die autoradiographische Methode im zellulären Bereich das gleiche wie der Zähler in der Biochemie.

Bei genügend dicker Emulsion erzeugen einzelne β-Teilchen von C-14 oder S-35 u.a.m. deutlich sichtbare Spuren von Silberkörnchen, die leicht ausgezählt werden können. Diese Methode der „*Spuren-Autoradiographie*" wurde vor allem von Ficq und Brachet ausgebildet. Die notwendigen Aktivitäten sind etwa 100mal geringer, was auch zu einer geringeren Strahlenbelastung der biologischen Objekte führt.

Ein besonderer Fall liegt bei der *autoradiographischen Verwendung von K-Strahlern* vor. Bei dem Zerfall eines K-Strahlers entsteht primär keine Teilchenstrahlung, sondern es wird primär ein Elektron der Elektronenhülle vom Kern absorbiert. Wegen des so entstandenen „Lochs" in der K-Schale kommt es dann zur Emission der charakteristischen Röntgen-K-Strahlung des Folge-Elementes. Bei mittleren und vor allem bei kleinen Ordnungszahlen wird diese Röntgenstrahlung in der eigenen Elektronenhülle des K-Strahlers fast vollständig absorbiert, wobei energiearme Elektronen entstehen. Ihre Energie entspricht derjenigen der H-3-β-Strahlung bzw. ist noch kleiner. Mit K-Strahlern kann deshalb ein mit H-3 vergleichbares, sehr großes Auflösungsvermögen erzielt werden, wie es z.B. bei J-125 und Cr-51 bekannt ist.

Wegen der Zufallsnatur des radioaktiven Zerfalls ist die Zahl der Silberkörner oder der „β-Spuren" z.B. über einem Kern mit einem *statistischen Fehler* behaftet. Für seine Berechnung gilt das gleiche wie bei Messungen mit einem Zähler. Bei der Berechnung des statistischen Fehlers kommt es immer auf die Zahl der erfaßten Zerfallsprozesse an. Wenn pro β-Teilchen u.U. mehr als 1 Korn erzeugt wird, so muß das berücksichtigt werden. Bei der Auswertung von Spuren-Autoradiogrammen, bei denen die Zahl der Spuren z.B. pro Kern im allgemeinen klein ist, sind die rein statistisch bedingten Schwankungen der Zahl der β-Spuren besonders groß, und es können einzelne Objekte trotz einer vorhandenen Markierung rein aus Zufall keine Spuren zeigen.

Wegen der notwendigen Feinkörnigkeit der autoradiographischen Emulsionen und der damit verbundenen geringen Empfindlichkeit muß im allgemeinen mit relativ großen Aktivitäten gearbeitet werden. Auf der anderen Seite darf die eingesetzte Menge einen bestimmten Wert nicht überschreiten, wenn die physiologischen Verhältnisse nicht gestört werden sollen. Das hat zur Folge, daß die markierten Ausgangspräparate eine sehr große *„spezifische Aktivität"* haben müssen. Sie muß wesentlich größer sein als bei rein biochemischen Versuchen. Die notwendigen spezifischen Aktivitäten liegen zwischen einigen Hundert und einigen Tausend mC pro mMol und mehr. Die Herstellung der markierten Ausgangspräparate stellt oft die — unerwartete — Hauptaufgabe eines Versuches dar. Die größten spezifischen Aktivitäten erhält man durch rein chemische oder auch biologische Syntheseverfahren. Die Methode von WILZBACH liefert im allgemeinen geringere spezifische Aktivitäten.

Zu autoradiographischen Untersuchungen können alle ebenen Präparate wie Schnitte, Ausstriche von Zellen, Deckglaskulturen, Quetschpräparate u.a.m. verwandt werden.

Von besonderer Bedeutung ist die Frage, in welcher *chemischen Bindung die Radioisotope im Material eines Schnittes* vorliegen. Bei der biochemischen Aufarbeitung von Gewebsproben kann diese Frage mit den üblichen Verfahren der Chemie sehr weitgehend gelöst werden. Demgegenüber sind einer chemischen Bearbeitung von Schnitten Grenzen gesetzt, wenn die Struktur des Schnitts nicht zerstört werden soll. Von großem Wert sind deshalb solche markierte Vorläufer, welche selektiv nur in eine einzige Substanz eingebaut werden. Ein Beispiel dafür ist die selektive Markierung der DNS durch Verwendung von H-3- oder C-14-markiertem Thymidin. Markierte RNS kann durch eine Behandlung des Schnitts mit RNase als solche erkannt werden. In diesem und anderen Fällen sind derartige Differenzmethoden aber notwendigerweise relativ ungenau. Die Bestimmung der chemischen Bindung der Radioisotope im Schnitt stellt eine von Fall zu Fall sehr unterschiedliche Aufgabe dar.

Nach diesen mehr allgemeinen, methodischen Bemerkungen zur Autoradiographie soll im folgenden ein gedrängter *Überblick über Anwendungen der Autoradiographie* gegeben werden. Vor einigen Jahren wäre eine solche Aufgabe noch relativ einfach gewesen. Wegen der Fülle des heute vorliegenden Stoffs kann hier nur eine Auswahl getroffen werden. Dabei sollen vor allem solche Anwendungen besprochen werden, welche die besonderen Aussagemöglichkeiten der Autoradiographie und ihre Grenzen kennzeichnen.

Einen breiten Raum nehmen in der Literatur Untersuchungen des *Eiweiß-stoffwechsels* mit markierten Aminosäuren ein. Die von verschiedenen Autoren mit H-3, C-14- und S-35-markierten Aminosäuren an Mäusen, Ratten, Kaninchen und anderen Tieren durchgeführten autoradiographischen Untersuchungen zeigten, daß bestimmte Zellarten wie die Hauptzellen des Magens, die exokrinen Pankreas-Epithelien u.a.m. sowie auch die Ganglienzellen gleich welcher Lokalisation einen sehr großen Eiweißstoffwechsel haben, während derjenige von Muskel-, Binde- und Stützgewebe sowie Glia und Mark sehr viel kleiner ist.

Die Deutung der Silberkorndichte der Autoradiogramme als ein Maß für die Größe des Eiweißstoffwechsels ist aber an eine wesentliche Voraussetzung gebunden. Die Aktivität, welche nach der Injektion einer markierten Aminosäure von einer Zelle inkorporiert wird, ist nicht ohne weiteres ein Maß für die Umsatzrate des Eiweißes dieser Zelle. Sie hängt vielmehr neben der gesuchten Umsatzrate auch von der *spezifischen Aktivität der freien Aminosäure* (Aktivität pro Mengeneinheit) am Ort der Eiweißsynthese ab. Die spezifische Aktivität der freien Aminosäure hängt ihrerseits mit der injizierten Aktivität, der Konzentration der freien Aminosäure am interessierenden Ort und mit der mehr oder weniger schnellen Durchmischung der markierten Aminosäure-Moleküle mit denjenigen des Aminosäure-„pool's" zusammen. Die inkorporierte Aktivität ist gleich

$$\text{Inkorporierte Aktivität zwischen Injektion und der Zeit } T = A_0 \cdot s_m \cdot T.$$

Dabei ist A_0 die gesuchte Umsatzrate der Aminosäure in Menge pro Zeiteinheit, s_m die mittlere spezifische Aktivität der freien Aminosäure für das Zeitintervall (T) zwischen Injektion und Tötung und T die Zeitdauer des Versuches. Die inkorporierte Aktivität und damit auch die autoradiographische Silberkornzahl ist also nur dann der gesuchten Umsatzrate ohne weiteres proportional, wenn die spezifische Aktivität der freien Aminosäure in den zu vergleichenden Zellen gleich ist. Das scheint in ein und demselben Organismus bei einer Reihe von Aminosäuren auch angenähert der Fall zu sein. Mit der Genauigkeit, mit der das der Fall ist, ist die Silberkorndichte dann tatsächlich ein relatives Maß für die Aminosäure-Inkorporation bzw. den Eiweißstoffwechsel der verschiedenen Zellarten.

Sehr viel schwieriger wird aber die Deutung von Autoradiogrammen, wenn der zeitliche Verlauf der Eiweißsynthese z.B. nach Hepatektomie in der regenerierenden Restleber untersucht werden soll. Solange in einem solchen Fall nicht bekannt ist, ob sich bei Injektion der gleichen Aminosäureaktivität zu verschiedenen Zeiten im Organismus auch immer die gleiche mittlere spezifische Aktivität der freien Aminosäure einstellt, sind solche Versuche prinzipiell nicht deutbar. Dabei ist ein Einfluß einer Hepatektomie auf die Konzentration der freien Aminosäure von vornherein durchaus denkbar. Ähnlich liegen die Dinge, wenn z.B. die Altersabhängigkeit des Eiweißstoffwechsels unter Verwendung von jungen und alten Tieren untersucht werden soll oder wenn normal ernährte Tiere mit eiweißarm ernährten Tieren u.a.m. verglichen werden. In solchen Fällen bedeutet eine Änderung der Silberkorndichte nicht ohne weiteres auch eine Änderung des Eiweißstoffwechsels. Das ist nur dann der Fall, wenn zusätzlich nachgewiesen wird, daß in den verglichenen Fällen die spezifische Aktivität der freien Aminosäure gleich war. Messungen dieser Art sind ein notwendiger Bestandteil der Deutung solcher Autoradiogramme, was in der Literatur oft übersehen wird.

Es liegt hier eine Situation vor, wie sie ganz allgemein für die Messung von Stoff-wechselraten typisch ist. Von der inkorporierten Aktivität kann also nicht ohne weiteres auf den Mengenumsatz geschlossen werden. Darauf kann nicht ein-dringlich genug hingewiesen werden.

Bei Versuchen dieser Art liefert die Autoradiographie also „*Umsatzraten*", während die cytophotometrischen und interferenzmikroskopischen Methoden der Histochemie die „*Menge*" von Substanzen einer Zelle zu messen gestatten. In dieser Hinsicht sind beide Methoden verschieden. Erst beide Methoden zusammen liefern die für den dynamischen Stoffwechsel der Zelle typischen Größen von „Umsatz" und „Menge" eines Stoffes.

Der *RNS-Stoffwechsel* ist vor allem mit H-3-Cytidin und H-3-Uridin und mit anderen H-3- und C-14-markierten Nukleosiden von vielen Seiten untersucht worden. Dabei findet man bekanntlich bei kurzzeitigen Versuchen zunächst nur eine Markierung des Nukleolus und des Kernchromatins. Bei größeren Zeiten erscheint dann die Markierung mehr und mehr im Cytoplasma, was allgemein als eine Migration von markierter RNS (oder ähnlichem) aufgefaßt wird. Es ist auch der Versuch gemacht worden, den quantitativen Übertritt der im Kern gebildeten markierten RNS ins Cytoplasma zu bestätigen.

Solchen Versuchen einer quantitativen Erfassung von H-3-Aktivitäten in den drei Zellstrukturen stehen aber, wie heute bekannt, zwei Schwierigkeiten im Wege. Einmal konnte gezeigt werden, daß bei der Fixation ein erheblicher Teil der markierten RNS in Lösung gehen kann, und zwar je nach Fixationsmedium zu einem sehr unterschiedlichen Prozentsatz. Zum anderen spielt die β-Selbst-absorption bei der Bestimmung der H-3-Aktivität in Nukleolus, Kernchromatin und Cytoplasma eine entscheidende Rolle. Bei kurzzeitigen Versuchen mit H-3-Nukleosiden liefert eine quantitative Auswertung der Autoradiogramme durch Silberkornauszählung ohne Berücksichtigung der β-Selbstabsorption, daß der Anteil des Nukleolus an der Inkorporation des ganzen Kerns nur 25% beträgt. Eine Berücksichtigung der in Nukleolus und Kernchromatin sehr verschiedenen β-Selbstabsorption führt aber auf einen Anteil des Nukleolus von 60% (!), d.h. auf einen 2,5mal größeren Anteil. Das zeigt sehr deutlich, daß bei einer quantita-tiven H-3-Autoradiographie die Berücksichtigung der β-Selbstabsorption im bio-logischen Material unerläßlich ist.

Eine große Zahl von autoradiographischen Arbeiten der letzten Jahre befaßt sich mit der *DNS-Synthese* unter verschiedensten Aspekten. Die Möglichkeit einer selektiven und dauerhaften Markierung der DNS mit markiertem Thymidin hat viele neue Möglichkeiten eröffnet. Es muß aber erwähnt werden, daß hier noch einige Fragen offen sind, welche damit zusammenhängen, daß Thymidin kein physiologisches Zwischenprodukt der DNS-Synthese ist.

Mit markiertem Thymidin konnte zum ersten Mal die Dauer der DNS-Syn-these bei vielen tierischen und pflanzlichen Zellen und die Lage der Verdopplungs-phase innerhalb des Generationszyklus bestimmt werden. Daraus lassen sich auch Werte für die Generationszeit und die anderen Phasen des Zellzyklus ableiten. Dabei fand sich bei fetalen Zellarten eine gut konstante DNS-Verdopplungszeit von 5—6 Std und bei erwachsenen Mäusen und Ratten — mit Ausnahme weniger Zellarten — eine konstante Verdopplungszeit von ca. 7,5 Std. Die unterschied-liche Generationszeit der verschiedenen Zellarten wird im wesentlichen durch die

Dauer der G_1-Phase (=Ende einer Mitose bis Beginn der DNS-Synthese) bestimmt. Die Zeitdauer vom Beginn der DNS-Synthese bis zum Ende der folgenden Mitose ist dagegen überraschend konstant.

Die Möglichkeit einer dauerhaften Markierung der DNS gestattet eine Untersuchung von proliferierenden Zell-Populationen wie bei Darm, Magen, Haut, Knochenmark u.a.m. Auch Fragen der Regeneration, der Cancerisierung von Gewebe, der Entwicklung des Gehirns u.a.m. sind bearbeitet worden. In diesem Zusammenhang ist von besonderem Interesse, ob mit einer Wiederverwendung von markierter DNS, welche beim Tod einer Zelle frei wird, gerechnet werden muß. Die bisher vorliegenden Arbeiten sprechen in dieser Richtung, wenn im einzelnen auch noch Fragen offen bleiben.

Ähnlich wie in der Biochemie mehren sich auch in der Autoradiographie die Arbeiten, bei denen in einem biologischen Objekt *gleichzeitig das Verhalten einer H-3- und einer C-14-markierten Substanz* untersucht wird. Es kann sich dabei um chemisch andersartige Substanzen oder um zwei verschiedene Markierungen derselben Substanz handeln. Diese Methode ist sehr wandlungsfähig und gestattet die Bearbeitung ganz neuartiger Fragestellungen. So kann die Dauer der DNS-Synthese von Zellen durch eine Kombination von H-3-Thymidin und C-14-Thymidin bei vielen — sonst nicht ausmeßbaren — Zellarten bestimmt werden. Weiterhin kann etwa der DNS-Stoffwechsel mit C-14-Thymidin und gleichzeitig der Eiweiß- oder RNS-Stoffwechsel mit einer H-3-Aminosäure oder H-3-Cytidin im gleichen Tier untersucht werden. Autoradiographisch kann zwischen H-3 und C-14 unterschieden werden, weil die extrem energiearmen H-3-β-Teilchen nur in einer 1—$2\,\mu$ dicken Schicht der Emulsion Silberkörner erzeugen, während die C-14-β-Teilchen wegen ihrer größeren Reichweite in einer dicken Emulsion sichtbare Bahnspuren von Silberkörnern hinterlassen.

Eine Fortentwicklung dieser Technik ist die sog. *Doppelschicht-Autoradiographie.* Hierbei wird auf den Schnitt zunächst eine erste relativ dünne Emulsion aufgebracht, welche bei entsprechender Dosierung der H-3- und C-14-Aktivität, nur eine der H-3-Aktivität entsprechende Zahl von Körnern ergibt. Dann wird nach photographischer Entwicklung dieser 1. Emulsion nochmals eine 2. Emulsion aufgebracht. Diese registriert dann bei langer Exposition die — möglichst schwache — C-14-Aktivität. Man kann dann deutlich zwischen einer H-3-Markierung (1. Emulsion) und einer C-14-Markierung (2. Emulsion) unterscheiden. Leider ist die autoradiographische Auflösung in der 2. Emulsion gering, so daß bei dicht nebeneinander liegenden C-14-markierten Kernen oder Zellen die Markierung nicht sicher zugeordnet werden kann.

Auf die Vielzahl der Arbeiten zum Stoffwechsel der Schilddrüse, der Mukopolysaccharide (S-35), des Knochens u.a.m. kann hier nicht eingegangen werden.

Ein Gebiet, das noch in Entwicklung begriffen ist, ist die *autoradiographische Darstellung wasserlöslicher Verbindungen.* Hierher gehören vor allem viele pharmakologisch interessante Substanzen. Während beim Eiweiß-, RNS- und DNS-Stoffwechsel die Herstellung von Autoradiogrammen wegen der Verwendbarkeit von Paraffinschnitten u.a.m. relativ komplikationslos ist, muß der ganze autoradiographische Prozeß bei der Darstellung wasserlöslicher Substanzen mit wasserfreien Flüssigkeiten und unter dem Gefrierpunkt durchgeführt werden. Nur unter diesen Bedingungen ist ein Verlust an markierter Substanz und eine Verwaschung der Lokalisation im Gewebe durch Diffusion vermeidbar. Es sind eine Reihe von Techniken beschrieben worden. Die Entwicklung ist aber noch im Gange.

Es ist erstaunlich, daß eine Autoradiographie des Stoffwechsels der *Fette* eigentlich völlig fehlt. Bei der Herstellung der Autoradiogramme können natürlich keine organischen Lösungsmittel verwandt werden. Auf diesem Gebiet dürften lohnende neue Fragestellungen liegen.

In letzter Zeit ist eine aussichtsreiche *Kombination von Autoradiographie und Elektronenmikroskop* entwickelt worden. Hierbei wird die photographische Emulsion auf einen dünnen Schnitt aufgebracht und dann, wie üblich, nach einer gewissen Expositionszeit photographisch entwickelt. Im Unterschied zur sonst üblichen Technik wird das Autoradiogramm dann im Elektronenmikroskop (bei geringer Vergrößerung) abgebildet. Durch eine nur sehr kurze photographische „Unter"-Entwicklung der Emulsion kann man erreichen, daß keine Silberkörner entstehen, welche das ganze Silber eines AgBr-Kristalls enthalten, sondern daß nur sehr kurze Fäden von Ag-Atomen entstehen, und zwar von einer optisch nicht mehr erkennbaren Größe. Diese Fäden sind aber im Elektronenmikroskop gut sichtbar. Hiermit — und nur hiermit — hängt die Vergrößerung des autoradiographischen Auflösungsvermögens zusammen. Diese Technik hat natürlich nur bei Verwendung von H-3-markierten Substanzen einen Sinn. Die Begrenzung des Auflösungsvermögens hängt auch hier mit der Größe der AgBr-Kristalle in der Emulsion zusammen. Ein latentes Bild, welches z.B. in der Mitte eines AgBr-Kristalls entsteht, kann im Kristall etwa bis zur Grenzfläche mit der Gelatine wandern. Der Silberfaden liegt dann nicht genau an der Stelle des primär entstandenen latenten Bildes. Die Feinkörnigkeit einer Emulsion kann aber auch nicht beliebig weit getrieben werden, weil dann die Empfindlichkeit immer kleiner wird. Die Empfindlichkeit der Emulsion ist aber bei diesem Verfahren besonders wichtig, weil die notwendigerweise dünnen Schnitte entsprechend weniger Aktivität enthalten. Es dürfte eine Verbesserung des autoradiographischen Verfahrens bis herab zu $0,1\,\mu$ erreichbar sein.

Eine große Zahl von autoradiographischen Arbeiten befaßt sich mit der Frage nach dem zeitlichen und örtlichen Verhalten der verschiedensten markierten Stoffe im Organismus *(Verteilungsautoradiographie)*. Die Aufnahme, Verteilung und Ausscheidung insbesondere von radiobiologisch interessanten Stoffen ist so untersucht worden, und zwar mit dem Ziel einer Beurteilung der lokalen Strahlenbelastung. Es sei hier nur erwähnt, daß die lokale Strahlenbelastung nach Inkorporation von markierten Substanzen auch direkt autoradiographisch gemessen werden kann, wenn man Schnitte nimmt, deren Dicke der maximalen Reichweite des verwandten Isotops entspricht. Die lokale Silberkorndichte ist dann direkt ein Maß für die lokale Strahlendosis im Gewebe.

Professor Dr. W. MAURER,
Institut für med. Isotopenforschung der Universität Köln,
5 Köln-Lindenthal (Deutschland), Kerpener Str. 15

University of Chicago

Nucleic Acids in Interphase Growth

By

HEWSON SWIFT

I have been asked to review the role of nucleic acids in interphase — a complex subject for a brief talk. I will limit myself to three specific aspects of the subject: (1) patterns of asynchrony in DNA synthesis, (2) heterogeneity in chromosomal RNA, and (3) some suggestions concerning nuclear-cytoplasmic exchange based on nucleic acid staining for the electron microscope. Most of our observations have been made on the salivary glands of *Drosophila virilis* larvae.

1) DNA synthesis in many interphase nuclei is asynchronous. Certain late labeling chromosomes or chromosome regions may still show H³ thymidine incorporation after DNA synthesis in the remainder of the nucleus is completed [2, 8, 9]. TAYLOR has suggested that DNA synthesis is under genetic control of the chromosome regions themselves and that gene balance is thus maintained during replication. In *Drosophila* salivary glands, Feulgen photometry indicates that DNA undergoes stepwise synthesis in the formation of the giant chromosomes. In thymidine radioautographs many nuclei show a uniform label over all chromosome arms. In some cells, however, only certain specific areas are labeled while other regions are unlabeled. These are often regions of high DNA concentration, and almost always include the chromocenter. We have interpreted this localized pattern as late labeling asynchrony. PLAUT [4] on the other hand has considered local labeling in *Drosophila* to represent waves of DNA synthesis passing down the chromosome. We find it difficult to reconcile this interpretation with our finding that localized labeling consistently involves certain chromosome regions. We would like to suggest that asynchrony of DNA synthesis in *Drosophila* is, at least in part, a function of late labeling, possible due to steric inhibition related to interphase condensation. Since the extent of chromatin condensation may vary with the metabolic condition of the nucleus, we may expect some accompanying variation in DNA synthesis patterns, in different stages of larval growth. In spite of some variability in detail of labeling patterns, however, all DNA synthesis in *Drosophila* can probably be attributed to chromosome replication. No evidence for "metabolic" interphase turnover has been found (but the DNA puffs of Sciarid flies form an interesting and unexplained exception [1, 5, 7]).

2) Synthesis of RNA on the chromosome is a heterogeneous process. The activation of a specific locus on a polytene *Drosophila* chromosome involves the accumulation of RNA within the DNA structure of the chromosome (puff formation). The RNA/DNA ratio within a puff is often much larger than unity. Most of the puff RNA is bound to non-histone protein in the form of irregular particles or filaments. Electron microscopy demonstrates that puffs contain ribonucleoprotein particles of size and appearance characteristic of the particular locus, with diameters ranging from 10 to 400 mμ. Radioautography with H³ cytidine shows that puffs may be rapidly labeled during their formation, but once formed may label much more slowly. These observations suggest that

chromosomal RNA shortly after its formation, is "captured" as ribonucleoprotein, possibly to permit additional RNA synthesis. For a period during puff formation the RNP accumulates faster than it is lost, but regression of a puff involves total loss of RNP from the locus.

The nucleolus in *Drosophila* salivary glands has a characteristic homogeneous portion, usually central, with areas containing ribosome-like particles at its periphery. The RNA/protein ratio is higher in the particulate areas. The homogeneous region contains a few spherical bodies up to $0.5\,\mu$ across, containing DNA, frequently with adjacent clear vesicles. Nucleolar particles are usually readily distinguishable from those of the chromosome, by their smaller and more uniform size, and dense packing. In radioautographs with H^3 cytidine, the nucleolus shows incorporation ahead of the chromosomes, except at those periods when new puffs are forming. Both the fine structure and labeling patterns suggest that nucleolar RNA is at least in part an independent fraction synthesized at the organizer locus (cf. [3]).

3) I would like to conclude with a few electron micrographs that suggest the polysome complex of ribosome and "messenger" RNA may be assembled in the region of the annuli of the nuclear envelope. In special cases this may also take place near the annuli of annulate lamellae, specific modifications of the endoplasmic reticulum that, at least in some cases contain RNA, but no ribosomes [6]. It seems possible that messenger RNA may become membrane bound in the nuclear envelope or in annulate lamellae, for the later attachment of ribosomes in the production of polysomal clusters or rosettes.

Literature

[1] FICQ, A., and C. PAVAN: Nature (Lond.) **180**, 983 (1957).
[2] LIMA-DE-FARIA, A.: J. biophys. biochem. Cytol. **6**, 457 (1959).
[3] MCMASTER-KAYE, R.: J. biophys. biochem. Cytol. **8**, 365 (1960).
[4] PLAUT, W.: J. molec. Biol. **7**, 632 (1963).
[5] RUDKIN, G., and S. CORLETTE: Proc. nat. Acad. Sci. (Wash.) **43**, 964 (1957).
[6] SWIFT, H.: J. biophys. biochem. Cytol. **2** (Suppl.), 415 (1956).
[7] — From: Molecular control of cellular activity, p. 73. ED. J. ALLEN: Mc Graw Hill Book Co. 1962.
[8] TAYLOR, J. H.: J. biophys. biochem. Cytol. **7**, 455 (1960).
[9] WOODARD, J., E. RASCH, and H. SWIFT: J. biophys. biochem. Cytol. **9**, 445 (1961).

Dr. H. SWIFT, The University of Chicago, Dept. of Zoology, 915. East 57th Street, Chicago/Illinois, USA

Les protéines du noyau en interphase

Par

R. Vendrely

Avec 3 figures dans le texte

I. Introduction

Les protéines essentielles des noyaux cellulaires les histones et les protamines furent découvertes et étudiées à la fin du siècle dernier. Mais après les beaux travaux des pionniers de l'école Allemande (Miescher 1868, Kossel 1884) l'étude de ces protéines fut un peu délaissée. Le développement magnifique des travaux sur les acides nucléiques, la découverte du code génétique et du mécanisme de la synthèse des protéines par l'intermédiaire de l'ARN ont relégué dans l'ombre les autres composants du noyau.

Cependant, un regain d'intérêt pour ces protéines semble se dessiner à l'heure actuelle, il est évident que ces substances plus ou moins liées à l'ADN, même si elles ne font pas partie du matériel génétique lui même, doivent avoir un rôle dans le fonctionnement, l'expression des gènes et la régulation de leur activité. Leur étude dans des tissus métaboliquement actifs, dans des tissus en voie de développement et dans le cancer doit conduire à de précieuses indications sur la nature de leur activité dans l'économie de la cellule vivante. Des travaux récents (Huang et Bonner 1962) ont donné des arguments de poids en faveur du rôle des histones dans le contrôle de l'activité des gènes (tel quel l'avaient proposé les Stedmann en 1950), en montrant que l'histone est capable d'inhiber dans la chromatine la synthèse d'ARN dépendant de l'ADN (ARN messager).

Il est donc important d'essayer de définir les différentes protéines nucléaires, de connaître leur composition et leur comportement métabolique dans le noyau des cellules normales ou pathologiques.

II. Les différentes protéines nucléaires

1. Le fractionnement des noyaux

La séparation des différents organites du noyau (chromosomes, nucléoles, suc cellulaire, membrane) n'a pas encore été résolue de façon satisfaisante. Les chromosomes interphasiques séparés par Mirsky et Ris (1947) renferment certes la substance des chromosomes mais elle est probablement contaminée par des débris de noyaux entiers. L'isolement de chromosomes métaphasiques bien reconnaissables semble être actuellement possible (Bloch 1962a, Somers et al. 1963, Chorazy et al. 1963) et cette nouvelle direction de recherche permettra sans doute des progrès dans la connaissance des protéines chromosomiques. L'isolement des nucléoles réussie pour la première fois par Vincent (1952) chez l'étoile de mer a permis à cet auteur de montrer que ces organites ne contiennent pas d'histone mais surtout des phosphoprotéines. L'histone trouvée dans les nucléoles de foie de rat par Monty et al. (1956) provient manifestement de la chromatine associée au nucléole qui est entraînée avec le nucléole. Muramatsu et al. (1963), Desjardin et al. (1963) constatent que les nucléoles de la tumeur de Walker ont un contenu en protéine plus élevé (7,8 picogrammes par nucléole) que celui des nucléoles de foie de rat (5 picogrammes par nucléole) ce qui est en accord avec des données cytochimiques plus anciennes (CASPERSSON et SANTESSON 1942).

Cependant ces travaux ne sont pas encore très développés et pour l'instant, la plupart des études biochimiques sur les protéines du noyau de la cellule se font par des méthodes de fractionnement qui soustraient successivement des noyaux des composants chimiques dont l'origine à partir de telle ou telle structure morphologique du noyau n'est ni simple, ni évidente. Le fractionnement utilisé par ZBARSKY et GEORGIEV (1959) sur des noyaux de foie de rat s'inspire de la méthode classique de MIRSKY et POLLISTER (1946) et parait une des plus satisfaisante. Ces auteurs ont tenté d'établir un bilan des différentes fractions et, en examinant au microscope électronique les noyaux résiduels au cours des extractions successives ZBARSKY et al. (1962) ont pu montrer une correlation entre les différentes fractions détachées du noyau et leur substratum morphologique.

Pour ce fractionnement il importe de s'adresser à des noyaux parfaitement isolés. La méthode au sucrose de CHAUVEAU et coll. (1956) qui donne d'excellents résultats a été fréquemment utilisée. La méthode de ZALTA et al. (1962) plus récente fait appel à un agent tensio actif et permet un excellent rendement de noyaux encore capables d'incorporer des acides aminés. Cependant, l'application de ces méthodes à l'isolement des noyaux de tumeur reste toujours délicate.

Les noyaux étant isolés on entraîne tout d'abord (1) les protéines labiles du suc nucléaire ainsi que les ribosomes nucléaires par un ou plusieurs lavages dans une solution de ClNa 0,14 M, puis (2) la désoxyribonucléoprotéine des chromosomes est dissoute dans du ClNa 1 à 2 M, il s'agit là essentiellement d'un complexe d'ADN et d'histone avec une petite quantité de protéine non histone insoluble en solution acide diluée — (0,2—0,5 M) qui serait également liée à l'ADN. DOUNCE et SARKAR (1960) attribuent un rôle spécial à cette protéine non histone, elle attacherait bout à bout les fibres d'ADN par des liaisons covalentes alors que l'histone s'enroulerait tout au long de la gorge de la double hélice de l'ADN. (3) Les noyaux résiduels insolubles en NaCl 2 M sont traités par un alcali dilué qui solubilise les protéines acides provenant du nucléole, une partie des chromosomes résiduels (qui demeurent après extraction de leur nucléoprotéine) et des lipoprotéines. (4) Le résidu final représente principalement les membranes nucléaires (Fig. 1).

Un procédé un peu différent comporte l'extraction immédiate des histones par un acide dilué puis celle des protéines acides par un alcali, directement après le lavage des noyaux (STEDMANN 1947, DOUNCE 1955, BUTLER et col. 1963, STEELE et BUSCH 1963).

De toutes les protéines nucléaires, les histones et les protamines sont les plus étudiées néammoins les protéines non histones du noyau commencent à attirer également l'attention des biochimistes.

2. Les protéines non histones

Les protéines non histones représentent un ensemble complexe et mal connu. Elles s'opposent aux histones par leurs propriétés de solubilité, leur composition chimique (présence de tryptophane, proportion plus faible en acide aminés basiques) et leur comportement métabolique.

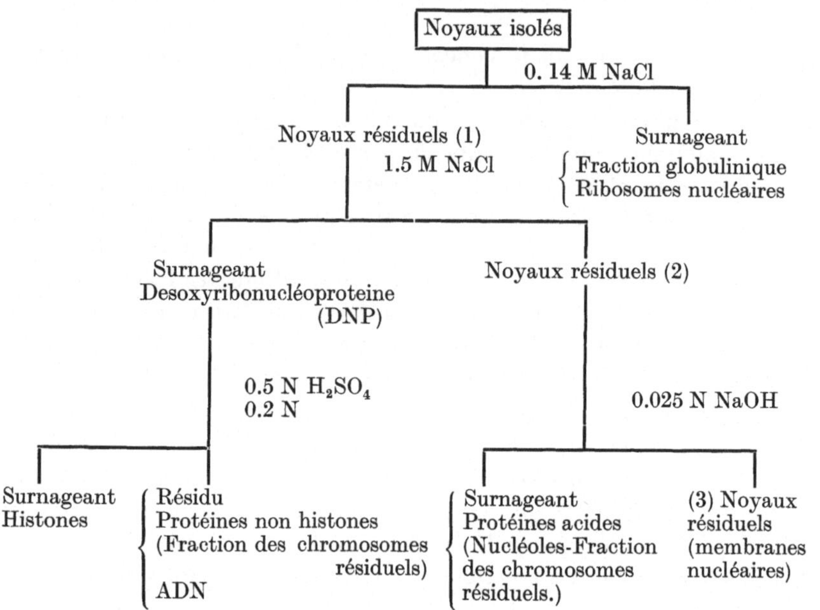

Fig. 1. Isolement de diverses fractions cellulaires (d'après Zbarsky et Georgiev)

C'est pourquoi on les considère souvent en bloc sans tenir compte de cette complexité. Les travaux ultérieurs devront certainement essayer de faire un tri dans ce mélange composite et d'analyser le comportement de ses différents composants.

Ces protéines non histones englobent en effet les protéines du suc nucléaire (globulines) celles des ribosomes dont la présence a été démontrée dans le noyau (Frenster et al. 1960, Wang 1960, Pogo et al. 1962), les protéines du nucléole qui sont encore mal connues, les protéines non histones des chromosomes et les constituants protéiques de la membrane du noyau, à cela s'ajoutent différents enzymes (Siebert 1963), qui interviennent dans des mécanismes de biosynthèse, notamment la biosynthèse de protéines, dans des réactions de transformation d'énergie ou dans des phénomènes de régulation. Parmi ces enzymes, la RNA polymérase doit être liée à la chromatine (Huang et Bonner 1962), de même que la DNA-polymérase (Keir et al. 1962), d'autres enzymes semblent plus particuliers aux nucléoles; nucléoside phosphorylase (Baltus 1954). On a pu mettre en évidence également dans les noyaux des enzymes glycolytiques, des enzymes du cycle de l'acide citrique, la DPN pyrophosphorylase etc.

La teneur en protéines non histone dans les noyaux varie considérablement selon le tissu: noyaux de foie de veau 48%, rein et rate 39%, thymus 13%, érythrocytes de poulet 12%, et quant aux noyaux de sperme chez la truite (Félix et al. 1956), ils ne seraient guère constitués que de nucléoprotamine.

Il apparaît donc que c'est la teneur en protéines non histone du noyau qui conditionne sa taille. La cytochimie quantitative (réaction de Millon) a permis de confirmer cette notion dans différents tissus (HIMES et al. 1955) et lors des variations de taille du noyau consécutives à une stimulation ou à une mise au repos de la cellule (ALFERT et al. 1955).

L'activité métabolique du noyau est due en grande majorité à celle de ses protéines non histone, un grand nombre de travaux biochimiques (DALY et al. 1952, BRUNISH et LUCK 1952, SMELLIE et al. 1953, ALFREY et al. 1955, HOLBROCK et al. 1960, STEELE et BUSCH 1963, OKUDA et al. 1963) l'ont montré sur un grand nombre de tissus.

Mais il serait évidemment intéressant de préciser la part prise dans ce métabolisme par chacune des fractions du noyau: chromosomes, nucléoles, suc cellulaire. La technique autoradiographique (FICQ 1955, 1959, DROZ 1958) doit permettre d'aborder ce problème ainsi que les progrès dans les techniques de fractionnement du noyau. Des travaux récents (BUSCH et al. 1963a) indiquent que dans les noyaux de foie c'est la fraction des noyaux résiduels qui incorpore le plus de lysine C^{14}, puis vient la fraction des protéines chromosomiques acides, la fraction des ribosomes nucléaires et en dernier la fraction des globulines du suc nucléaire.

Les protéines acides du noyau ont peut être un rôle très important dans la régulation de l'activité des gènes. Comme le fait remarquer BUSCH leur caractère acide et leur actif renouvellement dans le noyau suggèrent qu'elles pourraient intervenir dans le fonctionnement des histones en se liant momentanément à elles, pour libérer ainsi l'ADN et lui permettre d'assurer la synthèse d'ARN messager. Cette liaison de protéines acides à l'ADN avait été suggérée par ALFERT (1959) par des données histochimiques, la diminution de colorabilité par le vert rapide pouvant dans certains cas s'interpréter comme un masquage des histones par les autres protéines.

Citons pour terminer l'élégant travail de SEED (1963, 1964) qui, en combinant la photographie à intervalles réguliers de cellules en culture et l'autoradiographie à montré que dans les cellules normales la synthèse des protéines nucléaires commence 12 heures après la division, alors que dans les cellules tumorales elle est continue et d'importance égale au cours du cycle entier. Cette synthèse serait donc soumise normalement à un contrôle qui ferait défaut dans les cellules cancéreuses.

III. Histones et protamines

1. Définition

Les protéines les plus caractéristiques du noyau cellulaire sont les protéines basiques liées à l'ADN: histones dans tous les tissus somatiques, protamines dans les spermatozoïdes de certaines espèces.

a) Les plus simples sont les protamines qui sont des molécules de faible poids moléculaire, 2000 à 12000 (FÉLIX 1960) ne renfermant qu'un petit nombre d'acides aminés: 6 ou 7 en dehors de l'arginine qui est l'acide aminé les plus caractéristique de la protamine et atteint parfois plus de 90% de l'azote protéique. On ne rencontre jamais de protamine dans les tissus somatiques.

b) Les histones ont une complexité intermédiaire entre celle des protamines et celle des protéines plus élevées, leur masse moléculaire est d'environ 8000 à 30000 (BUTLER et DAVISON 1957). Elles renferment à peu près tous les acides aminés usuels avec une certaine prédominance des acides aminés basiques (arginine, lysine, histidine).

Le mode de préparation des histones n'a guère varié dans son principe depuis les travaux fondamentaux de Kossel.

Cependant comme le fait remarquer Butler et Davison (1957), ce mode de préparation par traitement acide laisse pourtant à désirer car il est assez brutal et risque de séparer artificiellement les parties les plus basiques d'une nucléoprotéine préexistante. On peut dont se demander si l'histone est véritablement une entité cellulaire ou une fraction basique réalisée par le traitement acide. La distinction entre l'histone et au moins une partie des protéines non histones des chromosomes est peut être, de ce fait un peu arbitraire.

Sur le plan cytochimique la coloration d'Alfert et Geschwind (1953) au vert rapide pour l'histone est la plus employée, dans une méthode plus récente, Black et Ansley (1964) utilisent le nitrate d'argent.

2. *Passage de l'histone à la protamine*

Le problème de la transformation de l'histone en protamine lors de la spermatogénèse chez certaines espèces animales est très intéressant sur le plan de la biologie générale et son étude peut donner des indications utiles sur le mode de liaison entre ADN et protéines basiques ainsi que sur l'importance de ces protéines dans l'activité de la cellule.

On a pensé quelquefois qu'au cours de la spermatogénèse la protamine dérivait de l'histone par amputation d'une partie de la molécule ne laissant subsister que la partie contenant l'arginine en abondance. Nous avons pu montrer (Vendrely et al. 1956, 1960a, 1960b) qu'en réalité il s'agit d'un remaniement plus complexe avec addition d'arginine à la molécule d'histone ou bien d'une démolition de l'histone et d'une synthèse de protamine. Pour cela nous avons considéré non plus l'histone ou la protamine isolée, mais le complexe nucléohistone dans son ensemble en rapportant les analyses effectuées sur la protéine à l'ADN qui lui est associée.

Nous avons pu montrer ainsi que les acides aminés basiques (arginine, lysine et histidine) de l'histone d'érythrocytes de poisson se trouvent en quantité suffisante pour saturer pratiquement tous les groupements phosphates de l'ADN (Tableau 1).

Tableau 1

Molécules d'arginine, de lysine et d'histidine pour 100 atomes de phosphore nucléique

	Fraction désoxyribonucléoprotéique originelle calculée du noyau d'érythrocyte et du spermatozoïde				Désoxyribonucléoprotéine isolée du même matériel			
	Arginine	Lysine	Histidine	Total	Arginine	Lysine	Histidine	Total
Carpe								
Erythrocyte (noyau)	33,2	64,2	5,0	102,4	25,8	35,2	5,0	66,0
Spermatoïde . . .	31,2	58,8	5,0	95,0	24,5	35,4	5,0	64,9
Brochet								
Erythrocyte (noyau)	33,2	68,7	5,0	106,9	23,4	31,5	5,0	59,9
Spermatoïde . . .	68,7	26,2	2,5	97,4	49,6	14,1	2,5	66,2
Truite								
Erythrocyte (noyau)	33,9	63,0	5,0	101,9	24,3	32,6	5,0	61,9
Spermatoïde . . .	103,0	0,0	0,0	103,0	74,9	0,0	0,0	74,9

En ce qui concerne les protamines, nous avons distingué trois types de spermatozoïdes, l'un à histone typique (Carpe), l'autre à protamine (Truite, Saumon) et le troisième intermédiaire (Brochet) présentant des teneurs en arginine de plus en plus élevée. Mais quel que soit le type étudié, la somme des acides aminés basiques sature toujours les groupements phosphate de l'ADN, l'arginine remplaçant progressivement la lysine à mesure que la protamine devient de plus en plus typique.

Cette évolution d'une espèce à l'autre peut donner une idée du changement qui doit se produire au cours de la maturation des spermatozoïdes dans les espèces animales à protamine et que l'on ne peut pas étudier directement tout au moins par des moyens biochimiques.

La mise en évidence de l'existence d'un type protéique intermédiaire dans un noyau de spermatozoïde est a rapprocher des résultats de BLOCH (1962 b) obtenus

Tableau 2. *Valeurs analytiques relatives à la Fig. 2.*

DPN de	Truite	Brochet	Carpe
	Mol/100 atomes DNA P		
Gl	0	3,6	30,9
Asp	0	2,4	18,4
Th	0	2,2	22,2
Se	11,0	17,1	20,5
Pr	14,7	17,1	22,9
Al	2,4	10,9	39,6
V	8,3	7,4	19,5
M	0	2,5	3,0
Isol	1,1	1,2	15,8
L	0,4	2,8	29,4
Ty	0	0,8	10,3
PA	0	0,8	8,4
Ly	0	14,0	42,0
A	74	49	26,0

par des méthodes histochimiques. Cet auteur, étudiant l'escargot (*Helix aspersa*) et le calmar (*Loligo opalescens*) a pu démontrer que le processus de spermatogénèse s'effectue chez ces animaux en plusieurs étapes: l'une antérieure à la formation des spermatozoïdes et qui se caractérise par le remplacement de l'histone typique par une histone plus riche en arginine, l'autre se produisant au moment de la maturation des spermatozoïdes et au cours de laquelle l'histone anormale est remplacée par une protamine. Ainsi le type intermédiaire que nous avons trouvé dans le spermatozoïde de certaines espèces pourrait être représenté dans certains cas à l'état transitoire au cours de la spermatogénèse.

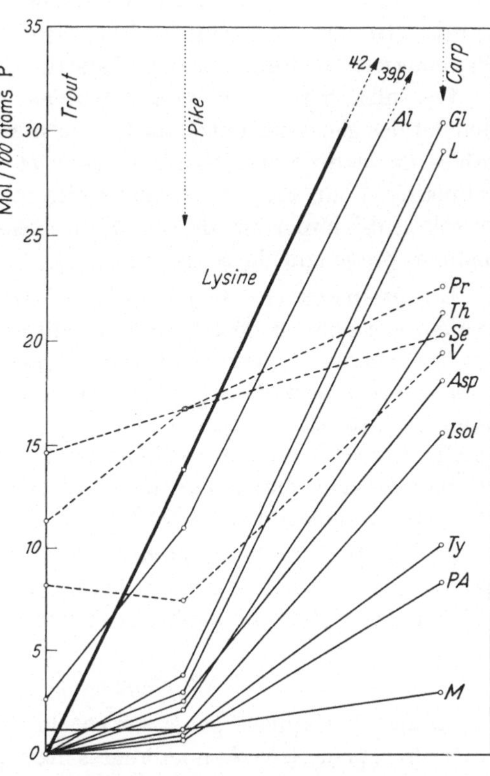

Fig. 2. Comparaison de la composition en acides aminés de nucléoprotéines isolées de 3 variétés de spermatozoïdes de poissons (Truite, Brochet, Carpe)

Reprenant nos trois espèces de poissons nous avons étudié ensuite l'évolution de tous les acides aminés lorsque l'on passe du type histone au type intermédiaire puis au type protamine. La Figure 2 et le Tableau 2 illustrent nos résultats et montrent que l'on assiste à une diminution notable des acides aminés basiques dans le type brochet, en rapport semblet-il avec la diminution de la lysine. Cinq des acides aminés neutres se comportent un peu différemment et subsistent en quantité appréciable dans la protamine typique: la proline, la sérine et la valine et dans une moindre mesure l'alanine et la méthionine. La conservation de ces acides aminés dans des noyaux où les gènes doivent être préservés mais n'ont pas à s'exprimer suggère leur importance dans la constitution d'une protéine de structure nécessaire à la protection des gènes par opposition à la protéine associée aux gènes en activité dans les cellules somatiques.

IV. Les histones

Un certain nombre de revues ont été consacrées aux histones ces dernières années (Butler et Davison 1957, Butler 1959, Moore 1959, Bloch 1961, Peacocke 1960, Philipps 1962, Bonner et Ts'o 1964).

Ces substances jouent vraisemblablement un rôle important dans le fonctionnement des gènes en contrôlant leur activité, elles seraient donc par le fait responsables de la différenciation des tissus et peut être du contrôle de la division cellulaire. Pour essayer de comprendre comment les histones pourraient remplir ce rôle il est important de savoir s'il existe des différences spécifiques entre les histones provenant de tissus différents.

Les Stedman (1950), Cruft et al. (1954), Mauritzen et Stedman (1959) avaient cru pouvoir déceler de telles différences et notamment entre histones de tissus normaux et histones de cancer, mais les méthodes d'extractions utilisées étaient fort imparfaites. Les travaux plus récents sur le dosage des acides aminés dans les histones de tissus différents (Crampton et al. 1957, Vendrely et al. 1958), d'espèces animales différentes (Vendrely et al. 1958) ou entre tissus normaux et cancéreux (Allfrey et al. 1955, Davison 1957) montrent au contraire une grande similarité entre ces histones.

Il importe donc de rechercher ces différences par des méthodes plus sensibles que celles d'une analyse globale de l'histone. C'est pourquoi le fractionnement de l'histone fait l'objet de nombreux travaux.

1. Fractionnement de l'histone

Ce que l'on appelle généralement histone est, en fait un mélange d'espèces différentes, peut être très nombreuses mais qui se groupent en 3 classes principales (Phillips 1962).

a) Les histones très riches en lysine avec un rapport lysine/arginine supérieur à 4.

b) Les histones modérément riches en lysine avec un rapport lysine/arginine entre 1 et 4.

c) Les histones riches en arginine avec un rapport lysine/arginine inférieur à 1.

Le fractionnement des histones démontré par le travail initial de Stedman (1951) a été réalisé depuis par l'emploi de très nombreuses techniques sur lesquelles il serait trop long de s'étendre ici. On opère soit directement en partant du tissu entier, soit plus fréquemment en partant des noyaux ou de la nucléo-protéine isolés. Les méthodes utilisées variant suivant les auteurs, il est parfois difficile de comparer dans le détail les résultats obtenus.

La caractérisation des diverses fractions est obtenue le plus parfaitement par électrophorèse sur gel d'amidon (SMITHIES 1955) ou par chromatographie sur résine. On a pu grâce à ces techniques séparer jusqu'à 16 fractions à partir de l'histone d'érythrocytes de poulet (NEELIN et CONNELL 1959).

Le problème est maintenant d'obtenir des fractions pures afin de les comparer entre elles. L'une des fractions du groupe des histones modérement riche en lysine a été obtenue sous une forme très pure et l'analyse chromatographique de ses peptides a été entreprise (HNILICA et al. 1963). Il apparaît déjà grâce à ces techniques que la composition des peptides dans les différentes fractions sont très différentes (PHILLIPS et SIMSON 1962).

S'il est vrai, comme on le pense actuellement, que les histones agissent comme suppresseurs de l'activité des gènes en les empêchant de fonctionner comme templates pour la production d'ARN messager, la séquence des acides aminés des différentes histones pourrait être très spécifique et correspondre à des séquences particulières de nucléotides de l'ADN. Il serait donc important d'arriver à fractionner le complexe de nucléohistone entier de façon à pouvoir envisager une comparaison entre la séquence des nucléotides de l'ADN et celle des acides aminés dans l'histone associée. Certains travaux déjà anciens (CHARGAFF et al. 1953,

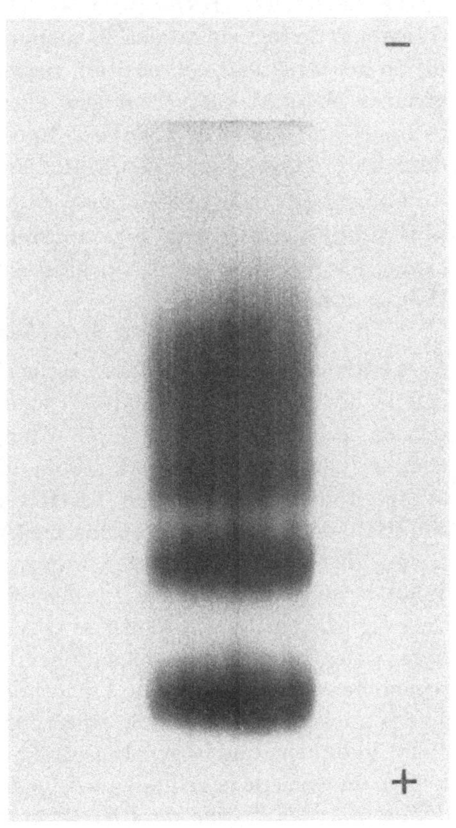

Fig. 3. Fractions nucléoprotéiques (précipitables à pH 4.7) obtenues par autolyse de chromatine de thymus de veau. Electrophorèse sur gel d'agarose-sephadex G. 200

LUCY et BUTLER 1954, 1955) montrent que ce fractionnement n'est pas possible sur une nucléoprotéine intacte, il faut qu'une dénaturation préalable de la protéine intervienne. C'est peut être le fait que la nucléoprotéine dénaturée est dans un état différent du matériel natif qui a découragé les chercheurs de poursuivre leur travaux.

Nous avons essayé d'aborder ce problème de façon nouvelle en étudiant l'autolyse aseptique de chromatine de thymus de veau (VENDRELY 1964). Cette action enzymatique très graduelle et plus physiologique aboutit à la formation de fragments de nucléohistone de faible masse moléculaire (8×10^4 à 10×10^4) insolubles à pH 4,7. L'analyse électrophorétique de ces produits d'autolyse sur gel d'amidon et gel de Sephadex (Vendrely et al. 1964), révèle l'existence d'au moins trois fractions migrant vers le pôle positif (Fig. 3). L'analyse des acides aminés de la protéine et des bases de l'acide nucléique pour chacune des fractions est en cours; cette nouvelle méthode permet donc d'espérer des résultats intéressants.

Quoi qu'il en soit, le fractionnement des histones n'a pas encore permis de trouver des différences significatives entre les fractions d'histones de différents tissus cancéreux. Hnilica et al. (1962a) comparant des fractions similaires d'histones de foie, de rate et de tumeur de Walker chez le rat n'ont trouvé aucune différence dans leur composition respective en acides aminés ainsi que dans leur groupes N terminaux. Peut être chez des espèces très éloignées (germe de blé comparé au thymus de veau) est-il possible de déceler une différence significative dans les fractions d'histone (Johns et Butler 1962).

Cependant, l'emploi des isotopes permet d'espérer la détection de différences plus subtiles concernant le comportement métaboliques d'histones de différents tissus, les résultats les plus positifs concernent ici le problème des histones du tissu cancéreux.

2. Métabolisme des histones

a) Histone et ADN. L'ADN est certainement la substance la plus stable de la cellule vivante (Vendrely 1955) et dans la majorité des cas, son métabolisme semble se limiter à sa duplication lors de la duplication des chromosomes. Or comme l'ont montré Alfert (1955), Bloch et Godman (1955) par des mesures cytophotométriques lorsque l'ADN double dans un noyau l'histone double simultanément. Le rapport histone/ADN se montre constant dans diférents tissus, au cours de la croissance normale ou pathologique (Rasch et Woodard 1959), et même après irradiation des cellules (Das et Alfert 1961, Ringertz 1963). Cela signifie-t-il que la synthèse d'ADN et celle d'histone sont rigoureusement liées ? Des travaux plus récents sembleraient indiquer qu'il n'en est rien. Des recherches biochimiques sur l'incorporation des précurseurs de protéines et des acides nucléiques dans le foie en régénération (Holbrook et al. 1962, Irvin et al. 1963) indiquent que la synthèse de l'histone semble s'étaler sur une portion plus longue du cycle de la cellule que celle de l'ADN. La phase de synthèse de l'histone débuterait avant celle de l'ADN et coinciderait avec une chute rapide de la synthèse de l'ARN. Ces résultats sont en accord avec l'idée d'une inhibition de la synthèse d'ARN par l'histone.

Enfin les travaux d'autoradiographie de Prescott et Bender (1963) sur des cellules de Hamster en culture montrent que la synthèse des protéines du chromosome ne se fait pas selon le mode semi conservatif qu'affecte l'ADN, au contraire, au cours de la replication les molécules d'histones semblent pouvoir se séparer d'une molécule d'ADN, s'associer à une autre et se renouveler continuellement.

Ainsi donc la constance du rapport histone/ADN n'est pas l'effet d'un métabolisme lié des deux substances, mais celui d'une association entre deux substances à renouvellement différent. D'ailleurs, ce rapport peut se trouver perturbé. Lindner (1963) montre que le fluorouracile provoque une augmentation considérable de l'histone par rapport à l'ADN dans les cellules d'ascite d'Ehrlich. Perugini et Soldati (1957), par ailleurs, signalent que la teneur en histone des lymphocytes de leucémie humaine évaluée par cytophométrie est de 15% supérieure à celle de lymphocytes normaux alors que la teneur en ADN est la même dans les deux cas.

b) Métabolisme propre de l'histone. Des nombreux travaux d'incorporation d'acides aminés dans les protéines du noyau que nous avons déjà cités, il ressort que le métabolisme des histones de tissus normaux, bien que faible comparé à celui des protéines non histone du noyau n'en est pas moins supérieur à ce que l'on pourrait attendre si le renouvellement de l'histone était uniquement lié à la duplication des chromosomes, BUSCH et al. (1963) indiquent que ce renouvellement et de 1 à 3 % par jour alors qu'il n'y a aucune synthèse d'ADN. De nombreux auteurs et, particulièrement l'école de Busch ont montré que d'une façon générale l'activité spécifique des histones de tumeurs étaient nettement plus élevée que celle des histones de tissus normaux et ceci dans un grand nombre de tumeurs (ROTHERHAM et al. 1957, BUSCH et DAVIS 1958, BUSCH et al. 1958, BUTLER et LAURENCE 1962, BUSCH et al. 1963a).

Il apparaît nettement que les différentes fractions d'histone ont des taux de renouvellement différents, les fractions riches en lysine ont l'activité spécifique la plus forte (HNILICA et al. 1962b). Ces constatations sont à rapprocher des travaux récents (HUANG et BONNER 1964) qui montrent que les différentes fractions d'histones de thymus de veau ont des activités différentes en ce qui concerne l'inhibition de la synthèse d'ARN dépendant de l'ADN dans des systèmes in vitro. Les histones ayant le rapport arginine/lysine le plus bas sembleraient inhiber le plus cette synthèse. Le comportement métabolique des différentes fractions d'histone doit donc avoir une grande signification fonctionnelle.

D'autre part (BUSCH et al. 1963a) il semble que dans les cellules cancéreuses des fractions riches en lysine aient tendance à se complexer avec des protéines acides du suc nucléaire. BUSCH voit dans ce phénomène un mécanisme possible d'explication de la multiplication incontrôlée des cellules cancéreuses par suppression de la fonction régulatrice possible de certaines histones sur la synthèse de l'ADN.

V. Conclusions

Nous avons pu constater au cours de ce bref exposé que l'étude biochimique des protéines nucléaires n'en est qu'à son début. Le perfectionnement récent des techniques d'isolement et de fractionnement des noyaux, l'utilisation des isotopes pour l'étude du métabolisme, l'amélioration des techniques de fractionnement des histones et l'analyse fine des fractions obtenues, enfin l'emploi des techniques modernes utilisant les systèmes de biosynthèse in vitro permettront sans aucun doute un développement considérable sur la nature et le rôle des protéines nucléaires. Ces notions établies par les biochimistes doivent orienter des recherches histochimiques sur ces substances, les techniques de microspectrophotométrie et d'interférométrie, l'autoradiographie ont déjà donné de précieux renseignement et peuvent permettre de résoudre des problèmes particuliers inaccessibles aux biochimistes.

Bibliographie

ALFERT, M.: Chemie der Genetik, S. 73—84. Berlin-Göttingen-Heidelberg: Springer 1959.
— H. A. BERN u. R. H. KAHN: Acta anat. (Basel) **23**, 185 (1955).
— et I. I. GESCHWIND: Proc. Nat. Acad. Sci. (Wash.) **39**, 991 (1953).

Allfrey, V. G., M. M. Daly and A. E. Mirsky: J. gen. Physiol. 38, 415 (1955).
Baltus, E.: Biochim. biophys. Acta (Amst.) 15, 263 (1954).
Black, M. M., and H. R. Ansley: Science 143, 693 (1964).
Bloch, D. P.: The cell in mitosis. New York: Acad. Press 1961.
— (a) Proc. nat. Acad. Sci. (Wash.) 48, 324 (1962).
— (b) J. Histochem. Cytochem. 10, 137 (1962).
—, and G. C. Godman: J. biophys. biochem. Cytol. 1, 17 (1955).
Bonner, J., and P. Ts'o: The Nucleohistones. San Francisco-London-Amsterdam:Holden-Day, Inc. 1964.
Brunish, R., and J. M. Luck: J. biol. Chem. 198, 621 (1952).
Busch, H., and J. R. Davis: Cancer Res. 18, 1241—1256 (1958).
— — and D. C. Anderson: Cancer Res. 18, 916 (1958).
— W. J. Steele, Cw. Hnilica, and H. Mavioglu: J. cell. comp. Physiol., Suppl. 1 to vol. 62, No 2, 99—110 (1963a).
Busch, H. B., P. Byvoet, and H. R. Adams: Exp. Cell Res., Suppl. 9, 376—386 (1963b).
Buttler, J. A. V.: Expos. ann. Biochim. méd., Ser. XXI, 41 (1959).
— Exp. Cell Res., Suppl. 9, 349 (1963).
—, and P. Cohn: Biochem. J. 87, 330 (1963).
—, and P. F. Davison: Advanc. Enzymol. 18, 161 (1957).
—, and D. J. R. Laurence: Brit. J. Cancer 14, 758 (1960).
Caspersson, T., et L. Santesson: Acta radiol. (Stockh.), Suppl. 46, 1 (1942).
Chargaff, E., C. F. Crampton, and R. Lipshitz: Nature (Lond.) 172, 289 (1953).
Chauveau, J., Y. Moule, and C. Rouiller: Exp. Cell Res. 11, 317 (1956).
Chorazy, M. R., A. Bendich, E. Borenfreund, and D. J. Hutchinson: J. Cell Biol. 19, 59 (1963).
Crampton, C. F., W. H. Stein, and S. Moore: J. biol. Chem. 225, 363 (1957).
Cruft, H. J., C. M. Mauritzen, and E. Stedman: Nature (Lond.) 174, 580 (1954).
Daly, M. M., V. G. Allfrey, and A. E. Mirsky: J. gen. Physiol 36. 173 (1962).
Das, N. K., and M. Alfert: Proc. nat. Acad. Sci. (Wash.) 47, 1 (1961).
Davison, P. F.: Biochem. J. 66, 703 (1957).
Desjardin, R., K. Smetana, W. J. Steele, and H. Busch: Cancer Res. 23, 1819 (1953).
Dounce, A. L.: The nucleic acids (Chargaff et Davidson ed.), II, p. 93. Acad. Press 1955.
—, and N. K. Sarkar: The cell nucleus, p. 206. Butterworth 1960.
Felix, K.: Advanc. Protein Chem. 15, 1 (1960).
— H. Fischer, and A. Krekels: Progr. Biophys. 6, 1 (1956).
Ficq, A.: Exp. Cell Res. 9, 286 (1955).
— Lab. Invest. 8, 237 (1959).
Frenster, J. H., V. G. Allfrey, and A. E. Mirsky: Proc. nat. Acad. Sci. (Wash.) 46, 432 (1960).
Hamer, D.: Brit. J. Cancer 5, 130 (1951).
Himes, M. H., A. W. Pollister, and B. C. Moore: J. Histochem. Cytochem. 3, 390 (1955).
Holbrook, D. J., J. H. Evans, and J. L. Irvin: Exp. Cell Res. 28, 120 (1962).
— J. L. Irvin, E. M. Irvin, and J. Rotherham: Cancer Res. 20, 1329 (1960).
Hnilica, L., E. W. Johns, and J. A. V. Butler: Biochem. J. 82, 123 (1962a).
— C. W. Taylor, and H. Busch: Fed. Proc. 21, 406 (1962b).
— —, H. Mavioglu, and H. Busch: Fed. Proc. 22, 658 (1963).
Huang, R. C., and J. Bonner: Proc. nat. Acad. Sci. (Wash.) 48, 1216 (1962).
Huang, R. C., J. Bonner, and K. Murray: J. molec. Biol. 8, 54—64 (1964).
Irvin, J. L., D. J. Holbrook, J. H. Evans, H. C. McAllister, and E. P. Stiles: Exp. Cell. Res., Suppl. 9, 359 (1963).
Johns, E. W., and J. A. V. Butler: Biochem. J. 84, 436 (1962).
Keir, H. M., R. M. S. Smellie, and G. Siebert: Nature (Lond.) 196, 752 (1962).
Kossel, A.: Physiol. Chem. 8, 511 (1884).
Lindner, A., T. Kutkam, Y. R. Sankaranara, R. Rucker, and J. Aradondo: Exp. Cell Res. Suppl. 9, 485 (1963).
Lucy, J. A., and J. A. V. Butler: Nature (Lond.) 174, 32 (1954).
— — Biochim. biophys. Acta (Amst.) 16, 431 (1955).

MAURITZEN, C. M., and E. STEDMAN: Proc. roy. Soc. B **150**, 299 (1959).

MIRSKY, A. E., and H. RIS: J. gen. Physiol. **31**, 1 (1947).

MONTY, K. J., M. LITT, E. R. M. KAY, and A. L. DOUNCE: J. biophys. biochem. Cytol. **2**, 127 (1956).

MOORE, S.: Rapp. Cons; Chim. Solvay **11**, 77 (1959).

MURAMATSU, M., K. SMETANA, and H. BUSCH: Cancer Res. **23**, 510 (1963).

NEELIN, J. M., et G. E. CONNELL: Biochim. biophys. Acta (Amst.) **31**, 539 (1959).

OKUDA, J., D. SZAFARZ, and Y. KHOUVINE: C.R. Acad. Sci. (Paris) **257**, 2904 (1963).

PEACOCKE, A. R.: Progr. Biophys. **10**, 55 (1960).

PERUGINI, S., U. TORELLI et M. SOLDATI: Experientia (Basel) **13**, 441 (1957).

PHILLIPS, D. M. P.: Prog. Biophys. **12**, 211 (1962).

—, and P. SIMSON: Biochem. J. **82**, 236 (1962).

POGO, A. O., B. G. T. POGO, V. C. LITTAU, V. G. ALLFREY, A. E. MIRSKY et M. G. HAMILTON: Biochem. biophys. Acta (Amst.) **55**, 849 (1962).

PRESCOTT, D. M., and M. A. BENDER: J. Cell comp. Physiol., Suppl. **1**, **62**, 175 (1963).

RASH, E., and S. W. WOODARD: J. biophys. biochem. Cytol. **6**, 263 (1959).

RINGERTZ, N. R.: Exp. Cell. Res. **32**, 401 (1963).

ROTHERHAM, J., J. L. IRVIN, E. M. IRVIN, and D. J. HOLBROOK: Proc. Soc. exp. Biol. (N.Y). **96**, 21 (1957).

SEED, J.: Nature (Lond.) **198**, 147 (1963).

— J. Cell Biol. **20**, 17 (1964).

SIEBERT, G.: Exp. Cell Res., Suppl. **9**, 389 (1963).

SMELLIE, R. M. S., W. M. McINDOE, et J. N. DAVIDSON: Biochim. biophys. Acta (Amst.) **11**, 559 (1953).

SMITHIES, O.: Biochem. J. **61**, 629 (1955).

SOMERS, C. E., A. COLE, and T. C. HSU: Exp. Cell Res., Suppl. **9**, 220 (1963).

STARBRUCK, W. C., and H. BUSCH: Cancer Res. **20** (6), 890 (1960).

STEDMAN, E., and E. STEDMAN: Symp. Soc. exp. Biol. **1**, 232 (1947).

— — Nature (Lond.) **166**, 780 (1950).

— — Phil. Trans. B **235**, 565 (1951).

STEELE, W. J., and H. BUSCH: Cancer Res. **23**, 1153 (1963).

VENDRELY, C., A. KNOBLOCH et R. VENDRELY: Biochim. biophys. Acta (Amst.) **19**, 472 (1956).

VENDRELY, R.: The Nucleic acids, vol. II, p. 155. New York: Acad. Press. 1955.

— In: The Nucleohistones, (J. BONNER, and P. Ts'o ed.), p. 307. San Francisco-London-Amsterdam: Holden-Day Inc. 1964.

— Y. COIRAULT et A. VANDERPLANCKE: C.R. Acad. Sci. (Paris), sous presse (1964).

— A. KNOBLOCH, and H. MATSUDAIRA: Nature (Lond.) **181**, 343 (1958).

— A. KNOBLOCH-MAZEN, and C. VENDRELY: The Cell Nucleus, p. 200. London: Butterworths Sci. publ. 1960a.

— — — Biochim. Pharmacol. **4**, 19 (1960b).

VINCENT, W. S.: Proc. nat. Acad. Sci. (Wash.) **38**, 138 (1952).

WANG, T. Y.: Biochim. biophys. Acta (Amst.) **45**, 8 (1960).

ZALTA, J. P., R. ROZENCWAJG, N. CARASSO et P. FAVARD: C.R. Acad. Sci. (Paris) **1**, 412 (1962).

ZBARSKY, I. B., i. G. P. GEORGIEV: Biokhimiya **24**, 192 (1959).

— N. P. DMITRIEVA, and L. P. YERMOLAEVA: Exp. Cell Res. **27**, 573 (1962).

Institut de Recherches Scientifiques sur le Cancer,
16, Avenue Vaillant-Couturier,
Villejuif (Seine), France

From the Department of Pharmacology and the Beaumont-May Institute of Neurology,
Washington University School of Medicine, St. Louis, Missouri

Microanalysis for Histochemical Purposes*

By

OLIVER H. LOWRY

With 4 Figures in the Text

Histochemical and cytochemical methods in general fall into two main classes. In the first class are methods which depend on assessment under the microscope of the amount of substance present. The substance itself may be estimated, as in the case of nucleic acid, or a product formed by the substance may be estimated, as in the case of an enzyme. The assessment may be made by eye or by photometer. We may call this *slide histochemistry;* most histochemical methods at present are of this variety. The other variety of histochemistry depends on direct quantitative analysis of a small fragment of tissue of known histological composition. This may be called *test-tube histochemistry.* It was developed and exploited by LINDERSTRØM-LANG and HOLTER in a series of brilliant investigations.

It is generally held that of the two methodologies slide histochemistry is more sensitive, gives far sharper localization, and is less difficult and time consuming; whereas test-tube histochemistry offers greater specificity and better quantitation. In this presentation I will try to show that test-tube histochemistry can probably be extended almost to the limits of the light microscope and that direct chemical analyses are not necessarily difficult.

There are three distinct steps in test-tube histochemical analysis, 1. isolation of the sample without change in the localization or concentration of the substance to be measured, 2. determination of the sample size, and 3. performance of the chemical analysis itself. These steps will be considered separately.

Isolation of Sample

There is much in favor of isolation by a) quick freezing, b) sectioning in a cryostat, c) drying in vacuum at -40^0 *after* sectioning, and d) separation of the sample by dissection under a microscope [1]. By following this sequence drying under vacuum is rapid and no embedding is required. To sample living tissue in the fresh state must be regarded as hazardous. Unicellular organisms or blood cells can often be manipulated without apparent damage but cells present in a solid tissue are easily injured during isolation. This would be particularly disturbing in the case of metabolically labile substances; in addition leakage even of proteins can occur. Sampling from frozen-dried sections avoids these dangers.

Quick freezing. Rapid freezing is necessary to keep ice crystals small; ice crystal artefacts are preserved by drying and make subsequent identification of structures difficult or impossible. (In the usual frozen section, which is not

* Supported in part by grants from the American Cancer Society (P-78), and the United States Public Health Service (NB-00434, and B-1352).

dried, ice crystal artefacts disappear when the section is allowed to thaw before fixation.) Manipulation of tissue prior to freezing is exceedingly critical if metabolites are to be measured, since many of these are profoundly affected by (for example) a few seconds of anoxia. Enzymes are presumably unaffected in concentration or distribution so long as the cells are still alive at the time of freezing. For some purposes even long periods after death may not effect the results. Thus ROBINS *et al.* [2] found that the activities and distribution of many enzymes in the three major cerebellar layers were unchanged by a delay of 6 hours after death before freezing. Once frozen, tissue can be preserved for months and possibly years at low temperature (-80^0) without detectable change in appearance of sections subsequently cut, or in the concentration or activity of labile metabolites or enzymes. This is not true of storage at higher temperature, -20^0 for example.

Sectioning. This is best carried out in a cryostat as originally described by LINDERSTRØM-LANG and MØGENSEN [3]. The thickness that is suitable will depend on the size of the structure to be analyzed. Sections as thin as 2 or 3 μ can be cut satisfactorily. The thinner the section the lower the temperature required. Too cold, sections will crack, too warm they will crumple or stick to the knife. A mechanical device to force the sections into a ribbon was originally described [3] but we have found it more generally useful to handle the sections one at a time using a camel's hair brush to keep each section flat.

Drying. Sections are conveniently dried and stored in special holders inside of a large glass tube fitted with a stop cock [1]. A suitable drying temperature is -40^0. At this temperature a vacuum of 0.01 mm Hg is sufficient; 8 or 10 hours is ordinarily ample time for drying. At temperatures above -30^0 some shrinkage can occur, at lower temperatures the vapor tension of ice falls so low as to make drying difficult. Failure to obtain satisfactory dry sections is usually the result of thawing and refreezing at some step in the procedure. After drying, sections can be stored for years at -20^0 without deterioration.

Dissection. The storage tube and contents are brought to room temperature before breaking the vacuum in order to prevent condensation of moisture which would be ruinous. Most enzymes and labile substrates in the dry sections will withstand at least a few hours at room temperature. This permits leisurely dissection. Storage tests have shown that at room temperature enzymes are much better preserved under vacuum than in air [4]. Therefore, when a tube containing sections is brought to room temperature it is kept evacuated at all times except for the brief intervals necessary to remove a few sections for dissection. It has also been found that certain metabolites in the sections can be altered if the air is very humid. Consequently the relative humidity during dissection is kept below 60%. A small amount of radium kept nearby eliminates static charges which may otherwise be very troublesome.

Splinters of razor blades mounted on a suitable handle make satisfactory knives for dissection. It is particularly advantageous to give flexibility to the blade by interposing a stiff hair or bristle between it and the holder (Fig. 1). Dissection is conveniently carried out on an opalescent plastic surface illuminated intensely from below. The plastic surface does not dull the knife edge, and the diffused light is particularly favorable for illuminating objects in the frozen-dried

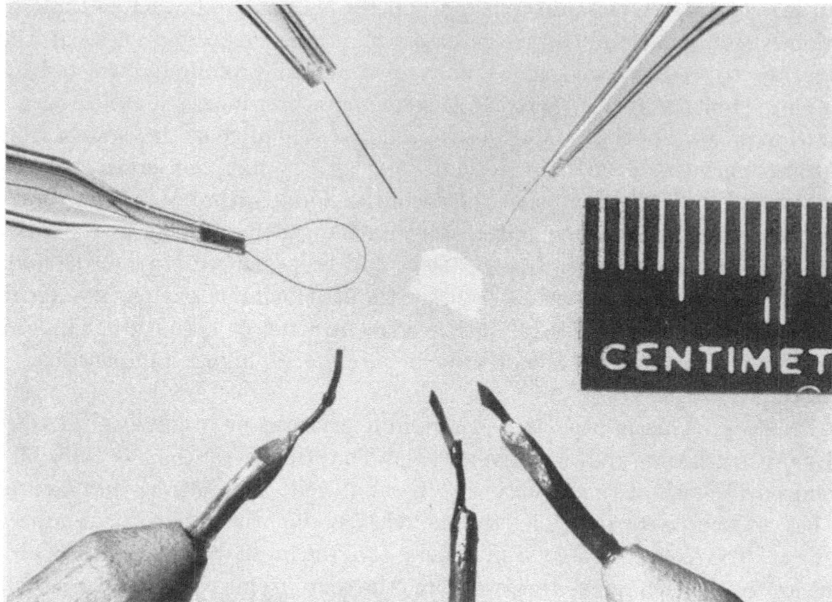

Fig. 1. Dissecting tools. — The knife blades are splinters of razor blade broken so as to give a suitable shaped edge. These are sealed to a copper wire which permits adjustment of angle with a pair of pliers. In 2 cases a piece of stiff hair or bristle has been interposed between blade and wire. Epoxy resin was used as the adhesive. Also shown are a hair loop, two "hair points" with which fragments of tissue can be picked up for transfer and a 20 μ section of cerebellum. The finer hair point is a sable hair from a small paint brush. On the end of this has been sealed a 2 mm piece of quartz fiber 3 or 4 μ in diameter. This hair point is suitable for picking up very small samples. (From reference [5])

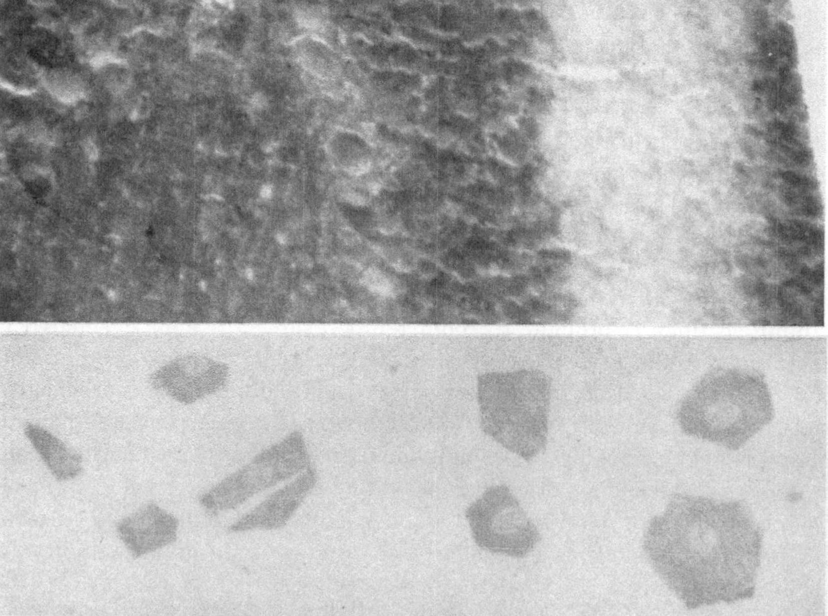

Fig. 2. Frozen-dried tissue as it appears during dissection. In the upper portion of the figure is shown rabbit cerebellum from a 20 μ section. In the lower portion are shown various stages in the isolation of nuclei from dorsal root ganglion cells. The samples are from an 8 μ section. (From reference [6])

material. If tremor of the hand is minimized by holding the knife as close as possible to the blade, free-hand dissection to within $10\,\mu$ is easily attainable. Even finer free-hand dissection is possible (to within 1 or $2\,\mu$) by the following manuever: A knife with flexible shaft is used. The point of the blade is touched to the plastic surface just ahead of the sample. This almost entirely eliminates tremor and creates a fulcrum. The blade is exactly positioned, without displacing the tip, by slight lateral pressure. The blade can now be brought down where desired by simply increasing downward pressure, since the flexible shaft permits the blade to rotate slightly.

Landmarks to guide dissection are easily visible in frozen dried material (Fig. 2). Small individual structures may be hard to see in thick sections. This fact will determine the most suitable section thickness. Large nuclei are easy to see and have been cleanly dissected for analysis ([7] and Fig. 2). It is believed that structures 1 or $2\,\mu$ in diameter could be visualized and, with the aid of micromanipulation, could be dissected out for analysis. After dissection, samples are picked up with "hair points" (Fig. 1). The diameter of the hair point tip should be much smaller than that of the sample.

Measurement of Sample Size

The most sensitive and satisfactory way found to measure the size of samples dissected is to weigh them. For this purpose a "fish pole" balance [1] is quite satisfactory. This is simply a quartz fiber mounted horizontally (Fig. 3). The

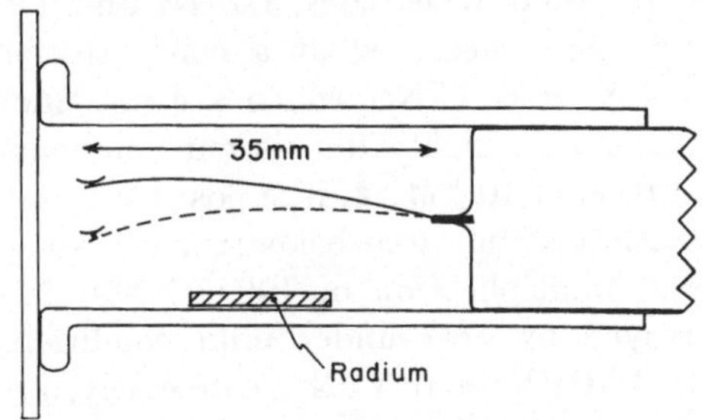

Fig. 3. Quartz fiber fishpole balance. The balance shown has a sensitivity of 0.002 µg (0.005 mm displacement of the tip) and a load limit of 1 µg (2.5 mm displacement). The displacement is measured on a mm scale in the ocular of a wide angle microscope (2 × objectives) mounted horizontally. The quartz fiber is about $10\,\mu$ diameter. The pan is made of very thin glass. The fiber is sealed with epoxy resin to a copper wire previously sealed to the plunger of a 20 ml syringe. The tip of the syringe is cut off and the plunger reversed. The angle of the fiber can be adjusted by bending the copper wire with pliers. The radium is used to eliminate static electrical charge. To increase the sensitivity 100-fold the fiber diameter is reduced to $1\,\mu$, the length to 8 mm. The pan is eliminated, since adhesive forces are sufficient to hold samples weighing less than 0.05 µg to the fiber tip. The syringe must he smaller (2 ml) to reduce air currents. (From reference [8])

sample is added under supervision through a wide angled microscope which is also used to measure the displacement. For satisfactory precision, a particular balance can cover only a 5-fold range of sample size. Therefore, a series of balances may be needed (see Fig. 3). This disadvantage is counterbalanced by the high sensitivity possible. The most sensitive balance to date was reproducible to ± 2 picograms i.e. to within 10% of the weight of one red blood cell. Balances 10 to 100 times this sensitivity can probably be made without great difficulty.

This would permit weighing dry samples of tissue 1 or $2\,\mu$ in diameter. Two disturbing factors have to be controlled, static electric charge and thermal air currents. A small amount of radium in the balance case eliminates static charges, a small balance case minimizes air currents. A balance with a sensitivity of 2 picograms is sufficiently stable in a chamber of 0.2 ml but would be useless in a 2 ml chamber. As balances become smaller illumination needs to be increased. When this becomes necessary filtered light of 500 to 600 mμ wavelength is used in order to give greatest visibility with least heat.

Chemical Analysis

A wide variety of tools and methods have been used for test tube histochemical analyses. Glick has described a great many of these in an excellent and detailed manner [9]. The suitability of a given method will depend among other things on the sensitivity required. A titrimetric procedure may be sufficiently sensitive to measure the dipeptidase in a $20\,\mu$ section 4 mm in diameter, but a method a million-fold more sensitive will be needed to measure the glucose in a single liver cell. It is not the purpose here to discuss the sensitivity of all the various analytical systems, but instead to describe one which has great flexibility, specificity and almost unlimited sensitivity.

This system is a fluorometric one which employs pyridine nucleotides in combination with purified enzymes. It is based on the following facts: 1. fluorometry is inherently very sensitive, 100- to 1000-fold more sensitive than colorimetry; 2. the reduced pyridine nucleotides, DPNH and TPNH, are themselves fluorescent and can be easily measured at a concentration of 10^{-7} M; 3. the oxidized nucleotides, DPN$^+$ and TPN$^+$, which are not fluorescent as such, are converted by strong alkali into highly fluorescent compounds [10] that can be measured at a concentration of 10^{-8} M; 4. it is possible to increase the sensitivity 10,000-fold by "enzymatic cycling" (see below); if necessary the process can be repeated to give overall multiplication of 100,000,000; 5. DPNH and TPNH are qualitatively destroyed by acid under mild conditions that leave DPN$^+$ and TPN$^+$ untouched; 6. DPN$^+$ and TPN$^+$, conversely, are quantitatively destroyed by weak alkali (pH 12) without development of fluorescence and without detectable loss of DPNH or TPNH; and 7. almost any substance in living cells, whether enzyme or substrate, can be induced with the aid of suitable auxiliary enzymes to oxidize or reduce a pyridine nucleotide.

The measurement of glucose in samples of various size will serve to illustrate the system. In this instance the primary enzyme reactions are:

$$\text{Glucose} + \text{ATP} \xrightarrow{\text{hexokinase}} \text{glucose-6-P} + \text{ADP} \tag{1}$$

$$\text{Glucose-6-P} + \text{TPN}^+ \xrightarrow{\text{dehydrogenase}} \text{6-P-gluconate} + \text{TPNH} \tag{2}$$

Case 1. To measure glucose in a frozen-dried tissue sample weighing 10 μg (10^{-10} moles of glucose). The sample after being heated briefly in acid to destroy enzymes is added to 1 ml of reagent in a fluorometer tube. The reagent contains both of the enzymes shown above plus excess ATP and TPN⁺. Glucose is completely converted to 6-P-gluconate with the formation of an equivalent amount of TPNH which is measured by the rise in fluorescence.

Case 2. To measure glucose in a 1 μg dry tissue sample (10^{-11} moles of glucose). The procedure is the same as in case 1 except that the volume of reagent is reduced to 20 μl and after time for the two enzyme reactions to be complete, a buffer of pH 12 is added. Brief heating at 60⁰ destroys TPN⁺ without making it fluorescent. The sample is now added to 1 ml of 6 N NaOH (containing a little H_2O_2 to convert the TPNH to TPN⁺) and the fluorescence is read an hour later.

Case 3. To measure glucose in 0.01 μg dry tissue (10^{-14} moles of glucose). Except for the last step, the analysis is conducted as in case 2 but with reduced volumes. After destroying excess TPN⁺, the sample is added to 100 μl of "cycling reagent". This reagent contains glucose-6-P dehydrogenase and its substrate glucose-6-P (G6P), together with glutamic dehydrogenase and its substrates α-ketoglutarate (αKG) and NH_4^+. As soon as TPNH is added the following reactions begin to take place over and over in rapid succession [*11*]:

$$\text{TPNH} + \alpha\text{KG} + \text{NH}_4^+ \longrightarrow \text{glutamate} + \text{TPN}^+ \tag{3}$$

$$\text{TPNH} + \text{6-P-gluconate} \longleftarrow \text{G6P} + \text{TPN}^+ \tag{4}$$

The overall reaction is:

$$\alpha\text{KG} + \text{NH}_4^+ + \text{G6P} \longrightarrow \text{glutamate} + \text{6-P-gluconate} \tag{5}$$

Since total TPN is rate limiting, product formation is proportional to the total amount of TPN present and to the time of incubation. With high levels of both enzymes the amount of each product formed in an hour can be 10,000 times the amount of TPN present.

After a fixed incubation period the reaction is stopped with heat and to the sample is added 1 ml of reagent containing 6-P-gluconate (6PG) dehydrogenase and excess TPN⁺ and the following reaction takes place:

$$\text{6PG} + \text{TPN}^+ \longrightarrow \text{ribulose-5-P} + \text{CO}_2 + \text{TPNH} \tag{6}$$

The TPNH (now 10^{-10} moles instead of 10^{-14} moles) is measured by its fluorescence.

Case 4. To measure glucose in 10^{-4} μg (100 picogram) dry tissue (10^{-16} moles of glucose). This has not yet been done in fact, but in principle would not be difficult. The first steps would be the same as in case 3 but with reduced volumes. After reaction 6 the sample would be heated with weak alkali to destroy excess TPN⁺. At this point there would be present approximately 10^{-12} moles of TPNH. The sample would now be added to fresh cycling reagent to yield, after an hour of cycling and repetition of reaction 6, 10^{-8} moles of TPNH, i.e. 100-fold more TPNH than needed for direct measurement.

Both single and double cycling give surprisingly reproducible results [*11*]. The cycling process introduces a degree of specificity which eliminates some of the adventitious errors and blanks.

The examples given are for measuring glucose, but reactions 1 and 2, or 2 alone, are equally useful for measuring ATP, glucose-6-P, hexokinase or glucose-6-P dehydrogenase, depending on which component is omitted. By the addition of creatine kinase and ADP the system has been used to measure P-creatine. This is only one of many comparable analytical enzyme sequences. It is a rare metabolite or enzyme which cannot be led with 2 or 3 auxiliary enzymes to a pyridine nucleotide. If the final step is conversion of TPNH to TPN⁺, the excess TPNH is destroyed with weak acid before proceding. If the final step involves DPN rather than TPN a different but analogous cycling reagent is used when the extra sensitivity of cycling is required.

At present the chief analytical limitation in test-tube histochemistry is not sensitivity but the availability of the necessary auxiliary enzymes. Fortunately the number of highly purified crystalline enzymes available commercially is

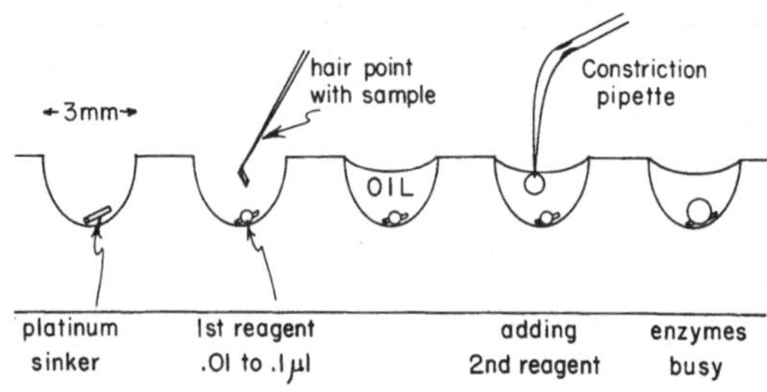

Fig. 4. Oil well technique. This represents a portion of a Teflon block in which holes have been drilled as shown. The platinum sinker is needed to anchor the droplet which might otherwise be trapped in the oil-air interface. (From reference [5])

increasing rapidly. The investigator should be warned however that trace contaminations with other enzymes are not uncommon and because of this, and because of enzyme instability, frequent tests of specificity and of reaction rates are necessary to avoid calamitous errors.

With double cycling it is theoretically possible to measure 10^{-18} moles of a given compound. This is approximately the amount of product that a single molecule of the average enzyme would produce in one hour. If test-tube chemistry can measure one enzyme molecule it becomes, as far as enzymes are concerned, as sensitive as any other kind of histochemistry could possibly be. The limiting factor for test-tube histochemistry therefore is not analytical sensitivity, but the capacity to isolate identifiable cytological structures. So far no one has tried to push the isolation process to the limit.

To utilize the potential analytical sensitivity introduces a problem that has not yet been mentioned. This concerns the troubles that arise when one tries to work at very high dilution . These troubles are contamination, side reactions, and blanks. For example, it is difficult to work with lactate at concentrations lower than 10^{-6} M or DPN at concentrations lower than 10^{-9} M because of contamination. To achieve a concentration of 10^{-6} M with 10^{-18} moles of material the volume has to be reduced to 0.001 μl. This is clearly too small a volume to handle in open test tubes. A solution to this problem appears to be to work with micro droplets under oil. Fig. 4 illustrates how this can be accomplished. The oil not only prevents evaporation of the micro drops, but decreases access of CO_2 which would otherwise disturb the pH. A series of reagents can be added without removing the droplet from the oil well. When necessary a rack of samples under oil can be heated to temperatures approaching 100^0 with little danger of evaporation. Cycling can be carried out directly in the oil well. Finally the sample is transferred from the oil well into 1 ml of reagent or diluting fluid in a standard macro fluorometer tube.

It was mentioned earlier that slide histochemistry today far outranks test-tube histochemistry in regard to volume of work being done. I wonder if it is not desirable to persuade more people to get into the test-tube field. Slide histochemistry goes faster, but in some cases it unquestionably gives erroneous results, and for small molecules it is rarely suitable at all. Resolution with test-tube chemistry to 10 μ is today very easy and has been accomplished to 2 or 3 μ. There seems no reason at all why resolution to less than 1 μ cannot be achieved. Test-tube histochemistry has much to offer. In the case of enzyme distribution studies, natural substrates can be employed, unstable enzymes can be measured, the enzyme concerned can always be made strictly rate limiting, diffusion arte-facts can be eli ninated. In the small molecule field test-tube chemistry is far ahead. Even compounds present in low concentration can be accurately localized. Both slide histochemistry and test-tube chemistry are needed, and can com-plement each other, but we do need a shift in the balance between the two.

Bibliography

[1] Lowry, O. H.: J. Histochem. Cytochem. 1, 420 (1953).
[2] Robins, E., D. E. Smith, G. E. Daesch, and K. E. Payne: J. Neurochem. 3, 19 (1958).
[3] Linderstrøm-Lang, K., and K. R. Møgensen: C. R. Trav. Lab. Carlsberg, Sér. chim. 23, 27 (1938).
[4] Lowry, O. H., N. R. Roberts, D. W. Schulz, J. E. Clow, and J. R. Clark: J. biol. Chem. 236, 2813 (1961).
[5] — Bull. N.Y. Acad. Med. 38, 789 (1962).
[6] — Harvey Lect., Series 58, 1 (1963).
[7] — J. V. Passonneau, and M. K. Rock: J. Histochem. Cytochem. 8, 335 (1960).
[8] — J. Mt Sinai Hosp. 30, 375 (1963).
[9] Glick, D.: Quantitative Chemical Techniques of Histo- and Cytochemistry, vols. I, II. New York and London: Interscience 1962, 1963.
[10] Kaplan, N. O., S. P. Colowick, and C. C. Barnes: J. biol. Chem. 191, 461 (1951).
[11] Lowry, O. H., J. V. Passonneau, D. W. Schulz and M. K. Rock: J. biol. Chem. 236, 2746 (1961).

Prof. O. H. Lowry,
Dept. of Pharmacology, Washington Univ. School of Medicine,
4577 McKinley Avn., St. Louis, Missouri 63110/USA

The Biological Institute of the Carlsberg Foundation, 16 Tagensvej, Copenhagen N, Denmark

Microgasometric Methods
Cartesian divers

By

Erik Zeuthen

With 6 Figures in the Text

I. Introduction

A general review of the microgasometric methods used in histo- and cyto-chemistry was recently published by Glick [*14*]. The review covers 166 pages and of these 138 pages are devoted to a description of the Cartesian divers, introduced into biology by Linderstrøm-Lang [*35*] in 1937. For micro measurements the Cartesian diver principle has uniquely demonstrated its versatility and applicability in the fields of cell chemistry and of cellular physiology. I shall therefore limit myself to a *discussion of some Cartesian diver methods*. I shall stress newer developments and at the end of my lecture show how the diver and a density gradient can be combined to give a new sort of diver, *the non-Cartesian Gradient Diver*, which has the special virtue that it permits continuous recordings over days.

II. The Standard Diver

The Cartesian diver (Fig. 1) is a reaction chamber which floats submerged. Continued buoyancy of the diver signals constancy of the diver's gas volume. Reactions (in a) which result in uptake or evolution of gas from or into the diver's gas phase (d) are compensated by and read as pressure changes necessary to maintain buoyancy; see Fig. 1, left. Like the Warburg-respirometer the diver is a constant volume, changing pressure gasometer.

Once the idea of using the Cartesian diver as a reaction chamber came into Linderstrøm-Lang's mind, the first problem was how to introduce the sample. At the time both Linderstrøm-Lang and his close associate Heinz Holter were quantitative histochemists and mostly used tissue sections which for enzyme determinations were incubated with substrate in a micro test tube. The splitting of the substrate was measured by titration. It was quite natural therefore that also with the potentially much more sensitive gasometric method the decision was to pipette a fixed volume of medium with a single large cell (diameter 300 μ or so), or a reaction medium with a tissue section, into a reaction chamber, which was made from a $\sim$1 mm wide capillary tube. The one end was sealed and, if suitable, blown up to accomodate the sample while the other end remained open to permit introduction of the pipette. However, the Cartesian diver when floated in *water* showed itself to be leaky to air, even after the introduction of a sizeable diffusion path, represented by a column of water [the mouth seal (Fig. 1, c)] in the diver's mouth. From this observation followed the necessity of taking three steps, one leading to the other, before control divers could be had in which the equilibrium pressure remained constant (within 1 mm H_2O/hour). The three steps were 1. the use of an almost saturated solution of $NaNO_3$, $NaCl$

as the flotation medium. In this medium the solubility of gases is very low, and the diffusivity of the gases in the mouth seal reduces to an acceptable minimum. To prevent water to distil from the low salinity reaction medium (Fig. 1, *a*) to the high salinity flotation medium in the mouth (*c*), 2. a separate oil seal (*b*) was introduced between the reaction medium and the mouth fluid. Finally, now having eliminated the possibility for fast absorption of respiratory CO_2 in

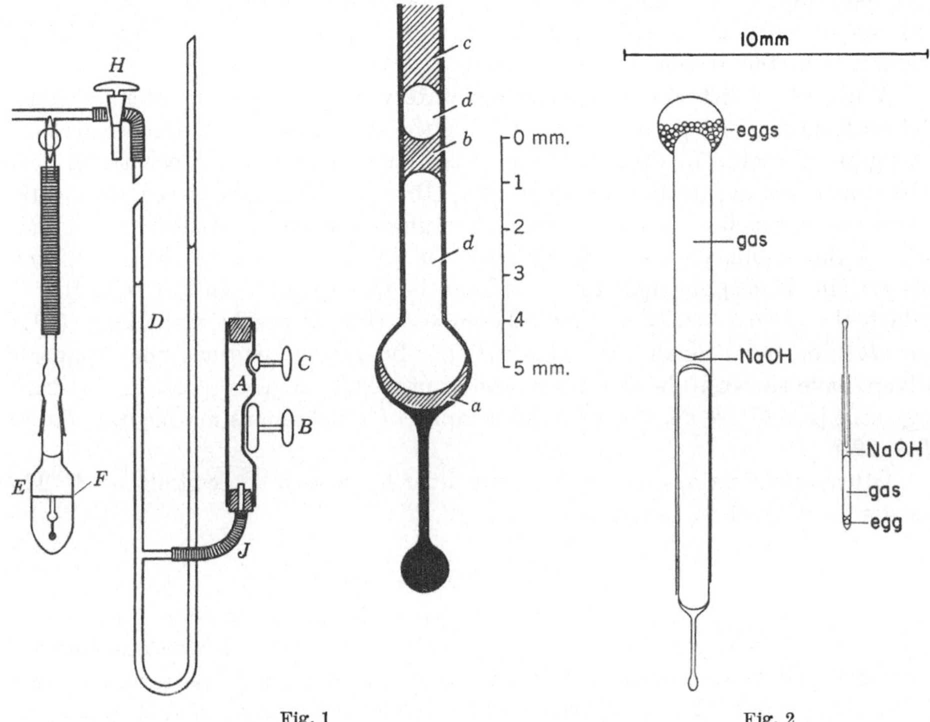

Fig. 1   Fig. 2

Fig. 1. Standard Cartesian diver, with manometer and flotation vessel (HOLTER [21])

Fig. 2. Stoppered divers for many (FRYDENBERG and ZEUTHEN [10]) and single sea urchin eggs (ZEUTHEN [67])

an alkaline flotation medium, 3. for measurements of respiration a separate alkaline neck seal (not shown in Fig. 1) was introduced between the oil seal, *b*, and the reaction mixture, *a*. The *Standard Cartesian Diver* is therefore charged by the successive pipetting of 3 or 4 accurately measured volumes of fluid into a dry reaction chamber. This, of course, requires accurate and special pipetting techniques. Below the oil seal the order of the seals can be changed at will and (diver siliconed on the inside [59, 60]) more seals or small side droplets can be introduced [5]. A neck seal and a side droplet beneath it or above it can be mixed when at any time excess or reduced pressure is applied to the system. As stressed by HOLTER [22] the system is versatile enough to permit on a microscale (error: with 10 μl-divers smaller than 10^{-3} μl/h) most of the measurements which in the Warburg-gasometer are performed on a several thousand times larger scale (error: larger than 1 μl/h). Like the Warburg apparatus the standard diver of LINDERSTRØM-LANG [36] and HOLTER [21] can thus be applied for cell physiological and for cell chemical studies, or at the same time for both.

Modifications. Working with the *Drosophila* egg, which is kept dry in the diver, NIGON and FOURCHE [51] used a narrow capillary tube, 0.5 mm wide and about 15 mm long, as the diver. The capillary is sealed in one end, and here the egg is deposited. Because there is no danger that the terrestrial object will dry up during the experiment, the authors could omit the seals and set the capillary afloat in Holter's concentrated saline medium which fills the outer 10 mm or so of the capillary, satisfying the conditions for buoyancy and for tightness of the system to gas.

With cells which are suspended in a watery medium loss of water by distillation from the biological medium holding the specimen can be omitted without using an oil seal. The diver floats in an isotonic medium which penetrates deep (10 mm or more) into the diver's mouth. Into the mouth, however leaving the meniscus untouched, is inserted a hollow glass stopper which leaves less than 5% of the capillary's cross section open for diffusion of gases. Such *Stoppered Divers* [10, 67, 69] are tight to air and can be made tight even to CO_2 [10] with which the diver is easily charged. The divers (Fig. 2) can be made large (10 μl air [10]) or quite small (0.05 μl air [69]). In various editions the stoppered divers have shown to be useful for studies of cellular objects, particularly single egg cells [11, 15, 16, 64, 69], or small samples of synchronous marine eggs (latest [10, 29]).

Other modifications of the standard diver have been successfully used. The reader is referred to earlier reviews [4, 22, 23, 37].

III. The Diver Balance

I have tried to point out that the use made by LINDERSTRØM-LANG and HOLTER of a technical gadget was determined by the type of investigation they were engaged in at the time. However, a choice, when first made, leaves other possibilities open. I myself have explored the possibilities arising when one decides *not* to use the diver as a reaction chamber, but as a balance [65] by the use of which reference measures can be obtained of the amount of material used in gasometric studies with the standard diver. When the Cartesian diver serves as a balance the biological object is placed in a cup attached to the outside of the diver (Fig. 4) which itself floats in a medium which is optimal for cellular functions. The diver's inside connects to the outside by a very narrow channel, the tail. In the diver is an air bubble. It is surrounded with fluid which is continuous, through the tail, with the outside medium. By virtue of its narrowness the tail represents the necessary efficient brake for diffusion of gases. At the same time it is wide enough to permit pressure equilibration. The equilibrium pressures of the diver are measured before and after loading. The difference between the two pressures is a measure of the RW of the load. The manometer and flotation vessel are — with slight modifications — as shown in Fig. 1. The diver is calibrated with small polystyrene beads [65] or other [41] standards of known RW.

The submerged, or reduced weight (RW) of a sample represents the difference between the weight (volume × density) of the cell and of the displaced medium.

$$RW = V \cdot \varphi_c - V \cdot \varphi_m = V\,(\varphi_c - \varphi_m).$$

The RW is therefore a function of cell volume (V), of cell density (φ_c) and of the density (φ_m) of the suspension medium.

When the flotation medium is water, the cell water does not contribute reduced weight. One measurement of the RW of a cell in water (or in a low salinity medium such as Ringer) is therefore [28, 71] a good measure of the cell's dry matter (rather: lipid-free dry matter since mixed lipids have low RW, their compound density being close to that of water). With suitable, large, non-ciliated and non-flagellated cells measurements can be repeated on the same individual cell at intervals over weeks so that the changes by growth [38, 53, 54, 58] or degrowth during starvation [28, 66] in dry matter can be followed.

If colloids in significant amounts are present in the flotation medium, thus making it heavier than water, the cell-water acquires buoyancy, and the reduced weight of the cell becomes smaller than in water [71]. With increasing outside colloid concentration the RW of the cell becomes first nil [42], then negative. From pairs of measurements in two isotonic media with different densities [42, 66] a number of cellular parameters can be followed in time: RW, density, volume and wet weight. As a colloid added to one of the media we have used Zulkowski soluble starch [42, 66]. Ficoll, a synthetic dextran polymer, is now recommended [26].

Preliminary unpublished experiments suggest that pairs of weighings in water, and in a H_2O—D_2O mixture after full exchange of the cell water with the surrounding medium, might give similar information which refer, not to the cell as a whole, but to the dry matter of the living and growing or degrowing cell. These experiences were had when the method was in this earliest stage of development [65] (appendix) and they ought to be followed up.

Starving cells combust their own matter and thus lose RW. In single large amoebae starving for weeks [28] and in developing eggs [44—48, 50] the loss in RW has been compared with e.g. [28, 44, 45, 48] the intensity of the combustions as measured by O_2-uptake with the standard diver, and [66] with the change in cell volume (amoebae) measured by paired weighings as described. In the starving amoebae the body substances are combusted in the order heavy-light-heavy which means that lipids are used mostly in mid-starvation. Cell water is pumped out in parallel with the loss by combustion of the heavy matter (carbohydrates, proteins) in the cell. In the eggs the sequence in which substrates (carbohydrate, lipid, protein) were combusted was analyzed.

The changing RW of a cell in which the cell water is in the process of equilibrating with outside heavy water (10% D_2O in H_2O) has been used for many studies of cell permeability [43, 49, 52, 55—57] and of water volume in live single cells [57].

From what I have said it will appear that the *Cartesian Diver Balance* is an instrument for the cell physiologist. It is useful also for the cell chemist who wants to know the amount of cellular material he analyses [1—3, 6, 17, 24, 25, 27, 34]. In one case [7] it has been used for following a histochemical reaction: the rate of deposition of copper-thiocholine formed in the Koelle-Friedenwald test for cholinesterase. The object is a single mouse gastrocnemius motor-end plate which rests on the cup of the balance during the experiment (Fig. 4).

More studies with the diver balance are in [71], which is a review.

Errors in work with Cartesian diver balances range from $\pm 10^{-8}$ mg RW with the smallest to $\pm 3 \times 10^{-2}$ mg with the largest divers used thus far, the objects studied from single (cytologically fixed) ciliate protozoa (*Tetrahymena* [38]) to 20—30 live frog-eggs [44]. There is no upper limit to the method.

Diver balances (error $\pm 10^{-5}$ mg RW for larger cells ([65] such as *Chaos chaos*) are easily made and handled. It is difficult to make divers for much smaller single objects [7, 38, 53] and I feel that it will not be easy to refine this simple instrument further.

New developments, by three Yugoslav visitors to our laboratory, Drs. BRZIN, KOVIČ and OMAN, are in progress. They claim that a *Magnetic Diver Balance*, only a piece of plastic with a small permanent magnet (AlNiCo) instead of the air bubble, is as sensitive as is the Cartesian diver balance. The diver is floated by a strong permanent magnet above it. The position of the magnet can be adjusted to the nearest 0.01 mm, and so can the position of the floating diver. This defines the magnetic force that acts on the diver excellently. The permanent magnet carries most of the load represented by the empty diver. The fine regulation, first of the empty then of the loaded diver, is by electric current through a coil around the flotation vessel. The current is read on a galvanometer. The magnetic diver is hardly easier to make than is the Cartesian diver balance. The advantages of this instrument is that the equilibrium situation is more stable in time and that the flotation vessel is open to the atmosphere, facilitating pipetting of the load. If the load is a non-Cartesian micro-gasometer ("ampulla", see later) the changes in gas volume (O_2 or CO_2) can be read (or recorded) as reduced weight changes. For such measurements the flotation vessel must be closed to the atmosphere. Finally, and above all, there is the hope that a very small balance, itself a tiny plastic-covered AlNiCo bead, can be incorporated by phagocytosis into growing cells. If this hope comes through it will be a simple matter to follow the growth e.g. from division to division, say of single suspended tissue cells.

IV. The Ampulla-diver Gasometer

The more promising recent developments with the Cartesian diver gasometer are with the simplest, the *Ampulla-diver*. I shall use the rest of my time for a discussion of this instrument used as a Cartesian (this section) and as a non-Cartesian diver (Sect. V).

The diffusion problem of the Cartesian diver respirometer is solved in different ways in the standard diver and in the ampulla-diver. With the ampulla-diver the flotation medium is isotonic with the reaction mixture in the diver; diffusion of gas between the diver and the flotation medium is minimized because the diver's inside connects to the outside through a long narrow tail as described for the diver balance.

In the original Cartesian diver gasometer [35] the diver is charged by accurate pipettings. With the ampulla diver a narrow capillary serves first as a pipette then as a diver. This is so because the outer part of the pipette (the "ampulla") when broken off serves as the diver.

In making the diver a narrow capillary of correct wall thickness [68] is pulled with a fine tip and an even finer "shaft". An "ampulla" is between the tip and the shaft. After the cell has been taken into the diver through the tip (Fig. 3), the tip is sealed with a microflame or with a heated mixture of wax and resin [62, 63]. The shaft is then broken so that the diver gets a "tail", just the same as the diver balance. After the filling, air is removed stepwise from the diver and replaced by isotonic flotation medium (alkaline for respiration measurements, acid when CO_2 should not be absorbed) until the diver is buoyant. In the diver are therefore only two fluid compartments: reaction medium and flotation medium. The gas volume of the floating diver is calculated from the known densities of the glass, of the media used in the diver and for floating the diver, and from

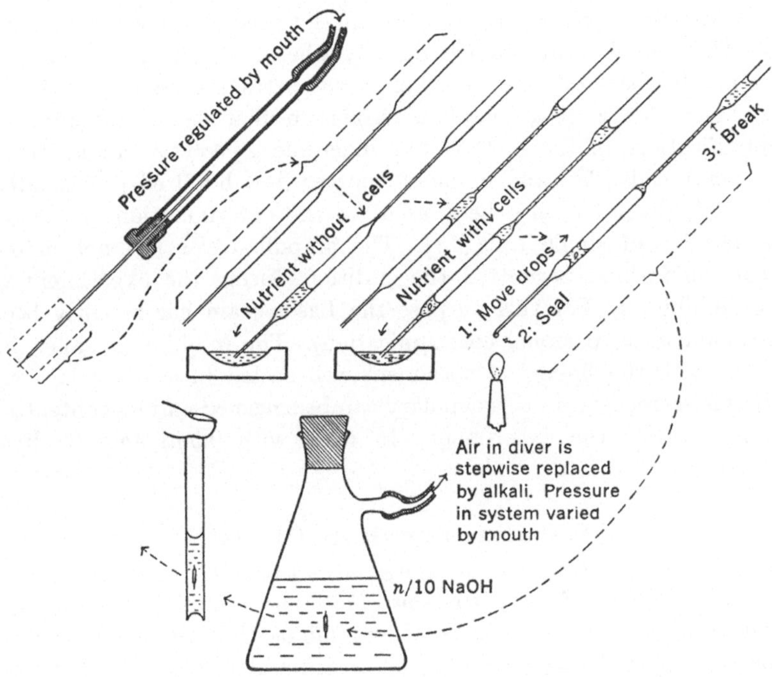

Fig. 3. Ampulla-diver. Filling and balancing of diver is shown. Diver is to be transferred to sensitive burette-manometer (ZEUTHEN [68]). Gassing of diver and manometer cannot be demonstrated here

the dry weight of the diver. The ampulla-diver can be charged with single objects up to 100 μ, the size of a sea urchin egg, or with a number of mammalian cells low enough (200—600) to permit counting of the cells actually transferred to the diver [9]. This solves an old sampling problem. When necessary a single diver can be reused, or refilled, e.g. with the same cell in new medium [7, 61, 62]. Sterile conditions are easily obtained. The ampulla-diver is therefore an excellent growth chamber for axenic cultures started from single cells [38, 68].

The ampulla-diver is used for estimating oxidative enzymes (cytochrome oxidase, succinodehydrogenase) in single nerve cells [18, 30—33], also for measuring α-ketoglutarate and glutamate oxidation [18, 19].

The ampulla-diver and the manometer can be easily charged with any gas mixture, e.g. with 5% CO_2 in N_2 [7, 8, 20, 62]. This makes it possible to measure CO_2 which at a physiological pH is liberated in the diver from a HCO_3^- buffer due to the formation of fixed acids. For cholinesterase the substrate is 6×10^{-3} M acetylthiocholine in saline with HCO_3^-. With 5% CO_2 the pH is 7.4. The flotation medium is 0.0166 M $NaHCO_3$ bubbled with 5% CO_2 [7]. Cholinesterase can be accurately determined [62] in single megakaryocytes [61, 62], in single nerve cells [12], in fractions of a nerve cell [12] and in single motor end-plates [6, 7].

In [7] the cholinesterase activity of single motor end-plates was measured with acetylthiocholine as the substrate. The rate of liberation of the acetyl end of the substrate molecule was followed. I have already referred to diver balance measurements in which the rate of precipitation of the thiocholine end of the substrate molecule with copper ions in the medium was measured. Comparison of the two methods reveal that in the precipitate we can account for only 30—60% of the thiocholine actually liberated. The remainder stays in solution, maybe because one thiocholine competes successfully with Cu^{++} for another thiocholine so that soluble disulfides form in significant amounts. By slight modifications of the method the anaerobic glycolysis can be determined in single nerve cells [20], and in small samples (few hundreds) of isolated Mast cells [8]. This diver has also been used for the determination of carbonic anhydrase activity of single cells [13]. The ampulla-diver has not before in a non-disturbing fashion permitted the addition during the experiment of substrates or inhibitors. For this purpose the Laser-beam has recently been used (Dr. A. HAMBERGER, personal communication). The inhibitor is taken into the diver along with the biological specimen, well sealed in a 10 μ-wide capillary. Using the Laser microbeam this capillary can be smashed and its content released at any time during the experiment. In work with 0.5 μl ampulla-divers the error is $\pm 2 \times 10^{-5} \mu l/h$ [9].

V. Gradient Diver Micro-Gasometry

Already LINDERSTRØM-LANG pondered over the possibility of combining his Cartesian diver and his density gradient. With the availability of the ampulla-diver and with the clear realization in our minds that we should not seek an exact solution to this complex problem, LøVLIE and I [40] tackled the problem with empirical methods. What we hoped to and did obtain was 1. a control diver which at constant pressure over a density gradient, made by dissolving Ficoll or Na_2SO_4 in water, comes to a stable equilibrium position at the level in the density gradient where the compound density of the diver equals the density of the displaced medium, an 2. and experimental diver which at constant pressure continues to migrate down or up (consumption or evolution of gas)

Fig. 4. Diver balance, micro-edition 5×10^{-4} μl air, bubble diameter about 0.1 mm. The air bubble is in a glass chamber surrounded by water which communicates with the flotation medium through the capillary tail (2—4 mm) of which only the upper part is shown. Neither water nor glass shows. It is all coated with the material (paraffin, polyethylene 1:1) from which the cup is made. A piece of muscle fiber with a motor end-plate is falling into the cup. The end-plate carries a Cu-thiocholine precipitate (BRZIN and ZEUTHEN [7])

Fig. 5. The photographic plate has been moved across a slit to permit repeated photography of an ampulla-diver which is charged with a tiny piece of plant *(Ulva mutabilis)*. Shifts between light and dark (top of figure) are correlated with shifts between photosynthesis (shining diver moves up) and respiration (the air in diver appears dark, and diver moves down in gradient). Spheres are density standards. Read from right to left (LøVLIE [39])

Fig. 6. Egg of *Psammechinus miliaris* in ampulla-diver. Cleavage from 2 to 8 cells. Exposures every 5 minutes. Migration of diver: 1.5 mm in 40 minutes. Read from right to left (LøVLIE and ZEUTHEN, unpublished)

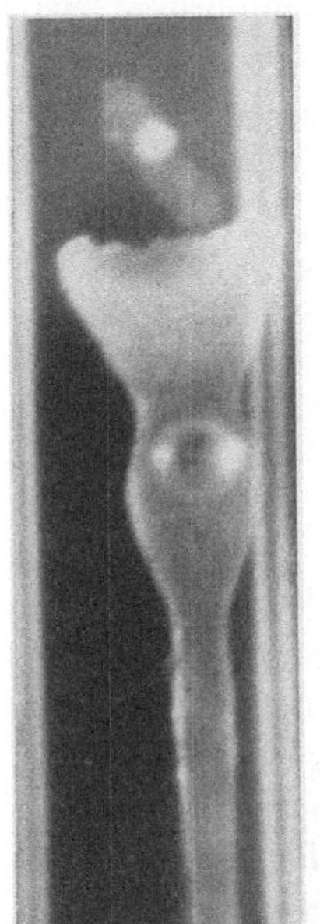

Fig. 4

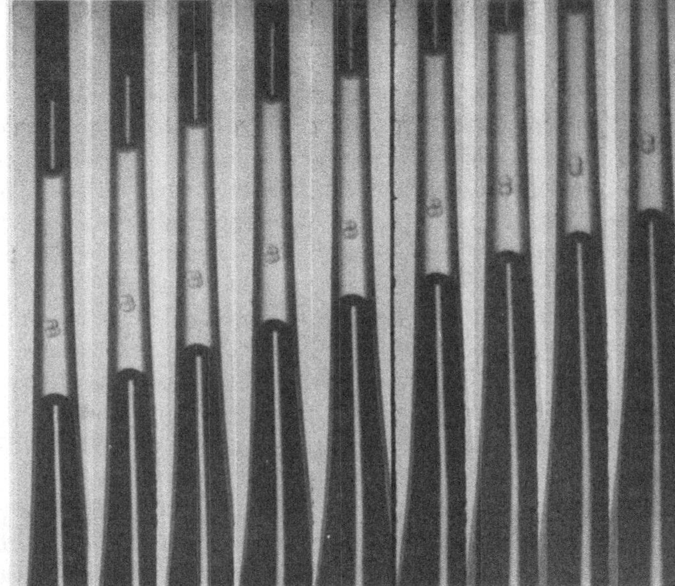

Fig. 6

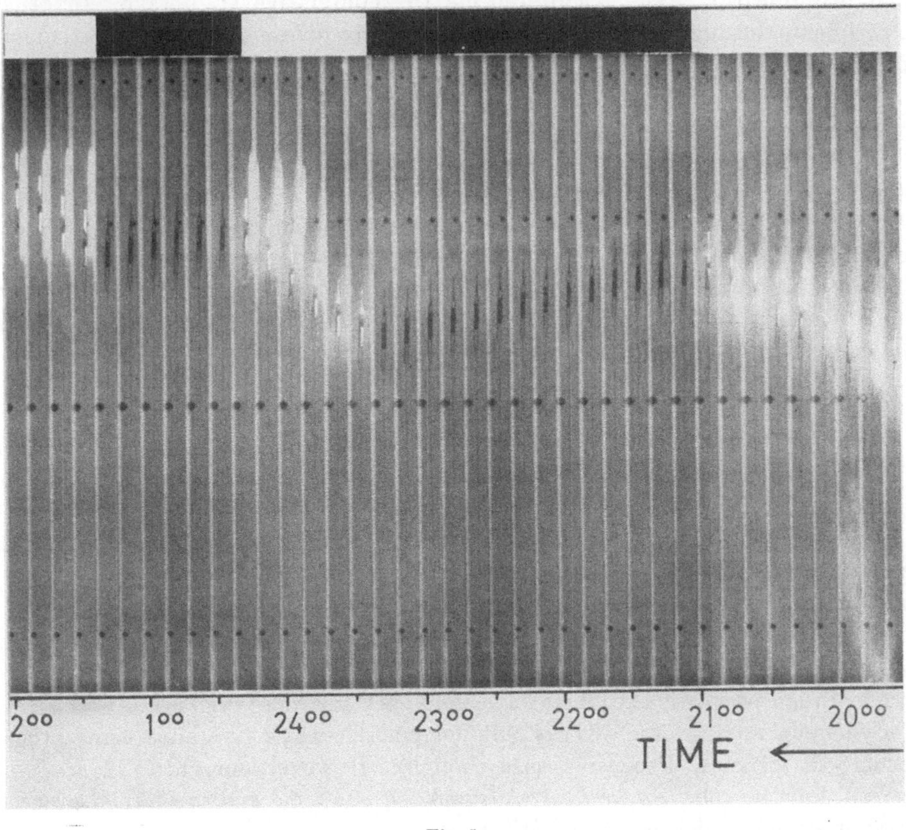

Fig. 5

always in close reach of the layer in which it is in density equilibrium. This gradient diver is non-Cartesian because we allow contraction or expansion of the gas. The migration is *recorded* automatically by photography (Figs. 5, 6). The system is calibrated using either the control or the experimental diver itself. The calibration is with the aid of an associated water manometer and in terms of the pressure changes necessary to move the calibration diver through the positions which at constant pressure are taken by the experimental diver.

We prefer inorganic (Na_2SO_4) density gradients in which 1 mm displacement of the control diver is produced by a change in pressure of one to a few mm H_2O. When selecting FICOLL [40] we planned to use gradients approximately isotonic with the biological medium. However, the diver's lowest part is initially filled with fluid isotonic with the droplet holding the biological sample. It is thus lighter than even the top gradient fluid. The diver floats tail down. The tail is tight to significant diffusion of gas and of salts, so even in Na_2SO_4 gradients [39] there is no indication of loss of water by evaporation from the biological medium in the diver. In Fig. 6 you see a record of a diver with a single egg of the sea urchin *Psammechinus miliaris*. While in the course of $^3/_4$ hours the egg goes from a 2-cell to an 8-cell stage the diver migrates $1^1/_2$ mm, roughly corresponding to a pressure change (calibration diver) of 2×10^{-4} atm. With this short migration, egg cleavage and migration can be recorded on the same plate and under rather high magnification. The record reflects the cumulative O_2 uptake. In other experiments the migration of the diver was directly followed using a cathetometer read to 0.001 mm. In agreement with the author's previous results [10, 29, 69, 70] a slight respiratory periodicity could be shown for the earlier divisions and a stronger periodicity after around the 64-cell stage.

LØVLIE [39] has used the diver-balance and the gradient diver in studies of growth, photosynthesis and cell division in small pieces of normal and mutant *Ulva* plants. His studies (Fig. 5) demonstrate that the gradient diver responds to rate changes with only a short lag, perhaps bad for some, but insignificant for many purposes. Such observations will be included in the final interpretation of the periodic respiration curves just mentioned. With 1.0 μl gradient divers the error is 10^{-5} μl/h [40].

Conclusions

A handy ampulla-diver carries a gas volume of around 0.5 μl. Handy diver balances are around 0.1 μl. When necessary both instruments can be made much smaller [7, 12, 38, 53, 54, 56]. When the smallest divers are selected for tightness to diffusion of salt (gradient diver) and/or air (see [9]) and used with special sensitive manometers (described in [10, 68, 69]) of which the gradient [40] can be dealt with as one, highest stability and sensitivity is obtained.

Even though the 27 years record of the divers in the field of cell chemistry is not bad, some would say fine, I feel to-day that the future of the divers — like much of their past — is in cellular physiology, perhaps more than in cell chemistry. For chemical analyses the diver is only one possible way out, among many others (usually a lot more expensive ones though). However, for chemical analyses we still kill the cell. So we deal with one cell one time and must then proceed to another cell which is always different.

This is where the divers offer, I would think unique, possibilities for many types of measurements. I have shown you some, both with the balance and with the gasometers. Using one suitable individual cell we can come back again and again measure, or we can continuously measure, vital functions such as changes in reduced weight, volume, density, wet weight, water content, respiratory and fermentative or photosynthetic rate. In short, in many respects we can deal with the cell as a costly individual.

References

[1] ANDRESEN, N., FR. ENGEL, and H. HOLTER: C. R. Lab. Carlsberg, Sér. chim. **27**, 408 (1951).

[2] —, and B. M. POLLOCK: C. R. Lab. Carlsberg, Sér. chim. **28**, 247 (1952).

[3] —, and C. W. MUSHETT: C. R. Lab. Carlsberg **33**, 265 (1963).

[4] BOELL, E. J.: The Cartesian diver technique in microrespirometry and enzyme assay. In: SASSER and JENKINS, Eds. *Nematology*, pp. 109—121. Chapell Hill: The University of North Carolina Press 1960.

[5] BORROW, A., and J. R. PENNEY: Exp. Cell Res. **11**, 188 (1951).

[6] BRZIN, M., and Z. MAJCEN-TKACEV: J. Cell Biol. **19**, 349 (1963).

[7] —, and E. ZEUTHEN: C. R. Lab. Carlsberg **32**, 139 (1961).

[8] CHAKRAVARTY, N.: C. R. Lab. Carlsberg (1964) (in press).

[9] —, and E. ZEUTHEN: C. R. Lab. Carlsberg (1964) (in press).

[10] FRYDENBERG, O., and E. ZEUTHEN: C. R. Lab. Carlsberg **31**, 423 (1960).

[11] GEILENKIRCHEN, W. L. M.: Effects of mono- and divalent cations on viability and oxygen uptake of eggs of *Limnaea stagnalis*. Diss. Utrecht 1961, Neerlandia.

[12] GIACOBINI, E.: Acta physiol. scand. **45**, Suppl. 156 (1959).

[13] — J. Neurochem. **9**, 169 (1962).

[14] GLICK, D.: Quantitative chemical techniques of histo- and cytochemistry, vol. I, pp. 123—289. New York and London: Interscience Publishers 1961.

[15] GONSE, P. H.: Exp. Cell Res. **8**, 550 (1955).

[16] — Biochim. biophys. Acta (Amst.) **24**, 267, 520 (1957).

[17] GREGG, J. R., and S. LØVTRUP: C. R. Lab. Carlsberg, Sér. chim. **27**, 307 (1950).

[18] HAMBERGER, A.: Acta physiol. scand. **58**, Suppl. 203 (1963).

[19] — J. Neurochem. **8**, 31 (1961).

[20] —, and H. HYDÉN: J. Cell Biol. **16**, 521 (1963).

[21] HOLTER, H.: C. R. Lab. Carlsberg, Sér. chim. **24**, 399 (1943).

[22] — The Cartesian diver. In: General cytochemical methods, vol. 2, Ed. J. F. DANIELLI, pp. 93—129. New York and London: Academic Press 1961.

[23] — K. LINDERSTRØM-LANG, and E. ZEUTHEN: Manometric techniques for single cells. In: Physical techniques in biological research, vol. 3, Eds. G. OSTER and A. W. POLLISTER, pp. 557—625. New York: Academic Press 1956.

[24] —, and B. A. LOWY: C. R. Lab. Carlsberg **31**, 105 (1959).

[25] —, and S. LØVTRUP: C. R. Lab. Carlsberg, Sér. chim. **27**, 27 (1949).

[26] —, and M. MØLLER: Exp. Cell Res. **15**, 631 (1958).

[27] —, and B. M. POLLOCK: C. R. Lab. Carlsberg, Sér. chim. **28**, 221 (1951).

[28] —, and E. ZEUTHEN: C. R. Lab. Carlsberg, Sér. chim. **26**, 277 (1948).

[29] — — Pubbl. Staz. Napoli **29**, 285 (1957).

[30] HYDÉN, H.: Nature (Lond.) **184**, 433 (1959).

[31] —, and P. W. LANGE: J. Cell Biol. **13**, 233 (1962).

[32] — S. LØVTRUP, and A. PIGON: J. Neurochem. **2**, 304 (1958).

[33] —, and A. PIGON: J. Neurochem. **6**, 57 (1960).

[34] KRUGELIS, E. J.: C. R. Lab. Carlsberg, Sér. chim. **27**, 273 (1950).

[35] LINDERSTRØM-LANG, K.: Nature (Lond.) **140**, 108 (1937).

[36] — C. R. Lab. Carlsberg, Sér. chim. **24**, 333 (1943).

[37] LØVLIE, A.: Kemisk Månedsblad Nr 3, Copenhagen [in Norwegian language] (1961).
[38] — C. R. Lab. Carlsberg **33**, 377 (1963).
[39] — C. R. Lab. Carlsberg. **34**, 77 (1964).
[40] —, and E. ZEUTHEN: C. R. Lab. Carlsberg **32**, 513 (1962).
[41] LØVTRUP, S.: C. R. Lab. Carlsberg, Sér. chim. **27**, 125 (1950).
[42] — C. R. Lab. Carlsberg, Sér. chim. **27**, 137 (1950).
[43] —, and A. PIGON: C. R. Lab. Carlsberg, Sér. chim. **28**, 1 (1951).
[44] — C. R. Lab. Carlsberg, Sér. chim. **28**, 371 (1953).
[45] — C. R. Lab. Carlsberg, Sér. chim. **28**, 400 (1953).
[46] — C. R. Lab. Carlsberg, Sér. chim. **28**, 426 (1953).
[47] — J. exp. Zool. **140**, 231 (1959).
[48] — J. exp. Zool. **140**, 383 (1959).
[49] — J. exp. Zool. **145**, 139 (1960).
[50] —, and B. WERDINIUS: J. exp. Zool. **135**, 203 (1957).
[51] NIGON, V., et J. FOURCHE: Bull. biol. Françe et Belg. **17**, 36 (1958).
[52] PIGON, A., u. E. ZEUTHEN: Experientia (Basel) **7**, 455 (1951).
[53] PRESCOTT, D. M.: Exp. Cell Res. **9**, 328 (1955).
[54] — Exp. Cell Res. **11**, 86 (1956).
[55] — J. cell. comp. Physiol. **45**, 1 (1955).
[56] —, and D. MAZIA: Exp. Cell Res. **6**, 177 (1954).
[57] —, and E. ZEUTHEN: Acta physiol. scand. **28**, 77 (1953).
[58] SATIR, P., and E. ZEUTHEN: C. R. Lab. Carlsberg **32**, 241 (1961).
[59] SCHWARTZ, S.: C. R. Lab. Carlsberg, Sér. chim. **27**, 79 (1949).
[60] WATERLOW, J. C., and A. BORROW: C. R. Lab. Carlsberg, Sér. chim. **27**, 93 (1949).
[61] ZAJICEK, J.: Acta physiol. scand. **40**, Suppl. 138 (1957).
[62] —, and E. ZEUTHEN: Exp. Cell Res. **11**, 568 (1957).
[63] — — Quantitative determination by a special "Ampulla-Diver" of cholinesterase activity in individual cells, with notes on other uses of the method. In: General cytochemical methods II, edit. by J. F. DANIELLI, p. 131. New York and London: Academic Press 1961.
[64] ZEUTHEN E.: C. R. Lab. Carlsberg, Sér. chim. **25**, 191 (1946).
[65] — C. R., Lab. Carlsberg, Sér. chim. **26**, 243 (1948).
[66] — C. R. Lab. Carlsberg, Sér. chim. **26**, 267 (1948).
[67] — Biol. Bull. **98**, 139 (1950).
[68] — J. Embryol. exp. Morph. **1**, 239 (1953).
[69] — Biol. Bull. **108**, 366 (1955).
[70] — Exp. Cell Res. **19**, 1 (1960).
[71] — The Cartesian diver balance. In: General cytochemical methods II, edit. by J .F. DANIELLI, p. 61. New York and London: Academic Press 1961.

Prof. ERIK ZEUTHEN,
The Biological Institute of the Carlsberg Foundation, 16 Tagensvej,
Copenhagen N, Denmark

Application of Histochemistry to Electron Microscopy

Moderator: R. J. BARRNETT, New Haven (USA).

RUSSELL J. BARRNETT: Relationship of Enzymatic Activity to Fine Structural Elements.

The combination of enzyme histochemistry and electron microscopy is a compromise discipline that must satisfy the stringent requirements of both fields. The most obvious requirements relate to the performance of *in vitro* histochemical tests so as to produce cytochemical results and to the maintenance of the fine structural details of cells and tissue. Although interesting results may be obtained by various preparative procedures prior to histochemical incubation, the most satisfactory involve the fixation of tissue in aldehyde fixatives, especially dialdehydes like glutaraldehyde or hydroxyadipaldehyde. These reagents not only preserve the fine structural details of cells and tissues in a satisfactory manner for electron microscopy, but also retain a wide variety of enzymatic activities, some of which have not survived fixation of any previous sort. This approach has allowed the investigation of the relationship of a variety of enzymatic activities to fine structural details of tissues and organs. — Thusly, the fine structural sites of activity of sites of various hydrolases will be demonstrated. In particular, the relationship of carboxylic acid esterase activity to ultrastructural elements will be discussed.

Dept. of Anatomy, Yale University School of Medicine New Haven, Connecticut (USA).

ELISABETH R. G. MÖLBERT: The Application of the Heavy Metal Method for Ultrastructural Enzyme Localization.

To show the presence of an enzyme at the cellular ultrastructure one follows — in most cases — the method of coupling one product of an enzymatic reaction with a strongly electron scattering partner. Heavy metal ions are to be preferred because of their high contrast. — The enzyme localization by means of the heavy metal method showed, that under certain circumstances the precipitates of heavy metal can imitate a specific enzyme localization. By using specific methods for preparing the tissues it was possible to coordinate light — and the corresponding ultrathin-sections and it was thereby possible to localize exactly the imitation. — This unspecific labelling of the ultrastructure was found in cases of insufficient fixation, pathologically altered tissues and by applying the "tissueblock method". — The reasons for those imitations will be discussed and explanations will follow of how to avoid these artefacts as far as possible by using well-known specific enzyme localizations.

Pathologisches Institut der Universität, 78 Freiburg i.Br. (Germany).

FRITZ MILLER: Enzymatic Activities in Renal Protein Absorption.

The digestive cycle following reabsorption of hemoglobin by cells of the proximal convoluted tubules in mouse kidney, and the uptake of ferritin by glomerular mesangial cells in the kidney of nephrotic rats were investigated by electron microscopical cytochemical procedures. — The tissues were fixed in glutaraldehyde, cut at $50\,\mu$ on a freezing microtome, incubated for acid phosphatase at pH 5,0 and thiolacetate esterase at pH 5,2, postfixed in OsO_4 and embedded in Epon 812. Satisfactory preservation of fine structure permitted the localization of the enzymatic reaction products on cell structures involved in uptake and digestion of exogenous proteins. It was found that lysosomal enzymic activities and incorporated exogenous proteins occur together in the same membrane bounded structures. In the cells of the proximal convolution, lytic activities become demonstrable within one hour after hemoglobin injection, appear first in apical vacuoles filled with hemoglobin, and persist in fully developed protein absorption droplets. At the end of the lytic cycle, the cells have an increased population of polymorphic bodies which exhibit lytic activities. In smaller numbers, identical bodies occur in controls. It is concluded that they represent remnants of previous digestive events. — In nephrotic rats ferritin was segregated in membrane bounded bodies in the mesangial and to a lesser extent in epithelial and endothelial cells. Most of these sites were marked by the reaction products of acid phosphatase and organophosphorus-resistant esterase and therefore identified as lysosomes connected with the digestion of incorporated exogenous proteins.

Elektronenmikroskopische Abteilung, Pathologisches Institut der Universität, 8 München (Deutschland).

ELIZABETH H. LEDUC and WILHELM BERNHARD: Water Soluble Methacrylates and Ultrastructural Cytochemistry.

Water soluble embedding media, particularly the methacrylates, hydroxyethyl and hydroxypropyl methacrylates, present a number of advantages for ultrastructural cytochemistry. Good preservation of morphological structures may be obtained in tissues or cultured cells fixed in the aldehydes, acrolein, formalin or glutaraldehyde, without post osmication. Ultrathin sections of these tissues exhibit low electron scattering but they can be stained in a number of ways to accentuate one or another cell component. In addition, such tissues may be specifically attacked by nucleases or proteases so that sites of RNA, DNA and protein may be localized in ultrathin sections. The efficacy of the enzyme depends on the choice of fixative, and on the final embedding mixture. Desoxyribonuclease digestion is possible only after formalin fixation. Ribonuclease digestion is slower after acrolein fixation than glutaraldehyde fixation but the final results are the same. Most rapid digestion is achieved when carbowax is added to the final embedding medium. Parallel 2 micron sections may be employed for light and fluorescence microscopy. Endogenous enzyme activity, acid phosphatase and esterase, for example, is preserved in tissues embedded in water soluble methacrylates and can be exhibited at the light microscope level.

Institut de Recherches sur le Cancer, Villejuif/Seine (France) and Dept. of Biology, Brown University, Providence, R.I. (USA).

MICHAEL L. WATSON and WILLIAM G. ALDRIDGE: Effects of Antibiotics on Cells as Revealed by Indium Staining of Nucleic Acids*.

Changes occur in nucleoli of rat hepatic cells within 15 minutes after injection of actinomycin-D into intact animals. The course of these changes with time after injection will be described in tissues processed by the indium procedure and also by conventional OsO_4 fixation. Briefly, nucleolar "ribosomes" in normal animals are rather homogeneously scattered throughout the nucleolus, but with some tendency for concentration near the periphery. After injection of actinomycin-D, there is progressive segregation of nucleolar protein and nucleolar "ribosomes" into two separate masses which are distinct in about 1 hour. Subsequent to this, the ribosomal mass becomes smaller and by 18 hours is difficult to make out. The protein mass is also reduced in size, but less so.

* These investigations were carried out under contract with the United States Atomic Energy Commission at the University of Rochester, Atomic Energy Project, Rochester, N.Y. and with the aid of funds provided by Research Grant CA 03589-06 from the National Institutes of Health, United States Public Health Service.

University of Rochester Medical Center, Rochester, N.Y. (USA).

HEWSON SWIFT: Nucleic Acids of Mitochondria and Chloroplasts.

We have studied mitochondria and chloroplasts of several cell types, utilizing cytochemical techniques for the electron microscope. Nucleic acids have been localized in micrographs after fixation in cold formalinphosphate, extraction with DNase or RNase followed by cold 5% trichloroacetic acid, embedding in Epon, and staining with uranyl acetate. Both mitochondria and chloroplasts frequently contain a filamentous component that binds uranyl ions and is removable with DNase. They also may possess a particulate ribosome-like component that binds uranyl ions and is removable with RNase. Changes in these filaments and particles with changes in cell states will be briefly discussed.

Whitman Laboratory, University of Chicago, Chicago, Illinois (USA).

JEAN PAUL REVEL: Application of Autoradiography to Electron Microscopy.

The biologist now has at his disposal a number of procedures for the preparation of autoradiographs suitable for examination in the electron microscope. These technical developments permit the study of cellular activity with greater resolution than heretofore possible, and autoradiography thus promises to become an even more important cytological tool than it is now. Autoradiography at the electron microscope level has been used for the localization of nucleic acids and proteins labeled with appropriate isotopic precursors. It is a useful tool for correlating the ultrastructural appearance of cells with their functional activity, in helping us to understand the respective role of various organelles and for clarification of interrelationships between various cellular compartments. Autoradiography can also provide direct confirmation of data obtained in cell fractionation studies, and may even help to extend these results, especially in the case of organelles which have, as yet, proved refractory to biochemical analysis. This presentation will describe methods suitable for the preparation of autoradiographs for electron microscopy, and will emphasize some of the results already obtained in the study of the function of various cell organelles in studies of cartilage synthesis, utilizing labeled proline, glucose and sulfate as matrix precursors.

Dept. of Anatomy, Harvard Medical School, Boston, Massachusetts (USA).

RICHARD A. RIFKIND, KONRAD C. HSU and COUNCILMAN MORGAN: Immunological Staining for Electron Microscopy.

Subsequent to the development and extensive exploitation of immunofluorescent staining, a variety of techniques has been proposed for the application of immunochemical stains in electron microscopy. The most widely used to date has been the ferritin-conjugated antibody technique originally described by SINGER (Nature *183*, 1523, 1959). Certain aspects of the conjugation procedure which appear likely to be responsible for a high yield of conjugated antibody will be discussed. These include: 1) the use of highly purified ferritin prepared by repeated crystallization in the cadmium sulfate; 2) the use of m-xylylene diisocyanate carefully protected from deterioration due to moisture; and 3) the use of close to equimolar ratios of globulin and ferritin. The results of immunologic analysis of the conjugates will be presented as well as methods for the purification of the product by ultracentrifugation and preparative electrophoresis. — Application of these coupled proteins as immunological stains for electron microscopy requires that tissues and cells be rendered permeable to the large ferritin-antibody conjugates while preserving as much as possible of their ultrastructural integrity. Preliminary fixation in dilute buffered formalin followed by either freeze-thawing, cryostat sectioning, or fine mincing has generally proved a satisfactory preparation for ferritin-staining. Subsequent fixation in glutaraldehyde and osmium tetroxide and embedment in epoxy resin results in fair to good structural preservation. By means of ferritin-conjugated antisera directed against a wide variety of antigens, including viruses, bacteria, intracellular proteins, and smaller molecules, it is demonstrated that the precision and resolution afforded by electron microscopy can be effective in localizing specific antigens.

Depts. of Medicine and Microbiology, Columbia University College of Physicians and Surgeons, New York, N.Y. (USA).

Physical- Optical Methods in Histochemistry: Instruments

Moderator: T. Caspersson, Stockholm (Sweden).
Co-Moderators: F. Ruch, Zürich (Switzerland),
PHILIP O. B. Montgomery, Dallas (USA).

T. Caspersson: Introduction and Plan for the Discussion.

In general the instrumentation used on the field is relatively complex. Many different types of techniques are being used. It is of importance for the future development that already now attention is focused on the fact that the combined use of different techniques often very greatly widens the field of applicability of quantitative cytochemistry to cell-physiological problems over that which individual techniques can give. Methods involving principles from very different realms of physical optics are being used and the concept is here taken in its widest sense to include also X-ray techniques and electron microscopic procedures. — It is not feasible within the time limits, even in a general way, to survey all instrumentation available at present. In accordance with what is said above the most important fields of approach have been selected and on each of these individual fields one or a small number of reports will present for the field typical working lines and furnish the starting point for discussion.

Institute for Medical Research, Karolinska Institutet, Stockholm 60 (Sweden).

A. Principle and Application of Different Methods and/or Instruments

F. Ruch: Photographische Mikrospektralphotometrie.

Die früher für spektrale Messungen im Mikroskop fast ausschließlich verwendete photographische Methode erweist sich auch heute noch für verschiedene Arbeiten als zweckmäßig. Der Umweg über die photographische Platte ergibt zwar eine geringe Einbuße an Genauigkeit und erfordert einen gewissen Zeitaufwand, bringt jedoch einige Vorteile mit sich. Besonders hervorzuheben ist das bleibende photographische Dokument als Auswertungsgrundlage, mit exakter Zuordnung von Objektstrukturen und Spektren. Dieselbe Aufnahme kann für weitere Auswertungen unter anderen Gesichtspunkten jederzeit erneut herangezogen werden. Sehr geringe Lichtintensitäten, wie z.B. bei Fluoreszenzuntersuchungen, können photographisch durch entsprechend lange Belichtungszeiten leicht registriert werden. Es sind Lichtquellen mit relativ geringer Leuchtdichte, wie z.B. die Wasserstofflampe verwendbar. Diese für Absorptionsmessungen im UV ideale Lichtquelle kann mit geeigneten Emulsionen bis zu Wellenlängen von 200 mμ benützt werden. — Die photographische Methode arbeitet nach dem Prinzip der Vergleichsmessung, d.h. auf jede Platte muß außer dem Objekt noch ein Referenzsystem (z.B. ein rotierender Sektor) aufgenommen werden, um Fehler durch den photographischen Prozeß ausschließen zu können. Zur Auswertung der Platten dient ein Mikrodensitometer. — Die spektrale Zerlegung des Lichtes erfolgt bei der photographischen Methode *hinter* dem Mikroskop. Dadurch können die Intensitäten verschiedener Wellenlängen gleichzeitig registriert werden. — Ein derartiger Mikrospektrograph für Absorptions- und Fluoreszenzmessungen im ultravioletten und sichtbaren Licht wird von Leitz hergestellt. An Stelle der photographischen Platte kann hier auch eine Fernseheinrichtung zur direkten Wiedergabe der Spektralkurven verwendet werden.

Photographic Microspectrophotometry. The photographic method — at one time used almost exclusively for spectral measurements with the microscope — still proves useful for various kinds of work. The disadvantage of the method is a certain loss of precision (caused by the photographic process) and its demand on time. It has, on the other hand, also certain advantages: Special attention is called to the fact that the photographic document will serve as a ready reference, which includes an exact co-ordination of the structures of the object and the spectra. The same photograph can be used for further analysis from a different aspect. Very low light-intensity, as encountered in fluorescent work, can easily be recorded by using appropriate exposure times. Light-sources with a relatively low light intensity — as for example the hydrogen lamp — can be used. If suitable emulsions are used, this lamp, which is ideal for measuring absorptions in the UV range, can be employed down to a wavelength of 200 mμ. — The principle of the photographic method is comparative measuring, i.e. each photograph must show not only the object but also a reference system (for example a rotating sector) so that faults, which are caused by the photographic process, can be excluded. A microdensitometer is used for the interpretation of the photographic plates. — As with the photographic method the spectral dispersion of the light takes place *behind* the microscope, the intensities of various wavelengths can be registered simultaneously. — A microspectrograph for absorption and fluorescence measurings in the UV and visible light range is produced by Leitz. The photographic plate can be substituted by a television set for the direct transmission of the spectral curves.

Institut für Allgemeine Botanik, Eidg. Technische Hochschule, Zürich (Schweiz).

P. O'B. MONTGOMERY: Flying Spot Microscopy.

The principle of visible light flying spot microscopy described by YOUNG and ROBERTS has been adapted to living cell studies in the ultraviolet by MONTGOMERY et al. (1956). The purpose of this adaptation was to produce a system which would permit the formation of ultraviolet absorption images of living cells for prolonged periods without introducing sufficient ultraviolet energy to produce detectable alterations in the specimen. Other uses of this technique have been described by BOX and FREUND (1959) and more recently by HOFFMAN (1962). Additional elaboration of these techniques by MONTGOMERY and BONNER (1958a), and BONNER (1962) have provided ultraviolet microbeam irradiation systems employing either the flying spot television microscope or flying spot scanning techniques in conjunction with visible light microscopy. Modifications of the flying spot television technique for ultraviolet microscopic irradiation studies have produced techniques for illuminating the specimen simultaneously or sequentially with different wavelengths of light (MONTGOMERY 1962). The principal purpose of this communication will be to present the technical aspects of these various scanning systems and to report some of the biological results achieved by their use.

The University of Texas, Southwestern Medical School, Dept. of Pathology, 5323 Harry Hines Boulevard, Dallas 35, Texas (USA).

T. CASPERSSON: Photoelectric Ultramicrospectrophotometry, Especially Rapid Procedures.

The large, scanning and recording ultramicrospectrophotometers permit work on a great variety of biological materials and can be used for the study of many different types of cellphysiological problems. They are especially well suited for work on small structures within the cell. On the other hand measurements of large inhomogeneous objects, as tissues or whole cells, are often quite timeconsuming in such instruments. During the last years the cytochemical variability between cells in tumor populations has attracted attention to an increasing degree. In this work it is often necessary to study large populations of optically very inhomogeneous cells or cell nuclei. The scanning and recording instruments are then not very suitable because of the time involved in the work. Because of this especially rapid ultramicrospectrophotometers have been developed for work as well in the visible as in the ultraviolet, in which such measurements can be carried through up to several hundred times faster than in the recording instruments. The instruments are of two types, a simpler one intended for work on whole cells only, and a more elaborate one which permits measurements also on selected cell parts.

Institute for Medical Research, Karolinska Institutet, Stockholm 60 (Sweden).

B. THORELL: Design and Performance of a Microspectrophotometer for the Determination of Respiratory Enzymes in situ.

Combination of highly sensitive recording circuits and microspectrophotometric technique will be described, which permit the assay of respiratory enzymes localized in defined subcellular areas. It has thereby been possible to detect, for instance, 10^{-20} mole or 6000 molecules of cytochrome b. — The noise level, observed in the recording of a single absorption spectrum with a $0.5\,\mu$ measured object area, is approximately 10^{-4} absorbancy units with a 2.5 second time constant. The average reproducability encountered when several curves are run of the same object is about 2×10^{-4} absorbancy units, which corresponds to 4% of the signal from the cytochrome absorption peak in the Soret region of a single (large) mitochondrion. — To achieve analysis of the multicomponent respiratory enzyme-system by the graphical solution of complex absorption spectra in the Soret region ($\sim$420 mμ), the microspectrophotometric record must have an "uncertainity" of a given curve not more than ± 2 mμ for the wavelength and $\pm 10\%$ for the peak heights. In the region of the α-bands ($\sim$550 mμ), however, a device to cool the cellular object down to the temperature of liquid nitrogen considerable increase the spectral resolving power.

Pathologisches Institut, Karolinska Institutet, Stockholm 60 (Sweden).

H. NAORA: Methodological Questions.

As reported previously, a microspectrophotometric apparatus was constructed in Japan 1955, avoiding Schwarzschild-Villiger effect and it has been widely used in studies on the nucleic acids and other compounds related to cell structures and on the physical and chemical properties of the microcrystals and fibers. — The apparatus is characterized by the use of two microscopic optical systems, one of which acting as a condensor by reversing its system. The main purpose of this apparatus is to eliminate Schwarzschild-Villiger effect completely and to facilitate the rapid measurement of the accurate extinction at the range of wave length from 2,400 Å to 22,000 Å. When the minute objects to be measured are spherical, such as the nuclei of small lymphocytes and isolated nucleoli, and even plane-parallel, very satisfactory results can be obtained. — In this communication, we shall describe briefly the optical properties of this microspectrophotometric apparatus and make some additional comments on the operation of this apparatus.

National Cancer Center, Research Institute, Tsukiji 5-Chome Chuo-Ku, Tokyo (Japan).

M. MENDELSOHN: Scanning Microscopy as Input to a Digital Computer.

Of the several approaches to the translation of images into a form suitable for computer analysis, one involves the measurement and recording of optical density point by point throughout the field of interest. The CYDAC system performs this task automatically for microscopical images. It consists of mechanical and flying spot microscopes with the appropriate electronic components for measuring optical density, converting the information to digital form, and recording the result in a strictly controlled format on magnetic tape. In approximately five seconds, 40,000 optical density measurements can be recorded from a $50 \times 50\,\mu$ field. Some of the technical and analytical difficulties in this work will be discussed, using illustrated examples of cell images which have been printed by a digital computer after being recorded by CYDAC.

Dept. of Radiology, Hospital of the University of Pennsylvania, Philadelphia, Pa. (USA).

B. Methods in Interference Microscopy

G. LOMAKKA: Hochauflösungsverfahren und Schnellmethoden.

Die manuelle Bestimmung des Trockengewichtes in biologischen Objekten mit dem Interferenzmikroskop ist ungenau und langsam. Durch Einführung einer automatischen Absuchungs- und Integrationsanordnung konnte die Genauigkeit und Schnelligkeit der Messung bedeutend verbessert werden. Es wird ein Instrument beschrieben, das das Bild im Interferenzmikroskop absucht und dabei die optischen Gangunterschiede mißt und kontinuierlich integriert. Es wird dazu ein Baker-Interferenzmikroskop benützt; die Richtung der Polarisationsebene im betreffenden Bildpunkt bestimmt die Phasenlage eines elektronischen Signals. Diese wird mit einem elektronischen Phasenmesser bestimmt und integriert. Der Integratorwert gibt das Oberflächenintegral des optischen Gangunterschieds an, das der Trockengewichtsmenge im abgesuchten Feld proportional ist.

High Resolution and Rapid Methods. The manual determination of the amount of dry mass in biological objects with the interference microscope is inaccurate and time consuming. By the introduction of an automatic device for scanning and integration the accuracy and speed for the measurement was considerably increased. An instrument will be described, that scans the image in the interference microscope and simultaneously measures the optical path difference and integrates it. The Baker interference microscope is used and the direction of the plane of polarization in the image point is transferred to a phase position in time for an electrical signal. The phase position is measured in an electronic phasemeter and the value integrated electronically. The integrator output gives the surface integral of the optical path difference, which is proportional to the amount of dry mass in the scanned field.

Institute for Medical Cell Research and Genetics, Karolinska Institutet, Stockholm 60 (Sweden).

C. Quantitative Microradiography

B. LINDSTRÖM: Quantitative Microradiography.

In biology and medicine total dry weight determinations and elementary analyses are the most important applications of quantitative microanalytical x-ray absorption methods. For total dry weight determination the specimen, such as a smear of isolated cells or a thin section $(3—10\,\mu)$ of solid tissue, is exposed to soft polychromatic x-rays, and the absorption image is registered on a fine-grain photographic emulsion together with a reference system consisting of a step wedge of nitrocellulose foils. The x-ray intensities transmitted through the specimen and through the step wedge are compared microphotometrically, and from this comparison the weight per unit area of cellular structures can be determined down to the range of dimension of $2—5\,\mu^2$. For elementary analyses the characteristic absorption edges for different elements are utilized. The specimen is then exposed to monochromatic x-rays with wavelengths on each side of a particular absorption edge. From differences in the two absorption images the weight per unit area of a particular element can be calculated. The method has been used for e.g. the analysis of calcium and phosphorus in calcified tissues and the analysis of sulfur in skin and cartilage during physiological and pathological conditions.

Dept. of Medical Physics, Karolinska Institutet, Stockholm 60 (Sweden).

D. Mass Determination with the Electron Microscope

G. F. BAHR: Mass Determination with the Electron Microscope.

The intentions in electron microscopy are similar to those in light microscopy, namely, to draw quantitative conclusions about the object from the contrast it causes in the recording system. The conditions of electron microscopy are more favorable, since the interaction of electrons with matter are less complicated than those with light. This entails, however, a loss of specific informations, which is compensated partially by the higher resolution of the electron microscope. Quantitative light and electron microscopy complement each other.

Armed Forces Institute of Pathology, Washington, D.C. (USA).

E. Fluorescence Methods

CH. LOESER: Perspectives Offered by Recent Experimentation with Television Fluorescence Microspectrophotometry*.

The use of a prism to disperse the light from the ocular of a fluorescence microscope, and a highly sensitive intensifier image orthicon television camera to scan the spectrum produced permits a dynamic study of interactions in living or fixed tissue. Corrected spectra result. For example, work with colleagues will be described in which fluorescence of weakly emitting neuropharmacologic and neurophysiologic entities has been followed intracellularly including chlorpromazine, serotonin, and reserpine. Work on the first mentioned has been carried out in neuron tissue culture in part. In addition, a) the spectrum of fluorescent immunochemical reactions, b) of carcinogens in interaction with living tissue, c) of *Euglena* autofluorescence, and d) catecholamines condensed with formaldehyde will be reported. — The effort, of course, is to obtain molecular information about the nature of the interaction; the compounds studied represent only a small part of the great number which the sensitivity of the method permits.

* Aided by grants from National Institutes of Health and the American Cancer Society.

Faculdad de Medicina, Universidad Nacional de Trujillo, Cátedra de Neuroanatomía, Trujillo (Peru).

R. RIGLER jr.: Quantitative Fluoreszenz-Mikrospektrophotometrie zur Bestimmung der Primär- und Sekundär-Fluoreszenz verschiedener Zellstrukturen*.

Zur Bestimmung der Fluoreszenz von Zellsubstanzen mit primärer Emission bei Erregung mit ultravioletter Strahlung (biogene Amine, Aminosäuren, Pyridinnukleotide) und mit sekundärer Emission nach Fluorochromierung (Nukleinsäure-Aminoakridinkomplexe) wurde ein Fluoreszenzmikrospektrograph entwickelt unter vorteilhafter Ausnutzung der gegenüber der Absorbimetrie bis zu 1000fach höheren Empfindlichkeit der Fluorimetrie und des durch die direkte Beziehung zwischen emittierter Lichtmenge und Gesamtmenge eines zur Fluoreszenz erregten Stoffes bedingten, sehr vereinfachten Meßverfahrens. Der schematische Aufbau ist folgender: Xenonbogen-Lampe (Osram XBO 450 W)-Quarzprismenmonochromator (Zeiss M4O III) mit automatischem Wellenlängenvorschub zur Zerlegung der Erregerstrahlung, achromatischer UV-Kondensor Zeiss $30\times$, NA $= 0,8$—$0,3$, bzw. Zeiss-Ultrafluar $100\times$ NA $= 0,85$, fluoreszenzfreies Spezialobjektiv Reichert $100\times$, NA $= 1,25$ bzw. $63\times$, NA $= 0,75$ (zur Registrierung der Emissionsstrahlung im ultravioletten Bereich: Spiegelobjektiv American Optical $50\times$, NA $= 0,56$) — Quarzprismenmonochromator (Zeiss M4Q II) mit automatischem Wellenlängenvorschub zur Auftrennung der Emissionsstrahlung — Photomultiplikator 1P28 mit Wechselstromverstärker. Die noch meßbare Änderung des unverstärkten Primärstromes liegt in der Größenordnung 10^{-15}—10^{-16} Amp. Die Kalibrierung der Erregerstrahlung erfolgt mit einem Uranylglasfilter (Schott GG 17). Durch Kombination von Iris- und Lochblenden bis zu $0,3\ \mu$ und von regulierbaren Spaltblenden (als Leuchtfeldblenden in die Objektebene projiziert) ist die Bestimmung der Intensität sowie der Energieverteilung der Primär- und Sekundärfluoreszenz fast jeder Zellstruktur, wie Chromosomen,

Nukleolus, Gesamtkern, möglich. Ein in das System eingebauter Polarisator und Analysator läßt auch die Bestimmung des Polarisationsgrades des emittierten Lichtes zu.

Quantitative Fluorescence-Microspectrophotometry for the Determination of Primary and Secondary Fluorescence of Different Cell Structures. For determination of the fluorescence of cell substances with primary emission by excitation with ultraviolett radiation (biogene amines, amino acids, pyridine nucleotides) and with secondary emission after fluorochromation (nucleic acid — amino-acridin-complexes) a fluorescence microspectrograph was developed using the advantage of an up to 1000 fold higher sensitivity in comparison with absorbimetry and of a very simplified measuring procedure (direct proportionality of emitted light and the total amount of the excited substance). Schematic setup: Xenon arc lamp (Osram XBO 450 W) — quartzprism monochromator (Zeiss M4O III) with automatic wave length registration for exciting radiation—achromatic UV-condensor Zeiss 30×, NA = 0.8—0.3, and Zeiss-Ultrafluar 100×, NA = 0.85—fluorescence free special objective Reichert 100×, NA = 1.25 and 63×, NA = 0.75 (for recording of emission in ultraviolett region: mirror objective American Optical 50×, NA = 0.56) — quartz prism monochromator (Zeiss M4Q II) with automatic wave length registration for emitted radiation—photomultiplier 1P28 RAC — AC amplifier. The still measurable variation of the unamplified primary photo current lies in the magnitude of 10^{-15}—10^{-16} amp. Calibration of exciting radiation by means of an uranyl glass filter (Schott GG 17). By combination of iris- and hole diaphragms down to 0.3 μ and of adjustable slit diaphragms (projected as light field diaphragm in the object plane) the intensity and energy distribution of primary and secondary fluorescence of almost every cell structure, such as chromosomes, nucleoli, whole nucleus, can be determined. A built in polarizer and analyzer allows the determination of the polarization degree of emitted light.

* In Zusammenarbeit mit T. CASPERSSON und G. LOMAKKA. Mit Unterstützung der Wallenberg-Stiftung Riksföreningen mot Cancer und Grant C-4716, U.S. Public Health Service.

Institute for Cell Research and Genetics, Karolinska Institutet, Stockholm 60 (Sweden).

Physikalisch-optische Methoden in der Histochemie:
Vergleich der Ergebnisse mit verschiedenen unabhängigen Methoden

Physical-Optical Methods in Histochemistry:
Comparison of Results with Various Independent Methods

Moderator: W. Sandritter, 63 Giessen (Deutschland).
Co-Moderator: F. Kasten, Pasadena/California (USA),
A. Krygier, Stettin (Polen).

A. Prologue

W. Sandritter: Einleitung.

Die physikalisch-optischen Methoden umfassen heute schon einen großen Teil des elektromagnetischen Spektrums: Röntgenhistoradiographie, Photometrie im ultravioletten und sichtbarem Licht, Interferenzmikroskopie, Infrarotphotometrie. Die quantitativen oder halbquantitativen Substanzbestimmungen erfolgen mittels photometrischer Methoden, wobei die Lichtabsorption mit Hilfe des Mikroskopes gemessen wird. Verschiedene Fehlerquellen können zur Verfälschung der Meßergebnisse führen: 1. *Meß- und Apparatefehler*, wie z.B. Schwarzschild-Villiger-Effekt, Größe der Meßblende, Höhe der Extinktionen u.a. 2. *Objektbedingte Fehler:* wie z.B. Streulicht durch Brechungsindexdifferenzen, Überlagerungen von Absorptionen verschiedener Substanzen u.a. 3. *Färbungsfehler:* wie z.B. ungenügende Stöchiometrie, Artefakte verschiedenster Art. — Diese kurze Aufzählung der Fehlerquellen zeigt, daß die Meßergebnisse mit anderen unabhängigen Methoden überprüft werden müssen.

Die Paneldiskussion soll zeigen, daß für den quantitativen Nachweis von DNS, RNS, Protein und Enzymen verschiedene physikalisch-optische Methoden eingesetzt werden können. Jede der Methoden hat Vor- und Nachteile und ist mit bestimmten, oft unvermeidbaren Fehlerquellen behaftet. Ein Vergleich der Meßergebnisse mit biochemischen Bestimmungen sollte deshalb angestrebt werden. Für vergleichende Messungen der DNS und RNS können die UV-Photometrie, die Feulgen-Photometrie, die Färbung mit basischen Farbstoffen, die Autoradiographie, Chromosomenzählungen sowie die Interferenzmikroskopie herangezogen werden. Die kombinierte Anwendung der cytophotometrischen Methoden und Trockengewichtsmessungen mit dem Interferenzmikroskop (oder der Röntgenhistoradiographie) mit Nukleinsäureextraktionen ist unseres Erachtens besonders zu empfehlen, da hierbei die Messungen mit verschiedenen Methoden an der gleichen Zelle ausgeführt werden können. Die Autoradiographie und die Chromosomenzählungen enthalten weitere bedeutsame Informationen, die in Beziehungen zu den cytophotometrischen und interferenzmikroskopischen Messungen gesetzt werden können. — Für quantitative Proteinbestimmungen stehen bekanntlich nur wenige Methoden zur Verfügung. Die UV-Photometrie ist ebenso wie die färberischen Reaktionen für Tyrosin und Tryptophan mit verschiedenen Fehlerquellen belastet. Die Interferenzmikroskopie kann auch hier mit Extraktionsmethoden zur qualitativen Differenzierung, z.B. von Zellkern-Eiweißkörpern herangezogen werden. Die neuerdings wieder so bedeutsame Klasse der Histon- und Protamineiweißkörper kann mit verschiedenen färberischen halbquantitativen Methoden erfaßt werden, deren Ausarbeitung noch weitgehend im Flusse ist.

Introductory remarks. Physical and optical methods include already a wide range of the electromagnetic spectrum: X-ray historadiography, photometry with UV- and visible light, interference microscopy, infra-red photometry. Quantitative and semiquantitative determinations are carried out by means of photometrical methods in which case the absorption of light is measured with the help of microscopes. Various sources of error are likely to affect the measurements. 1) *Measuring and instrument errors:* e.g. Schwarzschild-Villiger effect, the size of the slit, the intensity of the extinction, and others. 2) *Errors due to the objective:* as for example stray light caused by differences in the refraction index, superimposition of absorptions by different substances, and others. 3) *Staining errors:* as for example insufficient stoichiometry, artefacts of various kinds. — This brief enumeration of possible errors proves that the measuring results have to be checked by other independent methods.

The panel discussion will show that various physical and optical methods can be employed for the quantitative determination of DNA, RNA, protein and enzymes. Each method has its advantages and disadvantages and is subject to certain — often unavoidable —

deficiencies. A comparison of the measuring results with biochemical determinations is therefore desirable. For comparative studies on DNA and RNA, UV-photometry, Feulgen-photometry, staining with alkaline dyes, autoradiography, chromosome counts and interference microscopy can be employed. In our opinion it is especially advantageous to combine cytophotometrical methods and dry weight measurings (or X-ray historadiography) with extraction methods for nucleic acids. This combination makes it possible to employ different methods on the same cell. Autoradiography and chromosome counts give additional important informations which can be related to cytophotometrical and interference microscopic measuring results. — As is well known there are few methods only for the quantitative determination of proteins. UV-photometry has as many disadvantages as have the colour reactions for the determination of tyrosine and tryptophan. For this particular problem interference microscopy is an alternative method, as for example extraction methods can be used for the differentiation of nucleic proteins. Histones and protamines, a group of proteins which only recently gained new importance, can be demonstrated by several semi-quantitative staining methods, which, however, have still to be perfected.

Pathologisches Institut der Universität, 63 Giessen (Deutschland).

B. DNA

FREDERICK H. KASTEN: Deoxyribonucleic Acid. Comparison of Measuring Results: UV Photometry — Feulgen Photometry — Staining with Basic Dyes*.

Quantitative cytochemistry of DNA is carried out generally by Feulgen cytophotometry. Therefore, analytical results with this reaction are frequently used as a standard to compare data obtained by other less controllable cytochemical techniques. UV absorption analyses at 265 mμ allow the total nucleotides of both nucleic acids to be measured. The contribution by DNA is difficult to evaluate since nucleotide absorption is sensitive to pH changes and polymerization. Despite these limitations, there are published data which show reasonable agreement with Feulgen measurements. For example, growing cells *in vitro* show a continuous increase in UV absorption at 265 mμ in interphase nuclei. This increase is apparently due to DNA and is correlated with the progressive synthesis of DNA demonstrated by Feulgen cytophotometry. Other data will be presented summarizing agreement and disparity of published results obtained by UV and Feulgen cytophotometry. — Basic dyes are frequently employed as quantitative binding agents with cellular DNA. These include azure B, gallocyanin-chrome alum, methyl green, and methylene blue. Standardization of dye-binding methods requires judicious efforts to control interference from competing ions and to allow unequivocal interpretations to be made. The success with which this goal has been achieved will be evaluated and comparative data summarized.

* Supported in part by Public Health Service Research Grant CA 07017-01 from the National Cancer Institute and an Institutional Grant to the University of Southern California by the American Cancer Society.

Pasadena Foundation for Medical Research, Pasadena, California (USA) and University of Southern California School of Medicine, Los Angeles, California (USA).

W. SANDRITTER: Vergleich der Meßergebnisse: UV-Photometrie mit biochemischen Bestimmungen*.

Für UV-photometrische Bestimmungen des DNS-Gehaltes ergeben sich folgende Schwierigkeiten: 1. Welcher spezielle Extinktionskoeffizient (ε') soll für DNS gewählt werden? 2. Inwieweit gelingt es mit Ribonuklease, die gesamte RNS zu entfernen? 3. Inwieweit können andere Substanzen, die zur Absorption bei 260 mμ beitragen, ausgeschaltet werden (Tyrosin und Tryptophan, Streulicht)? 4. Es können nur Zellausstriche untersucht werden. — Der Vorteil der Methode liegt darin, daß die Mengenbestimmungen absolute Werte ergeben (10^{-12} g) und in Vergleich zu interferenzmikroskopischen Trockengewichtsbestimmungen gesetzt werden können. Damit kann, wie in biochemischen Messungen, der prozentuale DNS-Gehalt vom Trockengewicht angegeben werden. Eigene Untersuchungen an verschiedenen Zellpopulationen (Spermien, diploide Zellen: Thymus, kernhaltige Erythrocyten) zeigen, daß der nach Ribonukleasebehandlung gemessene DNS-Gehalt sehr gut mit biochemischen Werten übereinstimmt (ε' von 20, Streulichtkorrektur, Korrektion der Tyrosin- und Tryptophanabsorption bei 260 mμ).

Comparison of Measuring Results: UV-Photometry — Biochemical Determinations. Many difficulties arise when UV-photometrical methods are used for the determination of DNA. They can be expressed as follows: 1) What particular extinction coefficient (ε') is to be chosen for DNA? 2) How far is it possible to remove the total RNA with RN-ase? 3) To what extent can other substances be eliminated, which contribute to the absorption at 260 mμ (such as tyrosine and tryptophan, stray light) ? 4) Only cell smears can be examined. — The advantage of the method is, that quantitative determinations yield absolute values (10^{-12} g) which can be compared with interference microscopic dry weight determinations. Therewith the percentage dry weight DNA-content can be calculated as is possible with biochemical methods. Our own investigations of various cell populations (sperms, diploid cells: thymus, nucleus containing erythrocytes) show that the DNA content after DN-ase treatment corresponds very well with biochemical results (ε' 20, stray light correction, correction of tyrosine and tryptophan absorption at 260 mμ).

* Mit Unterstützung der Deutschen Forschungsgemeinschaft.

Pathologisches Institut der Universität, 63 Giessen (Deutschland).

W. OEHLERT: Vergleich der Ergebnisse der Feulgenphotometrie mit Autoradiographie.

Mit der Feulgenphotometrie werden Aussagen über relative DNS-Mengen erhalten, mit der Autoradiographie unter Verwendung von ³H-Thymidin dagegen Angaben über die Anzahl der Zellen innerhalb einer Zellpopulation, die sich während der Versuchszeit in der DNS-Synthesephase befinden, Angaben über Länge der DNS-Synthesezeit, die der prämitotischen Ruhephase und schließlich auch über die gesamte Generationszeit. Dagegen sind unter Verwendung der Autoradiographie keine Aussagen z.B. über das Ploidiemuster eines bestimmten Gewebes oder über Änderungen des Ploidiemusters während der Kanzerisierung möglich. Die mit beiden Methoden ermittelten Werte sind demnach verschieden und nicht ohne weiteres miteinander vergleichbar. — Wie die Ergebnisse autoradiographischer Versuche ergeben haben, dauert die DNS-Neubildung in den verschiedensten Zellen des Warmblüterorganismus und auch im Krebsgewebe 7—8 Std. Während dieser Zeit nimmt somit der DNS-Gehalt des Zellkerns kontinuierlich von einem „diploiden" zum „tetraploiden" Wert zu. Alle während dieser Zeit erfolgenden DNS-Messungen würden also „aneuploide" DNS-Werte ergeben. Während der prämitotischen Ruhephase, die im Karzinomgewebe auf mehrere Stunden verlängert sein kann, besitzt eine normalerweise diploide Zelle einen tetraploiden Chromosomensatz. Alle Zellen, welche während dieser Generationsphase feulgenphotometrisch untersucht werden, werden „tetraploide" DNS-Mengenwerte ergeben. Die Anzahl der „aneuploiden" Zellen innerhalb einer Population wird um so größer sein, je größer der Anteil ist, welchen die DNS-Synthesephase innerhalb der gesamten Generationszeit beansprucht. Vor allem bei schnell wachsenden Tumoren wird der größte Teil der gesamten Generationsphase von der DNS-Synthesephase eingenommen. Die in solchen Geweben ermittelten DNS-Werte für die Einzelzellen können also nicht das echte Ploidiemuster angeben. Die Kombination der Autoradiographie mit der Feulgenphotometrie würde über die Bestimmung der Generationszeit und der einzelnen Generationsphase den Aussagewert beider Methoden erheblich steigern.

A **Comparison of Results Obtained by Feulgen Photometry and Autoradiography.** The results of Feulgen photometric tests give information about relative quantities of DNA. Results obtained by autoradiographic methods (using ³H-thymidine) give information about the number of cells in a cell population which synthesize DNA at the time the investigation takes place, about the length of the DNA synthesis phase, the pre-mitotic phase, and finally about the length of the entire period of cell generation. It is not possible, however, to get any information about the chromosome pattern in certain tissues or about changes in this chromosome pattern during tumorigenesis by means of autoradiography. Data obtained by these two methods differ from each other and cannot be compared directly. Autoradiographic tests show that DNA synthesis in various cells of warm-blooded animals, including cancer cells, takes about 7—8 hours. Within this period the DNA content of the nucleus increases gradually from a "diploid" to a "tetraploid" standard. DNA estimations carried out during this period will therefore yield "aneuploid" results. During the pre-mitotic phase — which can last for several hours in neoplastic tissue — a normally diploid cell will have a haploid set of chromosomes. All cells tested during this period by Feulgen photometry will give "tetraploid" DNA results. The number of "aneuploid" cells in a cell population is directly proportional to the length of time which is required for the synthesis of DNA. This will especially be the case in rapidly growing tumours. DNA results obtained from individual cells in such tissue will not reflect the true chromosome pattern. A combination of autoradiography and Feulgen photometry would be of great value for the determination of the length and/or phase of the cell formation.

Pathologisches Institut der Universität, 78 Freiburg i.Br. (Deutschland).

H. G. SCHWARZACHER: Vergleich der Meßergebnisse: Feulgenphotometrie mit Chromosomenzählungen.

Die Konstanz von Zahl und Struktur der Chromosomen einer Spezies ist heute als gesichert zu betrachten. Ebenso haben Feulgen-photometrische Bestimmungen der DNS der Zellkerne für jede Spezies weitgehend charakteristische und konstante Werte ergeben. Die normalerweise vorkommenden Abweichungen der Chromosomenzahl verschiedener Zelltypen eines Individuums gehen einher mit gleichsinnigen Veränderungen der Feulgen-Werte der Interphasenkerne (haploide Geschlechtszellen, diploide und polyploide Zellen). In rasch wachsenden Zellpopulationen mit konstanter diploider Chromosomenzahl findet man entsprechend den Perioden des Zellzyklus Interphasenkerne mit Feulgen-Werten, die höher als „diploid" (Syntheseperiode) oder „tetraploid" (G2-Periode) sind. In transformierten aneuploiden Zellstämmen der Gewebekultur sind die Feulgen-Werte in guter Übereinstimmung mit der Chromosomenzahl. Beim Vergleich von Feulgen-Werten und Chromosomen verschiedener Spezies ergeben sich Korrelationen nur bei Berücksichtigung von Zahl und auch der Form (Armlängen) der Chromosomen. — Im allgemeinen sind innerhalb einer vorwiegend diploiden Zellpopulation die Abweichungen der Feulgen-Werte vom „diploiden" Mittelwert größer als die manchmal beobachtbaren Abweichungen der Chromosomenzahl vom konstanten diploiden Wert, die als Zählfehler oder präparatorisch bedingte Verluste von Chromosomen gedeutet werden können. Die Streuung der Photometerwerte ist meist größer als der einem bestimmten Gerät innewohnende Meßfehler. Die Frage, ob weitere Fehlerquellen der photometrischen Methode (Inkonstanz der Feulgen-Färbung, Inhomogenität etc.) oder tatsächlich DNS-Schwankungen der Interphasenkerne (bei konstanter Chromosomenzahl) vorkommen, scheint noch offen zu sein.

Comparison of Quantitative Results: Feulgen Photometry and Chromosome Counts. It can be considered a fact that the number and the structure of chromosomes in a species is constant. It is also a fact that Feulgen photometric determinations of nuclear DNA give largely characteristic and constant results for each species. Deviations in the number of chromosomes, which normally occur in different cell types of an individual, go parallel with variations of the Feulgen results on interphase nuclei (haploid germ cells, diploid and polyploid cells). In quickly growing cell populations with a constant number of diploid chromosomes, interphase nuclei which give a Feulgen result higher than "diploid" (period of synthesis) or "tetraploid" (G2-period) can be found, according to the period of the cell cycle. Transformed aneuploid cell lines in cell cultures show an almost complete correspondence between the number of chromosomes and the Feulgen results. If Feulgen results and chromosomes of different species are to be compared, a correlation will only be found if the number and the form of the chromosomes are taken into account. — In general the Feulgen result will —within a predominantly diploid cell population — deviate more from the "diploid" mean than does the number of chromosomes from the constant diploid rate. This latter fact can be regarded as the counting error of loss of chromosomes during the preparation. The deviation of the photometrical results is often higher than the measuring error of the apparatus used for the examination. The question whether there are genuine DNA fluctuations in interphase nuclei (while the number of chromosomes remains constant) or whether deficiencies of the photometrical method (inconsistency of the Feulgen method, inhomogeneity etc.) account for the results, remains unsettled.

Histologisch-Embryologisches Institut der Universität, Wien (Österreich).

H. KRAUS: Vergleich interferometrischer DNS-Bestimmungen mit anderen Methoden*.

Mit Hilfe des Interferenzmikroskopes können an der Einzelzelle verschiedene Zellkernfraktionen quantitativ bestimmt werden. In eigenen Untersuchungen wurde das Trockengewicht von Thymuslymphknoten und von in organischen Medien isolierten Kernen von Leber- und Nierenepithelien vor und nach Extraktion der Albumin-Globulin-Fraktion mit 0,14 m NaCl gemessen. Die erhaltenen Werte stimmen gut mit biochemischen Meßwerten überein. — Zur Bestimmung der DNS können Desoxyribonuklease, Trichloressigsäure und Detergentien benutzt werden. Bei zeitlich abgestufter Extraktion gelingt es, die Kinetik der Lösung der DNS zu verfolgen. Ein Vergleich mit cytophotometrischen Messungen ergibt einen übereinstimmenden Verlauf der Extraktionskurven. Der interferenzmikroskopisch gemessene Gewichtsabfall entspricht dem DNS-Gehalt, wie er aus biochemischen Messungen bekannt ist. Die Werte streuen allerdings stärker, so daß in manchen Fällen eine gleichzeitige Proteinextraktion nicht ausgeschlossen ist.

A Comparison of Interferometric DNA Determinations with Other Methods. The technique of interference microscopy makes it possible to determine various fractions of the nucleus on an individual cell. In the course of our own investigations the dry weight of thymus lymphocytes and nuclei of liver and kidney epithelium (isolated from organic media) has been measured before and after the extraction of the albumin/globulin fraction with 0.14 M NaCl. The results obtained are in agreement with biochemical findings. — For the determination of DNA, can be employed DN-ase, TCA, or detergents. If the extraction is carried out gradually over a period of time it is possible to observe the kinetics of the solution of DNA. A comparison with cytophotometrical results shows a corresponding shape of the extraction curves. The decrease of weight — measured by means of interference microscopy — is in accordance with the DNA content as it is known from biochemical findings. As the results spread over a relatively wide range a simultaneous protein extraction cannot always be excluded.

* Mit Unterstützung der Deutschen Forschungsgemeinschaft.

Pathologisches Institut der Universität, 63 Gießen (Germany).

C. RNA

G. KIEFER: Ribonukleinsäure. Vergleich der Meßergebnisse: UV-Photometrie — Färbung mit basischen Farbstoffen — biochemische Ergebnisse*.

Eine exakte mathematische Erfassung aller Bedingungen bei der Cytophotometrie ist nur begrenzt möglich. Daher spielen Messungen an Modellen und Vergleiche zwischen mikrophotometrischen und durch andere unabhängige Methoden gewonnene Ergebnisse für die Beurteilung der cytophotometrischen Methodik eine wesentliche Rolle. Es war daher für die quantitative Histochemie von großer Bedeutung, daß die von ihr zuerst erkannten Zusammenhänge zwischen der Ribonukleinsäuremenge und der Zellaktivität bzw. Proteinsynthese bald auch durch biochemische Untersuchungen bestätigt werden konnte. Ein direkter Vergleich von UV-photometrischen und biochemischen Bestimmungen bestätigte die Leistungsfähigkeit der Methode. Mit den heute zugänglichen verbesserten Apparaturen sollten solche vergleichenden Untersuchungen in verstärktem Maße durchgeführt werden, wobei neben der Biochemie auch Interferometrie, Historadiographie und Mikroextraktion (EDSTRÖM) zum Vergleich herangezogen werden könnten. Ebenso wünschenswert wäre ein Vergleich von UV-photometrischen Messungen und solchen im sichtbaren Spektralbereich nach Anfärbung mit basischen Farbstoffen. Die bisher in der Literatur vorliegenden Angaben sind spärlich (SWIFT, DI STEFANO), und es sollte ein größeres, möglichst verschiedenartiges Zellmaterial dabei untersucht werden. Dazu bieten sich eine Reihe von Farbstoffen an, für die ein konstantes quantitatives („Stöchiometrie") Verhältnis zur Nukleinsäure nachgewiesen wurde. Nach eigenen Erfahrungen spielt das begleitende Protein für die Färbbarkeit und für vergleichende Untersuchungen verschiedener Objekte eine wesentliche Rolle. Daher müssen die Bedingungen exakt konstant gehalten werden.

RNA. Comparison of Quantitative Results: UV-Photometry — Staining with Alkaline Dyes — Biochemical Methods. Only within certain limits is it possible to determine all conditions for cytophotometrical investigations in an exact mathematical manner. For the assessment of cytophotometrical methods it is of great importance to compare measurements obtained on models with the results of microphotometrical and other independent methods. In the field of the quantitative histochemistry it is therefore a matter of great importance that biochemical analyses confirm the correlation between RNA quantity, cell activity and/or protein synthesis. This fact was first discovered by histochemical means. A direct comparison of UV-photometrical and biochemical determinations confirms the efficiency of the method. With today's improved apparatus this kind of comparative study should be accomplished to a higher degree, employing interferometry, historadiography and microextraction (EDSTRÖM) in addition to biochemistry. It is desirable to compare UV-photometrical measurements with those obtained by staining with alkaline dyes in the visible spectral range. So far data in the literature are scarce (SWIFT, DI STEFANO) and a variety of cell materials ought to be investigated. For that purpose a whole range of dyes which react in constant quantitative proportions ("stoichiometry") with nucleic acids is available. Present investigations show that the concomitant protein is of great importance in staining and for the comparison of different objects. It is therefore important to work under exactly standardized conditions.

* Mit Unterstützung der Deutschen Forschungsgemeinschaft.

Pathologisches Institut der Universität, 63 Gießen (Deutschland).

D. Proteins

G. BENEKE: Eiweißkörper. Vergleich der UV-Photometrie und der Millon-Reaktion für den Nachweis von Tyrosin und Tryptophan*.

Bei dem Vergleich zwischen histochemischen Aminosäure-Reaktionen und der UV-Photometrie ergeben sich verschiedene Probleme und Schwierigkeiten. Nur zwei proteingebundene Aminosäuren sind diesem Vergleich überhaupt zugänglich: 1. Tyrosin, 2. Tryptophan. Beide sind mit histochemischen Farbreaktionen darstellbar. Beide absorbieren auch ungefärbt ultraviolettes Licht bei 280 mμ. Setzt man die Absorption durch das sichtbare Chromophor, das nach den Farbreaktionen gewonnen wurde, in Beziehung zur Absorption des Chromophors von Tyrosin und Tryptophan im ultravioletten Licht, so ergeben sich eine Reihe von Fragen: 1. Für den histochemischen Nachweis von Tyrosin hat sich bisher am besten die Millon-Reaktion bewährt, die in verschiedenen Varianten durchgeführt wird. Die Reaktion kann verschiedene, unerwünschte Nebenreaktionen auslösen, die natürlich die quantitative Auswertung beeinflussen. 2. Das gleiche gilt für den histochemischen Tryptophan-Nachweis. 3. Die UV-photometrische Auswertung erfaßt Tyrosin und Tryptophan gemeinsam. Ein Eiweißkörper, der diese beiden Aminosäuren enthält, liefert außer der echten Absorption auch noch eine sog. Streulichtabsorption, die in die Gesamtabsorption eingeht. Die Streulichtabsorption läßt sich auch am histologischen Schnitt, am besten mit einem Extrapolationsverfahren, eliminieren. In Lösungen läßt sich die gemeinsame Absorption von Tyrosin und Tryptophan bei 280mμ durch Ausnutzung der Rotverschiebung der Tyrosinabsorption im alkalischen Milieu trennen. Diesem Vorgehen stellen sich am histologischen Schnitt weitere Schwierigkeiten entgegen. — Unter Beachtung dieser Fehlermöglichkeiten und bei konstanten Versuchsbedingungen kann ein annähernder Vergleich der Meßergebnisse nach histochemischer Reaktion und im UV-Licht vorgenommen werden.

Proteins. A Comparison of UV-Photometry and Millon Reaction for the Determination of Tyrosine and Tryptophan. A comparison of histochemical reactions for amino-acids with UV-photometry results leads to certain problems and difficulties. Only two protein-bound amino-acids are accessible to comparison. 1. Tyrosine, 2. Tryptophan. Both are demonstrable with histochemical colour reactions and both absorb ultraviolet light at 280 mμ. If one compares the absorption by the visible chromophor — as obtained by colour reactions — with the absorption of the chromophor of tyrosine and tryptophan in the ultraviolet light, a number of questions arise: 1) For the histochemical determination of tyrosine the Millon reaction, which is used in several variations, proved to be the best method. The reaction can cause a number of undesirable secondary reactions which, of course, influence the quantitative evaluation. 2) The same applies to the histochemical determination of tryptophan. 3) UV-measuring results comprise both tyrosine *and* tryptophan. A protein which contains both these amino-acids gives, apart from its primary absorption, a so-called stray light absorption which enters into the total. The best method to eliminate this stray light absorption on histological sections is an extrapolation process. In solutions tyrosine and tryptophan can be distinguished by the fact that at an alkaline pH the absorption of tyrosine is shifted towards the red range of the spectrum. It is, however, difficult to carry out the same procedure on histological sections. — If these sources of error are born in mind, and the experimental conditions are kept constant, an approximate comparison of histochemical results with those obtained by UV-photometry can be accomplished.

* Mit Unterstützung der Deutschen Forschungsgemeinschaft.

Pathologisches Institut der Universität, 63 Gießen (Deutschland).

DAVID P. BLOCH: Comparison of Different Methods for the Demonstration of Histones and Protamines.

The standard cytological procedures for demonstrating histones and protamines are variants of the basic technique of ALFERT and GESCHWIND, in which histones are stained with acidic dyes (e.g. fast green) under alkaline conditions which prevent the staining of proteins with lower isoelectric points. Certain histones very rich in arginine, such as are found in many spermatids and sperms, stain even after deamination, in contrast to the great majority of histones whose staining is thereby abolished. Most protamines require special procedures for their identification, owing to their high solubility. Picric acid, substituted for trichloracetic acid during DNA hydrolysis, immobilizes protamines and prevents their loss. Eosin used for subsequent staining also forms a precipitate with protamine. Bromphenol blue will colorize the histones of the cleavage cells of some organisms which for unknown reasons do not otherwise stain. — A less empirical procedure can be used to distinguish various histones and also histones and protamines. This method employs double staining with fluorodinitrobenzene and Sakaguchi reagent. Protein bound arginine stains a bright pink, and lysine and tyrosine, an intense yellow. Most cell components stain various shades of orange. Microspectrophotometric determination of the shapes of the spectral absorption curves, or of the extinctions at 400 and 520 mμ, permit the characterization of histones and protamines in terms of their ratios of lysine plus tyrosine to arginine. This analysis is of course applicable to subcellular components.

Botany Dept., University of Texas, Austin, Texas (USA).

ARLINE D. DEITCH: Determination of Total Protein: Comparison Between Interference Microscopy and Different Staining Methods*.

Anhydrous mass, measured interferometrically, may be employed as a reference standard for protein content against which cytophotometric color reactions may be compared. The utility of the interferometric standard will depend upon the sensitivity of the interference determinations, the effectiveness of the fixation procedure and the completeness with which lipids, carbohydrates and nucleic acids can be extracted. It is possible to fractionate the protein mass by specific removal of histones or histone fractions. Mass measurements made before and after such extraction give the ratio of histone to non-histone protein in a cell type. — In histochemical color reactions the extent to which color will be formed will be influenced by the availability of reactive groups and the stability of the colored product. The specificity of the reaction may be determined on model substances and by appropriate blocking or extraction procedures. In order to obtain stoichiometry it may be necessary to remove competing groups. Fractionation of proteins by specific removal of proteins may be used in the cytophotometric estimation of protein as well as in the interferometric procedures. Data will be presented comparing the interferometric and cytophotometric determination of amounts of cellular proteins.

* Supported by grant AI-05708-01 VR from the USPHS.

Columbia University, College of Physicians and Surgeons, New York, N.Y. (USA).

B. THORELL: Microspectrophotometric Determination of Enzymes and Pigments.

The multicomponent metabolic systems in the cell, which generate the necessary energy for life, consist of enzymes some of which carry chromophoric groups and, therefore, lend themselves to spectrophotometric determination of quantity as well as of metabolic state. Such enzymes are the cytochromes and the pyridin-nucleotides. The absorption of radiant energy given by these enzymes in the layer of thickness of a single cell or cellular structure (10—0.1 μ), are in the order of 0.00001—0.001 absorbancy units. — Some special features of microspectrophotometric designs, capable of recording these very small absorbancies, will be described. — Examples will be given of microspectrophotometric measurements of respiratory enzymes within the single, living cell or intracellular structure under different, metabolic conditions. Thus, it has been possible to study, in a direct way, the metabolic states and dynamics of different intracellular compartments.

Dept. of Pathology, Karolinska Institutet, Stockholm 60 (Sweden).

Comparison and Criticism of Autoradiographic Methods

Moderator: S. R. PELC, London (England).
Co-Moderator: W. MAURER, Köln (Deutschland).

E. KOBURG: Differences in Grain Number Distribution due to Different Ways of Specimen Preparation.

It was repeatedly emphasized that autoradiography with tritium will be influenced to a high degree by absorption of tritium-β-radiation. That is why we assumed differences in the preparation of specimens to be responsible for different grain yields. — Sections and squashes of one and the same tissue and animal were used for stripping film autoradiography. — In squashes there are plenty cells with higher grain numbers than are in sections, but there are also comparatively more cells with 1 and 2 grains. The whole grain number distribution in squashes is rather broad and uncharacteristic compared with that of sections, whereas distributions derived from sections usually show two definite peaks: one in the vicinity of zero and one representing the mean incorporation rate. — The reasons of these differences and the advantages and disadvantages of the different preparations are discussed.

Städtische Krankenanstalten, Moorenstr. 5, 4 Düsseldorf (Deutschland).

A. FICQ: Track-Autoradiography.

The principle of the "track-method" (thick emulsions) will be presented. By means of several examples the advantages of this technique, in comparison with the "stripping film method", will be discussed. This comparison will also be presented in relation to the different radioelements labelling the molecules used as precursors. Finally, the renewal of interest in this method, due to the development of Molecular Biology, will be emphasized. (P^{32}-stars: viruses, phages; labelled DNA molecules: bacteria; double labelling experiments, etc.)

Lab. de Morphologie Expérimentelle, Université de Bruxelles (Belgium).

KAZIMIERZ OSTROWSKI and ERIC A. BARNARD: Use of ³H-Labelled Inhibitors for Localization and Estimation of Enzymes.

An alternative to the principle, widely used in histochemistry, of comparing enzyme activities using deposition of a product of the enzymatic reaction, has been proposed (OSTROWSKI, K. and BARNARD, E. A., Exp. Cell Res. *25*, 465, 1961; BARNARD, E. A. and OSTROWSKI, K., Biochemical J., *85*, 27P, 1962; OSTROWSKI, K., BARNARD, E. A., STOCKA, Z. and DURZYNKIEWICZ, Z., Exp. Cell Res. *31*, 89, 1963). This achieves, instead, the localization and the quantitative estimation of active centres of some enzymes by the use of ³H-labelled irreversible inhibitors. Application of the inhibitor, and washing, is followed by autoradiography. — Two examples will illustrate the usefulness of this principle: I. *Acetylcholinesterase* (AcChase) within the motor end-plates of mouse diaphragm after blockade by ³H-diisopropylphosphorofluoridate (DFP) gives sharp autoradiographs (ARGs). The specificity of the reaction was established and checked by several procedures: (i) Abolition of the colour reaction for AcChase (using acetylthiocholine as substrate), after blocking by ³H-DFP. (ii) Prevention of the ³H-DFP reaction by eserine (10^{-5} M). (iii) The sequence: DFP→ ³H-DFP gives negative ARGs. (iv) Protection by a specific substrate in the following sequence: Acetylcholine (10^{-2} M)→(Acetylcholine + unlabelled DFP)→buffer→³H-DFP gives sharp positive ARGs. (v) Treatment of blocked AcChase by the specific reactivator pyridine-2-aldoxime-methiodide (2-PAM), in the following sequence: DFP→2-PAM→³H- DFP, gives sharp ARGs at the end-plates. — Using grain counting, some quantitative estimations

	Mean Grain Count	Stand. Devn.	Number of unit areas counted
Glomerulus	2,7	1,6	1000
Prox. Conv. Tubule	7,4	2,3	1000
Thick part of Henle's loop .	2,1	1,1	500
Distal Conv. Tub.	3,3	1,4	1000
Background	0,2	0,5	1000

have been made to determine the influence of fixatives on the amount of active enzyme, and to study the kinetics of reaction of ^{3}H-DFP with AcChase *in situ* in unfixed diaphragms. — II. *Non-specific Esterases*. Relative amounts of the DFP-sensitive non-specific esterases in distinct parts of the mouse nephron *in situ* were estimated after blocking by ^{3}H-DFP. The Table gives the grain counts per unit area of 30,25 square micra.

Any small amounts of AcChase present were blocked by its specific substrates (acetyl-choline and acetyl-beta-methyl-choline), or by eserine. — Some classes of DFP-sensitive non-specific esterase can be measured using protection by particular substrates or their analogues (e.g. phenylacetic acid), in the sequence: substrate→(DFP + substrate)→buffer→ ^{3}H-DFP. Grain counting has been applied to determine classes in this way.

Histologic Dept., Medical Academy, Ul. Cherubinskiego 5, Warsaw (Poland) and *State University of New York, Buffalo (USA)*.

T. C. Appleton: Autoradiography of Soluble Compounds.

The technique of autoradiography is well established as a means of studying the incorporation of labelled compounds. The standard techniques, however, involve many solutions each of which may remove soluble or partly soluble materials. A new technique will be discussed for studying the uptake and distribution of applied labelled material whether soluble or insoluble, and for work on unextracted materials. No solutions that might leach out soluble compounds are used before the completion of photographic exposure. — Frozen sections are cut in a cryostat, kept in a darkroom, and are mounted directly onto cold (approx. —10° C) coverslips coated with a photographic emulsion. Melting of the sections is avoided by exposing at —20 to —30° C. The autoradiographs are given a quick fixation in Wolman's fixative prior to development. — The technique has proved both simple and reliable; the section is in secure and close contact with the film and gives excellent resolution. Experiments with tritiated Thymidine have shown nuclear and cytoplasmic labelling 15 minutes after injection and indicated that there was no slipping of autoradiograph; while experiments with labelled sugars have shown that highly soluble isotopes are retained. Control preparations with unlabelled material gave no autoradiograph or blackening to the film.

Medical Research Council, Biophysics Research Unit, King's College, London W. C. 2 (England).

P. Granboulan: Données techniques concernant la radioautographie au microscope électronique.

Depuis le travail princeps de Caro l'effort de tous les laboratoires s'est porté sur la recherche d'émulsions à grains plus fins. Il s'agit apparemment d'un problème de technique photographique plus intéressant pour l'émulsionneur que le praticien de la cytologie. En réalité deux ordres de difficultés peuvent se rencontrer: 1°) d'une part la sensibilité de ces émulsions sera-t-elle suffisante? On peut en effet admettre qu'elle décroit très vite avec la taille du grain vierge. 2°)D'autre part la proportion de gélatine ne sera-t-elle pas trop élevée? La gélatine se conduit comme si elle était insensible aux électrons, comme un véritable emballage du matériel radiosensible qui est le grain d'halogénure lui-même. Cependant on peut déjà espérer que les émulsionneurs pourront faire des émulsions à grains plus fins que 700 Å tout en respectant l'impératif classique des émulsions nucléaires: plus de 90% d'argent en poids. Les difficultés rencontrées au cours de ces essais seront discutées.

Technical Problems Concerning Electron Microscopical Radioautography. Since the basic work of Caro, the efforts of all laboratories have been directed towards the development of photographic emulsions with finer grains. The question is apparently more a problem of photographic technique, and thus more interesting for the manufacturer than for the cytologist. In practice two categories of difficulties are encountered: 1) On the one hand, will the sensitivity of these emulsions be sufficiently high? One has, in fact, to admit that it decreases rapidly with the size of the silver grain. 2) On the other hand, will the proportion of gelatine be too high? Gelatine behaves as if it were insensitive to electrons; it behaves like a real carrier for the radiosensitive material, the halogenous grain. It is to be hoped, however, that the manufacturers will be able to make emulsions with grains much smaller than 700 Å, taking into consideration the classical "must" for nuclear emulsions: a content of more than 90% of silver in weight. The difficulties encountered in the course of these experiments will be discussed.

Institut de Recherches sur le Cancer, Villejuif/Seine (France).

G. C. Budd: Semi-Quantitative Electron Autoradiography.

In electron microscope autoradiography many factors influence the relationship between the observed distribution of developed silver grains, and the site and amount of radioactive label within sections of cell organelles. Important among these factors are emulsion distribution and section thickness. — Some degree of reproducibility of results is possible when comparison of the grain distribution over different labelled structures in the same section is to be made. For this type of study the section thickness and emulsion distribution need not be very carefully controlled so long as the structures which are being compared are near to each other, and that large areas of labelled material are presented to the emulsion. — More careful control of section thickness and emulsion spreading is required if labelled regions in separate autoradiographic preparations from a series of specimens are to be compared. — By using serial sections from the same labelled specimen, the evenness of emulsion spreading has been examined under experimental conditions by observing the pattern of grains over each section in turn. — Similarly, serial sections provide a means of checking variations in the thickness of sections which occurs even when care is taken to collect sections which all show a similar interference colour. — Results obtained in checks of this type using the membrane method of electron microscope autoradiography will be discussed in relation to experiments in which the distribution of radioactive label in different specimens is compared.

Royal Free Hospital School of Medicine, London (England).

M. Chèvremont et S. Chèvremont-Comhaire: Etude cytologique et cytochimique des effects de la thymidine tritiée à haute radioactivité sur des fibroblastes cultivés in vitro.

Ces expériences sont entreprises dans un double but. D'une part, nous étudions les effets directs produits par un rayonnement émis par une substance se trouvant à l'intérieur même de la cellule et pendant des temps prolongés. D'autre part, les expériences sont susceptibles de fournir des indications sur les doses de tritium de la thymidine marquée à ne pas dépasser dans l'emploi de ce précurseur en technique histoautoradiographique. — Nous avons examiné les cellules vivantes à fort grossissement et en contraste de phase et leurs modifications éventuelles lors de l'interphase ou pendant la mitose. Le métabolisme et la synthèse des ADN ont aussi fait l'objet de recherches quantitatives dans des noyaux individuels, par cytophotométrie en lumière ordinaire après réaction de Feulgen. — La thymidine tritiée à haute radioactivité provoque d'importantes altérations chromosomiques. Celles-ci sont, dans leur ensemble, assez comparables à celles que l'on peut observer sous l'effet des Rayons X ou d'un agent alkylant comme le Myleran. L'activité mitotique peut être très inhibée. — Le métabolisme et la synthèse des ADN peuvent être profondément perturbés. On observe notamment une accumulation importante de noyaux à valeur «tétraploïde» en ADN, dans certaines conditions.

Cytological Effects of Highly Radioactive Tritiated Thymidine in Fibroblasts Cultivated in vitro. The experiments were undertaken for two reasons: First, to study the direct effect of radiation, emitted from a substance inside the cell, over a prolonged period. Secondly, the experiments were expected to supply information about the maximum amount of tritiated thymidine that can be used in historadiography. — We examined living cells, and their eventual modifications, at the time of the interphase or during mitosis, under very strong magnification and under phase contrast. Metabolism and DNA synthesis were the object of quantitative investigations of individual nuclei (Feulgen reaction followed by cytophotometry using visible light). Highly radioactive tritiated thymidine caused important alterations in the chromosomes. They were, on the whole, comparable with cells treated with X-rays or an alkylated agent like Myleran. The mitotic activity may be strongly inhibited. — Metabolism and DNA synthesis may be profoundly disturbed. Under certain conditions a considerable accumulation of nuclei, giving "tetraploid" DNA results, could be observed.

Institut d'Histologie et d'Embryologie, Université de Liège, 20 rue de Pitteurs, Liège (Belgique).

Leonard F. Bélanger: The Origin and Maintenance of Osteoclasts as Visualized by H³-Thymidine Radioautography.

In 3 weeks old chick tibiae, most primitive cells are labeled 2 hours after subcutaneous administration of H³-thymidine. Labeled osteoclasts are progressively abundant after 1 and 2 days. The number of labeled cells decrease thereafter. In Vitamin D-treated birds, the total osteoclast population is decreased but the proportion of labeled nuclei in individual cells is increased as compared to rachitic birds. It seems that osteoclasts during their unknown life span, renew their nucleus population by annexation of primitive cells as older nuclei degenerate.

Dept. of Histology and Embryology, Faculty of Medicine, University of Ottawa, Ottawa (Canada).

E. J. MOYNAHAN: Autoradiography in the Study of the Biology and Pathology of the Skin.

Autoradiography, at first sight, would seem to be eminently suited to the study of problems concerning the biology and pathology of the skin, yet a survey of the literature reveals a certain poverty of references to the use of the technique in these fields. This is all the more surprising since the technique is not unduly elaborate, and the procedures are now fairly standardised. Furthermore, useful results were obtained quite early on when the technique was applied to such studies, e.g. the work of FELL, PELC and others on the incorporation of sulphur compounds into the epidermal cells and into those of the dermal-connective tissues. Studies with labelled DNA have also given promising results, but little has been done to investigate DNA metabolism in disease, even though reasonably good quantitative methods are now available. It is true that autoradiography has been used by virologists and as a result we have a much clearer picture of events inside a cell infected with virus, but we still know virtually nothing about what happens in such common virus diseases of the skin as warts and herpes simplex. This is all the more remarkable since local applications of radio-active tracers are beginning to be used by clinicians in the United States and elsewhere. Why is there this hesitancy and reluctance to use the technique? This would seem to be due to several causes: (1) ignorance of the cytochemistry of the skin cells concerned, so that useful labelled compounds cannot yet be employed; (2) lingering doubts about the validity of the results obtained with available techniques, especially those using electron-track emulsions, which perhaps do not localise sufficiently precisely to give the information desired, and, (3) the limitations of use of radio-active substances in human work, possibly owing to an exaggerated fear of the genetic and remote consequences following their use.

Dept. of Dermatology, Guy's Hospital Medical School, London (England).

E. STÖCKER: Größe und synthetische Aktivität des Nukleolus. Autoradiographische Untersuchungen mit H^3-Cytidin, -Leucin, -Phenylalanin, -Thymidin.

Eine Volumenzunahme des Nukleolus erreichten wir experimentell einmal in den Zellen des exkretorischen Pankreas der Maus durch eine Applikation von Pilocarpin und zum anderen in den Leberparenchymzellen der Ratte durch eine Fütterung von Thioacetamid (TAA). Unter diesen Bedingungen zeigen autoradiographische Untersuchungen mit H^3-Cytidin sowie H^3-dl-Leucin und H^3-l-Phenylalanin eine gesteigerte Inkorporation der Vorläufer in das Kernkörperchen. Die außerordentlich starke Vergrößerung der Nukleolen nach TAA-Gabe ermöglicht sogar eine hinreichend sichere quantitative Aufschlüsselung des autoradiographischen Befundes. Dabei stellten wir fest, daß der Einbau von H^3-Cytidin und H^3-Phenylalanin und somit die Syntheserate von fixierbarer RNS und von Eiweiß proportional dem Volumen erfolgt. Die Größe des funktionell angeregten Nulkeolus kann demzufolge als Maß seiner synthetischen Aktivität für diese Substanzen gelten. Gleichzeitig steht die volumetrische Nukleolus-Karyoplasma-Relation in einem konstanten Verhältnis zur Neubildungsrate von RNS und Eiweiß dieser beiden Kernbestandteile. — In großen Nukleolen haben wir (ALTMANN, STÖCKER und THOENES) darüber hinaus licht- und elektronenmikroskopisch einen DNS-Gehalt und autoradiographisch nach Injektion von H^3-Thymidin eine DNS-Synthese nachgewiesen. Die DNS-Neubildung im Kernkörperchen war dabei zeitlich gegenüber der in den übrigen Chromosomen verschoben. Die DNS-Menge und -Synthese im Nukleolus ist jedoch — im Vergleich zu seiner starken Neubildungsrate für RNS und Eiweiß — relativ gering. Demzufolge sind wir der Ansicht, daß der Nukleolus wohl ein eigener Syntheseort von RNS und Eiweiß ist, daß man aber darüber hinaus zusätzlich eine Migration derartiger, ursprünglich im Karyoplasma neugebildeter Substanzen in den Nukleolus annehmen muß, welche dann im Kernkörperchen vervielfältigt werden. Möglicherweise steuert dabei der Nukleolus organizer den Umfang einer derartigen nukleolären „Verstärkerfunktion".

The Size and Synthetic Activity of the Nucleolus. Autoradiographic Investigations with H^3-Cytidine, -Leucine, -Phenyl Alanine and -Thymidine. An increase in the volume of the nucleolus was achieved experimentally under the following circumstances: 1) in the cells of the excretory pancreas of the mouse, following an injection of Pilocarpine. — 2) In the cells of the liver parenchyma of the rat, after administration of thioacetamide (TAA) with the food. — Under these conditions autoradiographic examinations with H^3-cytidine, H^3-dl-leucine and H^3-l-phenyl alanine showed an increased incorporation of the precursor into the nucleolus. As the enlargement of the nucleoli after TAA application was extraordinarily great, it was possible to make a sufficiently reliable interpretation of the autoradiographic findings. We have now established that the incorporation of H^3-cytidine and H^3-phenyl alanine, and consequently the rate of synthesis of fixed RNA and protein, occurred in proportion to the volume. The size of the functionally stimulated nucleolus can therefore be considered as an expression of its synthesizing activity in regard to the two substances in question. Moreover, the volumetric nucleolus/karyoplasm relation is in constant proportion to the rate at which RNA and protein are produced by these two nuclear components. — Beyond that we (ALTMANN, STÖCKER and THOENES) succeeded in demonstrating DNA in large nucleoli by light and electron microscopy. After the injection of H³-thymidine it was also possible to demonstrate DNA-synthesis in the nucleoli. The production of DNA in the nucleolus does not take place at the same time as it does in the chromosomes. The DNA-quantity and synthesis in the nucleolus is, moreover, relatively small if compared with the considerable production rate of RNA and protein. We therefore conclude that the nucleolus is certainly a site of RNA and protein synthesis. Beyond that, however, it has to be assumed that there is an additional migration of these substances — originally produced in the karyoplasm — into the nucleolus, where they are then increased. It is possible that the nucleolus organizer controls the activity of such nucleolar "intensifying function".

Pathologisches Institut der Universität, 87 Würzburg (Deutschland).

Croissance interphasique normale et pathologique
Normal and Pathologic Interphase Growth

Moderator: J. FAUTREZ, Gent (Belgium).
Co-Moderator: T. CASPERSSON, Stockholm (Sweden).

J. FAUTREZ: Introduction générale.

Le thème de la présente discussion se rapporte à la croissance, c'est-à-dire à un accroisse-ment de matière vivante qui peut se manifester par des augmentations volumétriques et pondérales de la cellule ou de certains de ses organites, comme par l'accroissement de la teneur en certains composants particuliers. Envisagé nécessairement sous l'angle de la cytochimie, cette étude entre donc dans le cadre de la cytochimie quantitative. Nous nous limiterons ici à la discussion de résultats, les méthodes étant étudiées dans d'autres sections de ce Congrès. Mais nous nous limiterons également à l'étude de la croissance interphasique, c'est-à-dire celle qui se fait en dehors de la division cellulaire. Une distinction nettement tranchée est cependant impossible. A côté de «triggers» plus spécifiques, la croissance interphasique peut être un facteur déterminant préparatoire à la division. Cette croissance peut en outre être correlée à certaines activités physiologiques. Nous pouvons cependant éliminer les réduplications d'ADN et de protéines nucléaires directe-ment liées à la préparation de la mitose. Au cours des phénomènes envisagés, il s'agira avant tout de synthèse d'acides nucléiques (ADN et ARN) et de protéines. Dans l'état actuel de nos connaissances ces composants paraissent les plus importants, peut-être parce que nous connaissons peu de chose des autres. On peut envisager ces synthèses dans divers organites, mais également dans divers matériels: unicellulaires, cultures de tissu, oöcytes, organes de métazoaires. Afin de ne pas exclure artificiellement les importantes interactions entre les divers éléments de la cellule nous proposons de classer les communications selon le matériel usagé, en réservant une dernière section à des observations de croissance patho-logique.

Introduction. The subject of the present discussion is growth, that is to say the growth of living matter that is manifested in increases of volume and weight in the cell itself or its constituents, as well as by the increase in the amount of certain special cell components. Though necessarily considered from a cytochemical point of view this study nevertheless extends to the domain of quantitative cytochemistry. Here we will confine ourselves to a discussion of the results, while the methods will be discussed in other sections of the Congress. We will also confine ourselves to the study of interphase growth, that is to say to the growth that takes place without cell division. A clear cut distinction is, however, impossible. In addition to more specific "triggers" the interphase may be a decisive preparatory factor for cell division. This growth may furthermore be correlated to certain physiological activities. We can, however, eliminate the reduplication of DNA and nuclear protein, which is directly linked with the preparation of the mitosis. There is, above all, the problem of the synthesis of nucleic acids (DNA and RNA) and proteins. In the present state of our knowledge these components seem to be the most important ones; this is perhaps because we know so little about the others. These syntheses can be observed in various cell-constituents and also in various materials: unicellular organisms, tissue cultures, oöcytes, metazoal organs. In order not to exclude artificially important interactions between the various cell elements we propose to classify the communications according to the material presented while the final section will be kept for the discussion of observations concerning pathological growth.

Laboratorium voor Menslijke Anatomie, Bojlokekaai 1, Gent (Belgique).

A. Unicellulaires
Unicellular Organisms

J. J. TAYLOR: Nitrogenous Substances in the Yeast-Like and in the Mycelial Forms of Blasto-myces Dermatitidis.

Analyses of both the yeast-like and the mycelial phases of *Blastomyces dermatitidis* strain 6046 were conducted for nucleic and amino acids. Yeast phase cells grown in an arti-ficial medium contained more ribonucleic acid (RNA) and total amino acids than mycelial phase cells from the. same medium. Concentrations of both substances decreased as the cells aged. Deoxyribonucleic acid concentration was similar in the two growth phases and varied little with age — When certain pyrimidine analogues were added to cultures, cells always assumed a yeast-like morphology, regardless of incubation temperature. The amounts of RNA and amino nitrogen in these cells were similar to those of yeast phase cells grown on normal media (although the former were slightly less), but significantly greater than concentrations in mycelial cells from either normal or analogue-containing media. — Cellular morphology could not be altered by cultivating cells in the presence of chlortetracycline, although amino nitrogen (but not RNA content) was significantly decreased under these conditions, and ninhydrin positive substances accumulated in the culture filtrates. — It is assumed that the biochemical reactions leading to polar cell wall synthesis (mycelial phase) are inhibited not only by elevated temperatures (leading to non-polar wall synthesis of the yeast phase), but also by the qualitative rather than the quantitative amino acid composition of the cells. It is also assumed that the synthesis of "nonsense" or non-functional RNA must account for this qualitative change, since the total amount of RNA is not altered appreciably.

Dept. of Microbiology and Public Health, Missoula, Montana (USA).

B. Cultures de tissu
Tissue Cultures

W. SANDRITTER: Interferenzmikroskopische Untersuchungen des Wachstums von Einzelzellen (HeLa-Zellen) in der Gewebekultur*.

Mit dem Interferenzmikroskop können die Fläche (Volumen), Masse pro μ^2 und das Trockengewicht von Einzelzellen in der Interphase und Mitose gemessen werden. Es wurden bisher 37 Einzelzellen (HeLa-Zellen) bis zu 32 Std in der prämitotischen und postmitotischen Phase im Abstand von 1 Std gemessen. (Für die Technik vergleiche Frankf. Zschr. Path. 70, 271, 1960.) Die Messungen ergaben: In der frühen Interphase, direkt nach der Telophase, nehmen nach dem Erscheinen des Nucleolus die Fläche und das Volumen von Kern (2fach) und Cytoplasma (6—8fach) sehr schnell zu (Wachstumskurve entspricht einer Potenzfunktion), während die Masse pro μ^2 um 50% abnimmt (Wasseraufnahme). Das Trockengewicht des Zellkerns bleibt lange Zeit unverändert, erst in der Präprophase (einige Stunden vor der Prophase) wird das Zellkerntrockengewicht verdoppelt. Das Cytoplasmatrockengewicht nimmt in der frühen Interphase um das Doppelte zu (linearer Trockengewichtsanstieg). In der Präprophase nehmen die Fläche und das Volumen von Kern und Cytoplasma um etwa 50% ab. Gleichzeitig steigt die Masse pro μ^2 um etwa 50% (Wasserverlust). In der Mitose wird das Trockengewicht von Kern (Chromosomen und Spindelapparat) und Cytoplasma halbiert und genau auf die beiden Tochterzellen verteilt. Daraus geht hervor, daß die Eiweißkörper des Interphasezellkernes in der Mitose den Spindelapparat bilden und nach der Mitose wieder als Kerneiweißkörper (Nichthistonprotein?) fungieren. — Der DNS-Gehalt des Interphasezellkernes beträgt etwa 10 % (Gewicht 100×10^{-12} g, DNS und RNS 10×10^{-12}g). Setzt man Gewichtsproportionalität von DNS und Histon voraus, so ergibt sich ein Nichthistonproteingehalt von 80%.

Interference Microscopic Investigations of the Growth of Individual Cells (HeLa-Cells) in Tissue Cultures. With the help of the interference microscope it is possible to measure the surface (volume), mass/μ^2 and the dry weight of individual cells during interphase or mitosis. 37 single cells (HeLa-cells) were measured during the pre- and postmitotic phase. Total time of observation: up to 32 hours; measuring carried out in intervals of one hour. (For details of the method see: Frankf. Zschr. Path. 70, 271, 1960.) The results were as follows: During early interphase (immediately after telophase) a rapid increase in the surface and volume of the nucleus (twofold) and the cytoplasm (6—8 fold) as well as a 50% decrease in the mass/μ^2 (due to the absorption of fluid) occurred as soon as the nucleolus became visible. The dry weight of the nucleus remained unchanged over a long period; only during the pre-prophase (a few hours prior to the prophase) the dry weight of the nucleus did double. The dry weight of the cytoplasm was doubled during the early interphase (linear increase of the dry weight). A 50% decrease in the surface and the volume of the nucleus and cytoplasm was observed during pre-prophase. At the same time a 50% increase in the mass/μ^2 (loss of fluid) was observed. During mitosis the dry weight of the nucleus (chromosomes and spindle) and the cytoplasm was divided into two equal parts, one being transmitted to each daughter cell. It can therefore be concluded that the proteins of the interphase nucleus form the spindle during mitosis. Afterwards they resume their position as nuclear proteins (non-histone protein?). — The DNA content of the interphase nucleus amounts to c. 10% (weight 100×10^{-12} g, DNA and RNA 10×10^{-12} g). Assuming a proportion of weight between DNA and histone the non-histone protein content is 80%.

* Mit Unterstützung der Deutschen Forschungsgemeinschaft.

Pathologisches Institut der Universität, 63 Gießen (Deutschland).

H. G. Schwarzacher: Asynchrone DNS-Synthese des X-Chromosoms und Sexchromatin-Bildung.

Radioautographische Untersuchungen mit ^{3}H-Thymidin an menschlichen Gewebekulturen (Fibroblastenkulturen und Leukozytenkulturen vom peripheren Blut) ergaben, daß die einzelnen Chromosomen einer Zelle nicht alle gleichzeitig DNS synthetisieren. Dabei wurde öfters festgestellt, daß in weiblichen diploiden Zellen eines der beiden X-Chromosomen später als die übrigen Chromosomen synthetisiert (z.B. German, 1962, Moorhead u. Defendi, 1963). Eines der beiden X-Chromosomen bildet auch das Sexchromatin in weiblichen Interphasenkernen (Ohno u. Makino, 1961). In polyploiden Zellen findet man, daß mit jeder Verdoppelung der DNS des Zellkernes sich auch der DNS-Gehalt des Sexchromatins verdoppelt. In Gewebekulturen menschlicher Fibroblasten ist festzustellen, daß der DNS-Gehalt (Feulgen) des Sexchromatins in Kernen der G2-Phase doppelt so hoch ist, wie in Kernen der G1-Phase. Diese Verdoppelung der DNS des Sexchromatins scheint sich aber über einen längeren Zeitabschnitt als nur den letzten Teil der Synthese-Phase zu erstrecken. In Fibroblastenkulturen von weiblichen Geweben finden sich regelmäßig 20—30% Zellkerne, die kein Sexchromatin erkennen lassen. DNS-Messungen (Feulgen) ergaben, daß solche Sexchromatin-negative Zellkerne sich in der G1—G2- oder S-Phase befinden können. Radioautographische Untersuchungen an Interphasenkernen deuten darauf hin, daß das spät synthetisierende X-Chromosom auch das Sexchromatin bildet. Beobachtungen des Sexchromatins in lebenden Zellen *in vitro* ergaben, daß Struktur und Größe des Sexchromatins während langer Zeiten der Interphase konstant bleiben.

The Asynchronous Synthesis of DNA by the X-Chromosomes and the Formation of Sex Chromatin. Radioautographic examinations (^{3}H-thymidine) of human tissue cultures (fibroblast cultures, and leucocyte cultures originating from peripheral blood) show, that the synthesis of DNA does not take place in all chromosomes synchronously. It has frequently been found that in female diploid cells one of the X-chromosomes synthesizes later than the other (German, 1962; Moorhead and Defendi, 1963). It is also one of the two X-chromosomes that forms the sex chromatin in female interphase nuclei (Ohno and Makino, 1961). In polyploid cells it can be shown that with every duplication of nuclear DNA the DNA content of the sex chromatin also doubles. Tests on cultured human fibroblasts confirm that the DNA content (Feulgen) of the sex chromatin is twice as high in G2-phase nuclei as in G1-phase nuclei. This duplication of the sex chromatin DNA seems to last for a rather long period and not only for the last part of the synthesis phase. 20—30% of the nuclei in fibroblast cultures of female tissues do not show the sex chromatin. DNA determinations (Feulgen) show that these sex chromatin negative nuclei are either in the G1—G2, or S-phase. Radioautographic investigations of interphase nuclei indicate that the sex chromatin is formed by the slowly synthesizing X-chromosome. *In vitro* observations of the sex chromatin of living cells show that during the interphase the structure and the size of the sex chromatin remains unchanged over long periods.

Histologisch-Embryologisches Institut der Universität Wien (Österreich).

C. Oocytes

J. Brachet: Lack of Biochemical Activity in the Interphase Nucleus of Unfertilized Sea Urchin Eggs.

Incorporation of amino acids into ribosomal proteins markedly increases after activation of sea urchin eggs. This increase cannot be explained by the production of messenger RNA by the female pronucleus of the activated eggs for the following reasons: 1) pulse experiments with ^{32}P or ^{3}H-cytidine have failed to demonstrate a measurable synthesis of nuclear RNA before the first cleavage; 2) pretreatment with actinomycin D has no effect upon protein synthesis at the time of fertilization and it does not stop early cleavage; 3) anucleate fragments of sea urchin eggs respond to activation by increased protein synthesis as well as nucleate fragments. It is concluded that either the unfertilized sea urchin eggs contain a store of masked messenger RNA's or that they can synthesize messenger RNA's on a cytoplasmic DNA template. Almost 50% of the latter is present in the anucleate halves.

Laboratoire de Morphologie animale, 1850, Chaussée de Wawre, Anderghem (Belgium).

W. S. Vincent: Interphase Growth of the Nucleolus*.

The nucleolus appears to increase in size during the early stages of oogenesis by the accumulation of fine granular material. These granules appear to be somewhat smaller than the ribosome particle. Early in the growth of the nucleolus, the content of nucleolar ribonucleic acid is high (16—20%); with increasing growth the proportion of RNA in the nucleolus decreases steadily so that the amount of RNA appears to be related to the surface area of the nucleolus rather than to nucleolar volume. — The apparent dry matter content (expressed as % volume) also decreases during nucleolar growth. This decrease appears to be related to an increase in vacuolar, or most likely nucleoplasmic space within the nucleolar substance. During this time there is an appearance of larger granules at the surface of the nucleolus which have dimensions similar to ribosomal particles. — The decrease in RNA concentration during nucleolar growth can be correlated with the appearance and accumulation of an unusual protein which, by the time the mature nucleolar size is reached, constitutes some 80% of the nucleolar materials. This protein is a neutral protein which is rich in the basic amino acid, lysine. Lysine is present at the expense of the other basic amino acids. This protein has been shown to combine with RNA in such a manner as to prevent its degradation by ribonuclease. Some evidence which suggests that this protein may associate with messenger RNA will be presented.

* Aided by grants from the National Science Foundation and the U.S. Public Health Service.

Dept. of Anatomy, University of Pittsburgh School of Medicine, Pittsburgh, Pa. (USA).

D. Tissues

S. R. PELC: Renewal of DNA.

After injection of H^3-thymidine in adult mice autoradiographs were found over nuclei of tissues showing no mitoses, such as brain and muscle and in tissues with very low mitotic indices, such as seminal vesicle. The number of labelled cells that divide was determined by counting mitoses with and without colchicine, counting of the number of labelled cells and grain counts of labelled nuclei at various times after injection. The results show that either none or only a small proportion of these cells divide. The results of biochemical and autoradiographic experiments showing instability of DNA will be reviewed. — It is suggested that differentiated cells periodically renew some or all of their DNA without going through division.

Medical Research Council Biophysics Research Unit, King's College, London W. C. 2 (England).

DAVID P. BLOCH: Comparison of the Chromatin Protein during Different Nuclear and Chromosomal States*.

The ratios of protein lysine plus tyrosine to arginine are determined for metabolically active and inactive nuclei, for nuclei prior to, during, and after DNA replication, and during mitosis. Comparison is made of the different chromosomes of a complement during division, and of normally condensed and heteropycnotic chromosomes. Genetic and "extragenetic" chromatin are also compared. The findings are discussed in relation to the question of histone variability and chromosome function.

* This work was supported by grants from the National Science Foundation (GB 442) and the United States Public Health Service (GB 09654).

Botany Dept., University of Texas, Austin, Texas (USA)..

A. K. SHARMA, A. SHARMA and A. CHATTERJEE: Tyrosine in Nucleo-Cytoplasmic Transfer.

Experiments were planned to localize the presence of protein bound amino acids in different parts of the nucleus and cytoplasm including the nuclear membrane and adjoining sectors. Modified Millon's reaction was adopted for the localization of tyrosine. — Evidences have been obtained to show that tyrosine plays a significant role in nucleo-cytoplasmic transfer. Distinct tyrosine positive strands have been obtained radiating from the nuclear membrane to the cytoplasm and specially towards the cell membrane at the interphase stage of the normal cell. In malignant cells on the other hand, this pattern is disrupted.

University College of Science, Botany Dept., 35, Ballygunge Circular Road, Calcutta 19 (India).

H. ROELS: Influence of Cell Activity on Nuclear Dry Weight.

The influence of cell activity on nuclear dry weight has been investigated by means of the scanning-integration interferometric method of SVENSSON (in different endocrine glands of the rat). Inhibition and stimulation of cell function are both followed by an increase of nuclear dry weight in the different layers of the adrenal cortex and in the adrenal medulla of the white rat. — In the thyroid however a decrease of nuclear dry weight has been observed in these circumstances. Accordingly hypophysectomy caused a decrease of nuclear dry weight in the thyroid of the white rat. It has no influence on the dry mass of the cell nuclei of the adrenal medulla. STH has no influence in the dry weight of the cell nuclei of the thyroid or adrenal medulla of hypophysectomized rats. Until now the contradictory results obtained in the adrenal and the thyroid remain unexplained. The observations in the adrenal show independent variations of dry weight and nucleic acid content of the nucleus. This suggests the possibility of synthesis of certain nuclear proteins without mediation of nucleic acids.

Laboratoire de Pathologie, Bijlokekaai 1, Gent (Belgique).

E. Croissance pathologique
Pathological Growth

E. GRUNDMANN: Das Interphasewachstum bei bösartigen Proliferationen.

In bösartigen Tumoren läuft die Interphase in prinzipiell gleicher Weise ab wie in normalen meristematischen Geweben. Die in einzelnen Tumoren gefundenen Variationen betreffen einmal die zeitlichen Korrelationen der Reduplikationsphasen. So kann die DNS-Verdoppelung relativ rasch nach der Telophase einsetzen, und das präkaryokinetische Intervall ist abnorm verlängert. In der Gewebekultur sind solche zeitlichen Verschiebungen durch Milieuänderungen zu erzielen; sie stellen also kein Charakteristikum der Tumorzellen dar. Weitere Besonderheiten sind in Zeitpunkt und Ausmaß der Proteinsynthesen festgestellt worden. Ob DNS und RNS zugleich an den gleichen Chromosomenstellen synthetisiert werden können, ist an Tumorzellen nicht sicher untersucht worden. — Das Prinzip der DNS-Verdoppelung vor Beginn der Kernteilung wird auch in der bösartigen Proliferation gewahrt. Allerdings ist in unterschiedlichem Ausmaß die der interphasischen Reduplikation vorgeschaltete Arbeitsphase der Zelle, die sog. Intermitose, reduziert, und in rasch proliferierenden Geweben kann sie ganz fehlen, so daß dann im Carcinom wie in der frühen Keimentwicklung eine Mitose unmittelbar auf die andere folgt ohne Zwischenschaltung einer Intermitose. Dementsprechend werden in der Interphase keine differentiellen Strukturen redupliziert, und die cytochemisch gefundenen Werte weichen von den Normalgeweben entsprechend ab. — Zu den Kennzeichen der bösartig proliferierenden Zellen gehören schließlich Dissoziationen zwischen Kern- und Zellteilungsmechanismus mit Bildung mehrkerniger Riesenzellen. Dies kann in bösartigen Tumoren ohne ersichtliche Ordnung eintreten. Auf die hierbei feststellbaren cytochemischen Besonderheiten wird hingewiesen.

Interphase Growth in Malignant Proliferations. In malignant tumours the interphase takes place in exactly the same way as in normal meristematic tissues. Variations, which are found in individual tumours, concern the temporal correlation of the reduplication phases. Thus the duplication of DNA can begin shortly after telophase and the prekaryokinetic interval is abnormally extended. In tissue cultures these temporary fluctuations can be induced by certain changes in the culture medium: they are therefore not characteristic of tumour cells. Other peculiarities found concern the moment and the extent of protein synthesis. So far it has not been demonstrated, in the case of tumour cells, whether RNA and DNA are being synthesized at the same chromosome sites. — The principle of DNA duplication prior to nuclear division is maintained in malignant proliferations. However, the so-called intermitosis (i.e. the working phase of the cell which takes place antecedent to the interphase reduplication) is reduced in varying degrees. In rapidly growing tissues it can be completely abolished, so that in cancerous tissue one mitosis is followed immediately by another and no intermediate intermitosis takes place. Correspondingly no differential structures are being reduplicated during the interphase. This accounts for the fact that the results of cytochemical determinations of tumour cells differ from the results given by normal tissues. — Another characteristic of malignant cells is the dissociation of the mechanism of nuclear and cell division, and the formation of polynuclear giant cells. This occurs in malignant tumours for no obvious reason. Cytochemical peculiarities which have been found in these cases will be discussed.

Pathologisches Institut der Bayer-Werke, 56 Wuppertal-Elberfeld (Deutschland).

C. VENDRELY, M. T. GRANGÉ, F. KASTEN, P. TOURNIER et R. WICKER: Etude de la teneur en ADN des noyaux de fibroblastes normaux et de fibroblastes transformés par le virus SV 40 et le virus polyoma.

Notre étude a porté sur une souche de fibroblastes de rat et sur deux autres lignées issues de la première et transformées respectivement par le virus SV 40 et par le virus du polyoma en souches présentant des caractères de malignité. — La teneur en ADN de la majorité des cellules de la souche transformée par le virus SV 40 se trouve être nettement supérieure à la teneur caractéristique de la souche témoin et suggère que la population est hypotétraploïde. Les résultats obtenus pour les cellules transformées par le virus du polyoma sont très dispersés et ne donnent pas l'image d'une population homogène. — L'étude des chromosomes des trois souches montre que dans le cas des cellules transformées par le virus SV 40, les nombres chromosomiques les plus fréquents se trouvent effectivement être hypotétraploïdes. Par contre, dans les cellules transformées par le virus du polyoma, l'étude du nombre chromosomique révèle une population homogène avec un nombre de chromosomes normal ou légèrement inférieur à la normale. — Une étude autoradiographique doit permettre de déterminer si le désaccord entre les résultats de la mesure cytophotométrique de l'ADN des noyaux et des numérations chromosomiques peut être attribué à un allongement particulier de la phase de synthèse de l'ADN dans la souche transformée par le polyoma.

A Study about the DNA Content of Normal Fibroblast Nuclei and Nuclei of Fibroblasts Transformed by SV 40 and Polyoma Virus. These experiments were carried out on a strain of normal rat fibroblasts and two sub-strains, transformed by SV 40 virus and polyoma virus, thus becoming malignant. The DNA content of the majority of cells, transformed by the SV 40 virus, was distinctly higher than the characteristic amount of the control strain; this suggested that the cell population was hypotetraploid. The results obtained from the cells transformed by the polyoma virus were disparate and did not give the impression of being a homogeneous population. Chromosome studies of the three strains showed that in the case of the cells transformed by the SV 40 virus, the most frequent number of chromosomes was in fact hypotetraploid. However, a study of the chromosome number of the cells transformed by polyoma virus, revealed a homogeneous population with either an normal or only slightly sub-normal chromosome number. An autoradiographic study is bound to determine whether the disagreement between the cytophotometric measurements of nuclear DNA and the chromosome number, in the strain transformed by polyoma virus, may be attributed to an extraordinary prolongation of the DNA synthesis phase.

Institut de Recherches sur le Cancer, 16, Avenue P. Vaillant Couturier, Villejuif/Seine (France).

Immunohistochemistry

Moderator: A. H. Coons, Boston (USA).
Co-Moderator: H. von Mayersbach, Nijmegen (The Netherlands).

ALBERT H. COONS: Introduction.

The panel will describe and discuss the methods of immunofluorescence and its potential usefulness in cytochemistry. Labeled antibody acts as a cytochemical fluorochrome with immunological specificity. It has a resolving power of about 0.1 μ when used with the light microscope and a sensitivity for antigen of about 10^5—10^6 molecules when they are closely packed together. Localization of small amounts of a macromolecular species (antigen) is therefore possible when it is present in high concentration. Enzymes and other proteins of cytochemical interest are usually immunologically specific for each species. For example, localization of amylase in cattle pancreas requires a supply of purified amylase from beef pancreas. Impurities in the amylase preparation may be represented by unwanted and misleading antibodies in the serum; if present they must be removed by specific neutralization or precipitation. — The panel will discuss this problem, and the methods and controls by which the specificity of the reaction can be established. Thereafter, the discussion will turn to examples of the successful use of labeled antisera in the localization of several products and structural components of cells. Examples of the use of contrasting colors for the localization of two different substances in the same preparation will be presented. — As a tool, immunofluorescence is simple in concept, but its successful use requires knowledge of immunology and skill in its application.

Dept. of Bacteriology and Immunology, Harvard Medical School, Boston, Massachusetts (USA).

F. PETUELY: Various Labels for Antibodies.

For marking antibodies with fluorescent dyes several conditions must be observed. 1) The marking substance must link up firmly with the antibody, 2) the antibody should not lose its immunologic activity and 3) the marking substance must be visible in very small quantities. After Coons has introduced fluorescein isocyanate in 1942 there were several other fluorescent-substances proposed for antibody marking. Mainly only 2 kinds of principles for chemical linkage of marking substances on antibody proteins are in use. One group uses the carbamido linkage corresponding to Coons' method; the other group uses sulphonyl chlorides which give stable sulphonamido linkages with the NH_2 groups of the antibody proteins. In practice only the original isocyanate proposed by Coons and its further development, the fluorescein isothiocyanat (Riggs) as well as the 1-Dimethylnaphthalene-sulfanilic acid (Mayersbach) and Lissamine-rhodamine (McEntegart) are in use. Every marking substance has its certain advantages and disadvantages from the point of view of its chemical, physical or biological behaviour. The chemical structure of the marking substances, and their optical behaviour as well as their action on antibodies will be discussed.

Bundesanstalt für Lebensmitteluntersuchung, Wien (Österreich).

GERALD GOLDSTEIN: Purification of Conjugates.

A rapid and simplified method of obtaining optimally conjugated antibodies has been devised. These antibodies are used without powder absorption and have no or negligible non-specific staining. The method is as follows: All procedures are at 4^0 C. Gamma globulins are obtained from serum by 1.6 M $(NH_4)_2SO_4$ precipitation, and DEAE-cellulose chromatography. The gamma globulin adjusted to pH 9.5 with 0.1 N NaOH is added to dry fluorescein isothiocyanate (8—25 μg/mg protein depending upon species). pH is maintained at 9.0—9.5 during the first hour. After 18—24 hours, the reaction mixture is diluted ten-fold with 0.01 M sodium phosphate, pH 7.5 and applied to a DEAE-cellulose column. Unreacted fluorescent materials and overcoupled globulins are very strongly bound to the column. Undercoupled globulins emerge with the starting buffer. Optimally coupled globulins are step eluted with 0.03 M sodium phosphate pH 7.2—7.5. The application of this method with rabbit, human, sheep, and goat antibodies will be described. Using tetramethyl rhodamine isothiocyanate, it is still necessary to remove unreacted fluorescent materials before fractionating fluorescent globulins on DEAE-cellulose.

Dept. of Microbiology, University of Virginia, School of Medicine, Charlottesville, Virginia (USA).

J. F. Ph. Hers: Preparation of Specimens.

Although the immunological disciplines, such as the technique for the purification of antigens and the preparation of high titered specific antisera in experimental animals are highly important, the use of fresh biological material and a correct handling of the tissue or cell specimens is likewise essential for good and reliable results. Generally speaking, tissue and/or cells from autopsy rooms are only suitable if specimens are taken as soon as possible after death. Exsudates and cells of patients have to reach the laboratory immediately from the bedside. The choice of the technique depends on the biological problem, the characteristics of the tissue concerned and the properties of the antigens under study. Sections of frozen or freeze-dried material have been successfully used and monolayers of tissue cultures may be treated. It has been also shown that smears of micro-organisms, sputum, nasal exsudate, blood, bone marrow, cell suspensions and acellular material can be used. Impression smears of tissue and preparations made with the aid of the sedimentation chamber have extended the results. — Although these techniques are well developed, most of the disadvantages of the fluorescent antibody technique occur owing to glassware, cold-techniques, storing, fixation procedures and the choice of buffers. Special attention will be given to these problems.

Academisch Ziekenhuis, Afdeling Inwendige Geneeskunde, Leiden (The Netherlands).

C. Grossi: Fixation and Buffers.

Fixation is important for immuno-histochemistry as many of the antigens to be traced are highly soluble and can be extracted very easily from tissues during incubation with labelled antibodies. As we have to deal with a very sensitive system (antigen-antibody) there is also the risk of denaturation. These two conditions are only apparently antipodes; in practice it is sometimes very easy to pass from the one to the other situation, that means a good quantitative preservation can be followed by a complete immunologic inactivation of the antigens. In a series of models the effect of several fixatives and fixing conditions has been studied and the different results will be given. — The choice of buffer during in-cubation and rinsing the tissues is also very important as several buffers tend to extract quite a lot of tissue components. This effect is again closely connected with the preparation of the tissues and the type of fixation. The buffers play further a very important role for unspecific colourations as they change the charge of the tissue proteins and/or modify the behaviour of the sections through extractions, both of which can enhance the intensity of the non-specific colourations.

Laboratorium voor Cyto-Histologie, Nijmegen (The Netherlands).

H. von Mayersbach: Non-Specific Reactions.

The occurrence of unspecific reactions is known since the early days of the method. As these unspecific reactions are normally quite strong they can be quite troublesome and cause faulty results. The strong colouring of normal tissues with labelled sera was the reason for using this in cancer diagnosis. "Unspecific reactions" are: 1) colouring through non-bound, free fluorescent markers, 2) colouring of tissues through the protein-dye complex. — The first kind of unspecific colouring is today no longer an actual problem as it is easily avoided by gel-filtration techniques. The colouring through the protein dye-complex is still an actual problem. The reason for this colouring is an electrostatic-non-immunologic inter-action between tissues and labelled proteins in which the protein dye-complex behaves like an acid dye e.g. eosin. The basic concept of this electrostatic binding could be established by many experiments changing the electric charge either of the tissues or the sera. Each treatment which increases the difference between the iso-electric points of these two com-ponents is followed by an enhancement of non-specific colouration, whilst decreasing the differences decreases this phenomenon. — Under usual conditions (gel-filtration and/or liver powder absorption of the sera) the unspecific reactions in fresh-frozen, fixed or unfixed sections can hardly be avoided. There exist some new techniques of tissue and serum treat-ment after which unspecific colouring can be efficiently lowered but strongly eosinophilic components are still coloured unspecifically. — In spite of these new techniques the proofs of specificity must be very seriously considered.

Laboratorium voor Cyto-Histologie, Nijmegen (The Netherlands).

Paul Klein: Significance and Specificity of Complement Staining.

Complement (C') can be bound to tissue elements in two different ways: either the Ag-Ab-Aggregates fix autologous C' *in vivo* during the disease process or the C'-fixation is artificially performed *in vitro* by exposing tissue sections to heterologous C'. In both cases staining of fixed C' components can be carried out by labelled Anti-C'. Demonstration of autologous C' involves difficult problems concerning the specificity of the Anti-C'. The *in vitro* fixation of heterologous C' allows the establishment of a rigorous control system concerning the well known fixation characteristics of C'. Both methods should be applied simultaneously in the search for Ag-Ab-complexes in tissue if high informative value of the staining procedure is desired.

Institut für Medizinische Mikrobiologie, Johannes Gutenberg-Universität, 65 Mainz (Germany).

A. Identification of Cell Products

KENJIRO YASUDA: Immunocytochemical Demonstration of Amylase in the Acinar Cells of the Pig Pancreas.

Antiserum against porcine pancreatic amylase was prepared in rabbits, and conjugated with fluorescein isothiocyanate. The serum gave only one line in two-dimensional gel diffusion (OUCHTERLONY) with purified porcine pancreatic amylase, and only one weaker line with an extract of pig pancreas. This line was shown to be due to amylase and anti-amylase by specific inhibition of enzyme activity. — When applied to formalin-fixed frozen sections of pig pancreas, the fluorescent antibody solution reacted with fine granules near the base of the cell. The majority of the zymogen granules were located near the apical region. This finding is consistent with those of others that amylase activity behaves differently in the centrifugal field from trypsinogen and chymotrypsinogen. — Attempts to localize trypsinogen and chymotrypsinogen, and to investigate the homogeneity or heterogeneity of zymogen granules, failed because of serious cross reactions in the antisera.

Dept. of Anatomy, Keio University, Tokyo (Japan).

PIERRE BURTIN: The Cellular Origin of Human Immune Globulins.

The cellular origin of the three immune globulins of human serum (γ globulins, β_2A globulin and β_2 macroglobulin) has been studied with monospecific antisera labelled with fluorescein or sulforhodamin. In normal organs (mainly spleen), the three immune globulins are found in plasma cells which often have a granular cytoplasmic structure. Each plasma cell reacts only with one monospecific antiserum, thus contains only one immune globulin. The same is seen in myeloma plasma cells. In Waldenström macroglobulinemia's lymph nodes, cells containing β_2 macroglobulin or γ globulins often have the morphology of lymphocytes. — Other studies are in progress on the origin of polypeptide chains of γ globulins: heavy and light chains from one part, types I and II light chains from the other part.

Immunochemistry Laboratory, Institut de Recherches Scientifiques sur le Cancer, Villejuif Seine (France).

B. Identification of Structural Components

HOWARD HOLTZER: Localization of Myosin and Actin in Striated Muscle.

Rabbit antisera are prepared against chick myosin, light- or heavy-meromyosin or actin. The gamma globulin fraction is conjugated to fluorescein-isocyanate or fluorescein-isothiocyanate. Several different kinds of experiments suggest that the cytological intracellular localization of myosin obtained by using the antimyosin and anti-light meromyosin are consistent with findings based on biochemical, physiological and cytological techniques. The status of the antisera against heavy-meromyosin or actin is as yet ambiguous. These antibodies, though specific for antigens associated with the myofibril, may be directed against components of the myofibril other than the primary contractile proteins. — The fluorescein-labelled antisera against myosin or light-meromyosin have been used (1) to "fix" selectively, or to render insoluble, myosin *in situ*, (2) to follow changes in the spatial distribution of myosin in various stages of contraction of the myofibrils, and (3) as a microtechnique based on spectrofluorimetry to estimate small quantities of the antigen. It has also been used to follow the synthesis of myosin and the differentiation of myofibrils in embryonic muscle cells.

Institute of Neurological Sciences, University of Pennsylvania, Philadelphia, Pennsylvania (USA).

A. E. SZULMAN: The ABH Antigens in Human Tissues.

The blood group antigens are widely distributed throughout the human body as constituents of cell-walls of the endothelium and various epithelia, and as water-soluble entities in mucous and other secretions. — The antigens have been investigated by immunofluorescence (employing frozen sections) in intra- and extrauterine life, starting with embryos of 5 mm (28 days). — The reagents employed were human hyperimmune anti-A, B, H sera and their fluorescein conjugates, the specificity of which was assured by direct testing for other known erythrocytic antigens and by obtaining clear control reagents by means of absorption with purified blood polysaccharides. Further controls were provided by obtaining negative results in heterologous staining. — The natural history of the antigens in embryogenesis reveals them as *cell-wall constituents* of endothelium and of the epithelia of the integument and the digestive tube, including their derivatives (e.g. anterior pituitary; thyroid). The antigens are permanent only in the endothelium and in stratified epithelia; they disappear from other cells by the end of the first trimester. — The water-soluble antigens first appear in 35 mm embryos in mucus glands. While the chemical identity of the mucus-bound substances has been much studied (KABAT, MORGAN and WATKINS) that of the cell-wall substances remains virtually unknown and presents a field for further histochemical endeavor.

Harvard Medical School, Boston, Massachusetts (USA).

Symposien · Symposia · Symposia

Comparison of Different Fixation Methods and their Significance for Histochemistry

Chairman: R. E. STOWELL, Washington D.C. (USA)

Co-Chairman: R. J. BARRNETT, New Haven (USA)

ROBERT E. STOWELL: Effects of Different Conditions of Freezing, Thawing and Storage on Tissues.

There have been many new developments in the freezing and storage of tissues of importance to histochemistry. Before frozen tissue banks for research and other purposes are established, we need to know more about the effects of various procedures upon the structural and chemical constituents. — In a series of experiments slices of mouse liver were frozen on dry ice (—79⁰ C), in isopentane (—155⁰ C), or in propane (—175⁰ C). Some tissues were freeze-substituted without thawing to study the ice crystal artifacts; others were thawed for biochemical analysis or fixed for general structural or electron microscopic studies. By electron microscopy of thawed liver, in tissue frozen on dry ice, and, to a lesser extent, in isopentane, the nuclear and cytoplasmic morphology was similar to control tissues except for the bile canaliculi and plasma membranes. After freezing with propane, there were substantial alterations in cytoplasmic organelles. Although there were some mitochondrial abnormalities after isopentane, after propane there were irregularities in shape, interruptions in continuity of membranes, pallor of matrix and absence of matrix granules. — Biochemical measurements of succinoxidase showed losses of 26, 29, and 37% after freezing on dry ice or in isopentane and propane respectively. Losses of succinic dehydrogenase ranged from 11% on dry ice to 23% in propane. Changes in glutamic dehydrogenase appeared to reflect alterations in mitochondrial structure. Freezing or sonic treatment produced a 6—7-fold increase in glutamic dehydrogenase released from disrupted mitochondria. The nuclear enzyme NAD-pyrophosphorylase was unaltered after freezing. — After freezing of liver on dry ice, the large ice crystals produced the most displacement of cells and their gross constituents. After thawing, tissues frozen in propane showed the greatest chemical alterations associated with ultrastructural changes especially of the mitochondria, and dry ice the least chemical or structural changes.

Armed Forces Institute of Pathology, Washington, D.C., 20305, Washington (USA)

K. OSTROWSKI: Freeze-Substitution (F/S) and Freeze-Drying (F/D) for Histochemical Purposes.

Table. *The content of N, P, and dry mass of rat liver after the substitution in methanol and acetone in the different temperatures. Results expressed as %-values of the control material*

	Control % of dry mass	mg N/100 mg*	μg P/100 mg*	Methanol —79⁰ C % of dry mass	% of content		—20⁰ C % of dry mass	% of content		+4⁰ C % of dry mass	% of content	
					N	P		N	P		N	P
Mean . . .	29.5	3.66	529.0	25.9	96.7	60.6	24.3	96.5	46.1	24.2	95.7	44.8
Standard deviation	±0.4	±0.07	±37.0	±0.8	±0.7	±3.3	±0.9	±1.6	±2.7	±0.4	±1.3	±1.8
Variation coefficient	1.4	1.9	7.0	3.1	0.7	5.4	3.5	1.7	5.9	1.6	1.1	4.0
				Acetone								
Mean . . .				30.9	100.9	97.5	31.0	101.2	98.5	30.7	100.3	94.1
Standard deviation				±1.4	±2.5	±2.0	±1.4	±2.4	±3.6	±1.0	±2.5	±0.7
Variation coefficient				4.5	2.5	2.1	4.4	2.3	3.7	3.3	2.5	0.7

* of fresh weight.

Since quenching, one important point in the F/S and F/D technic, is discussed in the paper by Dr. Stowell, in this presentation some other theoretical and practical details will be stressed: 1. F/S is in many ways the ideal technic for histochemistry provided that the proper choice is made of substituting medium and range of temperature for substitution of water. Examples of quantitative experiments concerning the losses of some substances in different media and different temperature ranges will be critically discussed. The table illustrates the type of such experiments.

2. The results of F/D depend to a great extent on the type of apparatus. New equipment now available will be briefly mentioned. A model of apparatus in which all parameters are strictly controlled and which does not possess the troublesome condenser (refilling the fluid gases, etc.) will be demonstrated. 3. Some recently published data concerning the application of low temperature technic in ultrastructure research will be discussed.

Histology Dept., Medical Academy of Warsaw (Poland).

Russell J. Barrnett: The Effects of Chemical Fixation in Various Aldehyde Reagents on Enzymatic Activity.

Fixation for enzyme histochemistry is performed for a variety of reasons, including: that morphologists are used to fixing tissues as a preservative method, that tissues once fixed can be prepared in routine and simple ways, that the specimens are more or less permanent, that fixation will hopefully preserve the morphology in an adequate state and finally, that the enzyme may be rendered insoluble *in vitro* without markedly effecting its activity. While the latter two reasons are the most significant, it should be borne in mind that some enzymes are firmly bound to membranous structure and can be demonstrated in disrupted tissue or in particulate fractions isolated from tissues. Other enzymes are less firmly bound to structure and may be rendered soluble and diffusable by disruptive pretreatment. Practically, therefore, the process of fixation is a compromise between retention of morphology and of adequate enzymatic activity. Toward this end, the choice of fixative is not only important, but also the pH of the fixing medium, the purity of the fixative, and the inclusion of various additives in the fixing medium, such as various cations, sucrose, or PVP, the latter for overall osmolar concentration. When fixative is used, generally, brief fixation in dilute and cold reagents minimizes the effect of the fixative on the enzyme molecule. — This presentation shall cover the use of the following fixatives: acrolein, acetaldehyde, crotonaldehyde, formaldehyde, hydroxyadipaldehyde, glyoxal, glutaraldehyde, methacrolein, and pyruvic aldehyde. The preservation of morphological detail in frozen sections shall be examined first and then the effects of tissue fixatives on the histochemical demonstration of a variety of enzymatic activities shall be covered. These include succinic dehydrogenase, DPNH and TPNH diaphorase, cytochrome oxidase, aliesterase, acetylcholinesterase, Cathepsin C, alkaline and acid phosphatase, various nucleoside mono, di and tri phosphatases and glucose-6-phosphatase.

Dept. of Anatomy, Yale University School of Medicine, New Haven, Connecticut (USA).

M. Wolman: Effects of Chemical Fixation on Membrane Constituents.

Fixation implies preservation of structure and depends therefore on conservation of membranes and of boundaries separating different phases. Best chemical fixation of membranes is obtained by the use of osmium tetroxide and of other oxidizing agents. These act on the hydroxyl, amino, and other oxidizable groups, in addition to their action on ethylenic links. By creating bridges between molecules these oxidants "freeze" the membranes and in most instances also other molecules which are attached to membrane constituents. It has recently been found in the writer's laboratory that divalent ions play an important role in stabilizing the structure of membranes which are mainly made of polar lipids. Calcium ions were shown to bind membrane proteins to the membrane lipids. Variations in the relative proportions of Ca and Na (or of other divalent to monovalent ions, but also of other lipophilic to hydrophilic influences) change the structure and spatial configuration of membranes. Such factors can change substrates which are not available for histochemical demonstration into stainable materials, but may also cause solution in the medium of substances which were primarily insoluble and stainable.

Dept. of Pathology, Government Hospital, Tel-Hashomer (Israel).

Histochemistry of Esterases and Peptidases

Chairman: EARL P. BENDITT, Seattle (USA)

Co-Chairman: A. G. E. PEARSE, London (Great Britain)

E. P. BENDITT: The Problem of Identification and Localization of Tissue Peptidases and Esterases*.

Hydrolases have been the enzymes most amenable to histochemical analysis in the 25 years since GOMORI and TAKAMATSU first demonstrated the alkaline phosphatases. The first substrate—Chloroacetyl Naphthol AS (synthesized by GOMORI) was classed among the "non specific" esterases. Later this substrate was shown by us to be selective for endoproteases resembling chymotrypsin. The development of a greater knowledge of the nature of enzyme action and specificity provides the histochemist with a wide range of new possibilities for making new selective and sensitive substrates for the analysis of tissue enzyme location. The coupling of this development with the recent advances in examination of enzyme kinetics *in situ* increase the defining power of the histochemical tools. Further refinement of gross tissue analysis by the zymogram technique should help in the definition of tissue enzymes. In the end our aim is the closest approximation possible to the true character and spatial distribution of enzymes in cells and tissues, since only with this information will the real function of these constituents be revealed. The aim of this panel will be to assess the current technical state in this area of endeavor.

* Supported in part by U.S.P.H.S. grant No. HE-03174.

Dept. of Pathology, Seattle, Washington (USA).

H. HANSON: Zum Vorkommen und zur biochemischen Differenzierung von Peptidasen tierischer und menschlicher Organe und Gewebe.

Unter Peptidasen werden Enzyme, die Peptidbindungen hydrolytisch spalten, verstanden (Peptidhydrolasen oder Proteasen). Bei ihrer Charakterisierung wurden bisher meist die Eigenschaften der Substratspezifität, wie sie für die im Verdauungskanal zur Wirkung gelangenden Proteasen bekannt sind, zugrunde gelegt. Diese Einteilung trägt dem viel mannigfaltigeren Vorkommen von proteolytischen oder peptidhydrolytischen Enzymen in den Geweben und Organen des tierischen und menschlichen Organismus nicht ausreichend Rechnung. Der biochemische Nachweis spezieller proteolytischer Enzyme in Organen und Geweben kann erfolgen 1. durch Einsatz enzymspezifischer, synthetischer Substrate, unter Umständen auch sog. chromogener Substrate, 2. durch Feststellung der Spaltungsorte an bestimmten in ihrer Aminosäuresequenz bekannten Peptiden oder Proteinen, 3. durch Herstellung bestimmter spezifischer Bedingungen (pH, Effektorenzugabe), 4. durch Lokalisation an bestimmten Zellbestandteilen (z. B. durch fraktionierte Zellbestandteilzentrifugation). Die Möglichkeiten 1—4 werden am Beispiel der Aminopeptidasen erörtert: die Leucinaminopeptidase aus Niere, Leber, Augenlinse usw. spaltet bevorzugt Leucinamid, viel schwächer L-Leucin-β-naphthylamid, besitzt ihr pH-Optimum bei pH 8,5, wird durch Mn^{2+} aktiviert und findet sich bei der Zellbestandteilfraktionierung vor allem im Überstand. Daneben kommt besonders in Niere, Placenta und anderen Organen ein Enzym vor, das L-Leucin-β-naphthylamid deutlich angreift, L-Leucinamid relativ schwächer, durch Mn^{2+} gehemmt wird, sein pH-Optimum bei 7,5 besitzt und vorwiegend in der Mikrosomenfraktion lokalisiert ist. Es weist Beziehungen zur Oxytocinase auf und ist vielleicht mit diesem Ferment identisch. L-Leucin-β-naphthylamid ist somit kein geeignetes Substrat zur Unterscheidung von Leucinaminopeptidase und anderen Aminopeptidasen. Entscheidend für die Zuordnung einer im Gewebe oder Organen nachgewiesenen Peptidase bleibt ihre Isolierung oder die Möglichkeit, ein Substrat einzusetzen, von dem bekannt ist, daß es nur durch diese Peptidase angegriffen wird.

The Presence of Peptidases in Animal and Human Tissues and Organs and Their Biochemical Differentiation. Peptidases are a group of enzymes which hydrolyse peptide linkages (peptide hydrolases or proteases). Hitherto their characterisation has been based on their property of reacting specifically with a substrate, as for example from the proteases of the digestive tract. This classification does not sufficiently take into consideration the multifarious presence of proteolytic or peptide hydrolytic enzymes in tissues and organs of the animal and human organism. Specific proteolytic enzymes can be demonstrated biochemically in organs and tissues 1. by employing enzyme specific synthetic substrates, in certain cases even so-called chromogenic substrates. 2. by identifying the site of the splitting, using certain peptides and proteins with known amino acid sequences. 3. by creating certain specific conditions (pH, addition of effectors). 4. by localizing them on certain cell components (e.g. fractionated centrifugation of the cell components). Taking aminopeptidases as an example these four possibilities are discussed as follows: Leucine aminopeptidase affects leucineamide strongly and less so L-leucyl-β-naphthylamide. It is optimally active at pH 8.5 and to be activated by Mn^{2+}. It is found mainly in the supernatant of the fractionated cell component centrifugation. In kidney, placenta and other organs, however, there is at the same time another enzyme present which has a strong effect on L-leucyl-β-naphthylamide and a relatively weaker effect on L-leucineamide. It is to be inactivated by Mn^{2+}, optimally active at pH 7.5, and its localisation is predominantly in the microsome fraction. It bears certain relationships to oxytocinase and is possibly identical with this enzyme. L-leucyl-β-naphthylamide is therefore not suitable for the differentiation of leucine aminopeptidases and other aminopeptidases. Determining factor for the classification of a peptidase, which has been demonstrated in tissues or organs, is therefore its isolation or the use of a substrate which is specifically affected by this enzyme.

Martin-Luther-Universität Halle-Wittenberg, Physiologisch-Chemisches Institut, Halle/Saale, Hollystr. 1 (Deutschland).

A. G. E. PEARSE: Esterase Activity of Cathepsins and Imidazoles, and its Relationship to the Nuclear Cycle.

E. 600-resistant esterase activity, whether demonstrated with indoxyl or other types of substrate, has been considered to be due to cathepsin-like enzymes. Present evidence supports this contention though complete proof is still lacking. It has been shown that imidazoles can hydrolyse aromatic esters and the results of the present work indicate that the indoxyl esterase activity of nucleoli is due to active sites in relation to histidine. It is suggested that positive copper phthalocyanin (Luxol Fast Blue) staining of such nucleoli, and of other lipid-free proteins, is due to complexing of the dye with histidine and not, as formerly considered, to base exchange with arginine.

Postgraduate Medical School, Duncan Road, London W 12 (Great Britain).

D. LAGUNOFF and E. P. BENDITT: Histochemical Demonstration of Proteolytic Enzymes — The Influence of the Naphtholic Moiety on Substrate Specificity.

The prime use of enzymatic histochemistry is the spatial localization of previously characterized enzymes. To find the locus of a particular hydrolytic enzyme in a tissue or within a cell, a substrate must be constructed that will be split by the enzyme and have the additional property that one of the products of hydrolysis can be captured and visualized at or near the site of the enzyme. Naphtholic esters and amides have commonly been utilized as histochemical substrates, and it has usually been assumed that the specificity of enzymes for potential substrates is little affected by the substitution of a naphtholic or naphthylamidic group for an aliphatic alcohol or amide. — Several proteolytic enzymes are readily amenable to histochemical demonstration using naphthol derivatives, and a study of the substrate specificity of one of these, chymotrypsin, has been carried out in some detail. A limited study of trypsin specificity has been conducted for comparative purposes. — Crystalline enzymes were used for kinetic studies which were performed utilizing a fluorometric assay. Apparent affinity constants (K_0) and maximum velocities (V_0) were determined from Eadie plots of the data. Semi-quantitative measurements were carried out on selected cell enzymes *in situ* by standard histochemical procedures. — The results indicate that although histochemically applicable substrates may be used to identify and distinguish enzymes with substrate specificities resembling chymotrypsin and trypsin, the presence of the naphtholic moiety in the substrate results in an unusual affinity of chymotrypsin for the substrates. Demonstrable specificity in this situation depends entirely on differences in V_0. An explanation of the effect in terms of a hydrophobic binding site on chymotrypsin is consistent with the data.

Dept. of Pathology, Seattle, Washington (USA).

GEORGE G. GLENNER: Characterization of Enzymes Hydrolyzing Acyl and α-Amino Acyl β-Naphthylamides.

By means of newly developed techniques it is now possible to characterize accurately several enzymes in histochemical systems and to compare directly their activity characteristics with those described biochemically. These methods permit a quantitative evaluation of (1) the kinetic characteristics of enzymes present in histochemical systems (2) the preferential capacity of these enzymes to hydrolyze non-histochemical substrates (3) the effect of diazonium salts, oxidants, metal ions, etc., on enzymic activity and (4) the relation between those enzymes found in homogenate and zymogram systems, and those found in the histochemical system. These and other methods including biochemical isolation techniques have permitted us to characterize a variety of enzymes (acylamidases) in mammalian tissues hydrolyzing acyl and α-amino acyl β-naphthylamides. Besides a γ-glutamyl transpeptidase activity and a trypsin-like activity, enzymes having a specificity for N-terminal aspartic and glutamic acid peptides (aminopeptidase A), enzymes (not leucinaminopeptidase) hydrolyzing leucyl β-naphthylamide, and a variety of other naphthylamide-cleaving activities have been characterized. The functional capabilities of several of these enzymes in mammalian tissues have been studied and their physiologic role assessed. The methods described can be easily modified to characterize other types of hydrolytic enzyme activities.

Section on Histochemistry, National Institute of Health, Bethesda, Maryland (USA).

ROBERT L. HUNTER, MARVIN S. BURSTONE and HISAKO O. YOKOYAMA: Zymogram Comparison of Esterases and Proteolytic Enzymes.

A wide variety of substrates have been used to study the relationship between esterases and enzymes hydrolyzing proteins and peptides in the tissues of albino Swiss mice. Substrates used to study esterases include α-naphthyl acetate, α-naphthyl propionate, α-naphthyl butyrate, naphthol AS acetate, naphthol AS-D acetate, naphthol AS-MX acetate, naphthol AS-D chloroacetate, naphthol AS-chloroacetate, indoxyl acetate, 5-bromoindoxyl acetate, and 5-bromo-6-chloroindoxyl acetate. Fast Blue RR was used as a coupling agent with the naphtholic substrates. — Substrates employed for the enzymes hydrolyzing peptides include L-leucyl-α-naphthylamide, L-leucyl-β-naphthylamide, DL alanyl-β-naphthylamide, chloroacetyl β-naphthylamide and N-benzoyl-DL-arginine-β-naphthylamide. Fast Garnet GBC was employed as coupling agent for these substrates. Gelatin in photographic film was used as a substrate for the localization of proteinases. — Multiple bands of proteolytic and esterolytic enzymes were observed. Some appeared to share an identical location on the zymogram, whereas others were clearly in different locations. The details will be presented and discussed.

Dept. of Anatomy, Stanford University, Stanford, California (USA) and National Cancer Institute of the National Institutes of Health, Bethesda, Maryland (USA).

OLAVI ERÄNKÖ: Multiplicity and Characterization of Histochemically Demonstrable Lyo- and Desmo-Esterases.

Naphtholic, indoxylic and thiocholine esters suitable for histochemical demonstration of esterase activity are to a varying extent split by a number of different enzymes. Esterases present in tissue extracts can be resolved into several active bands by starch gel electrophoresis, and these bands can be characterized using different substrate-inhibitor combinations. The majority of esterases in tissue extracts responsible for the electrophoretic band patterns are readily soluble (lyo-enzymes), and they are therefore not histochemically demonstrable in fresh tissue sections. In these, poorly soluble (desmo-) enzymes, which do not essentially contribute to the electrophoretic band pattern, are responsible for the histochemical reaction. Lyo-enzymes can be immobilized in the tissue by fixation, which makes it possible to study their distribution in sections, and desmo-enzymes can be mobilized from the tissue by repeated freezing and thawing, thus making them susceptible for analysis by electrophoresis. — It can be shown that lyo- and desmo-esterases produce different zymogram patterns suggesting truly different enzyme species. Fixation, besides immobilizing lyo-esterases, also inactivates some desmo-esterases, which can be helpful for their characterization. Comparison of histochemical reactions obtained with several substrate-inhibitor combinations in fresh and fixed sections, as well as in lyo- and desmo-zymograms sometimes make it possible to identify a localized histochemical reaction in a section with a specific band in the zymogram.

Dept. of Anatomy, University of Helsinki, Helsinki (Finland).

BENITO MONIS: Electrophoretic Evidence for Isozymes of Aminopeptidase (Aminoacyl Naphthylamidase) in the Rat. Serum, Renal and Urinary Enzymes Splitting D-L Alanyl-2-Naphthylamide and L-Leucyl-2-Naphthylamide*.

Bilateral nephrectomy elicited no changes of serum blood levels of aminopeptidase assayed with 2 chromogenic substrates (L-leucyl-2-naphthylamide and D-L-alanyl-2-naphthylamide) 2 days after the operation. Since rat urine contains similar enzymes it was postulated that the urinary enzyme is of renal origin and distinct from serum aminopeptidase. For this purpose, electrophoretic studies were undertaken. Starch block (for quantitative determinations) and starch gel for zymograms of aminopeptidase were used. It was shown that both procedures demonstrated the identity of urinary and renal aminopeptidase, which was distinct from the serum enzyme. The renal and urinary enzyme showed two distinct isozymes: one remaining at the origin and the other migrating somewhat less than the isozyme in serum. It is postulated that the isozyme remaining at the origin corresponds to a membrane bound form, whereas the electrophoretically mobile isozyme is found in the cellular supernatant. The aminopeptidase from serum was represented by a single enzyme which migrated farthest towards the anode.

* This investigation was supported by Grant A-3219 of the National Institutes of Health, Bethesda 14, Maryland.

Dept. of Pathology, Sinai Hospital, Baltimore, Maryland (USA), and Instituto de Patologia General, Fac. Medicina, Univ. del Nordeste, Corrientes (Argentina).

Histochemie der Haut, speziell der normalen und pathologischen Differenzierung der Epidermis

Histochemistry of the Skin, Especially of Normal and Pathological Epidermal Differentiation

Chairman: O. BRAUN-FALCO, Marburg (Deutschland).

O. BRAUN-FALCO: Einführung.

Es ist das Anliegen dieses Symposions, die Histochemie der normalen und pathologisch gestörten Differenzierung der Epidermis zu diskutieren. Eine der wichtigsten Funktionen der Epidermis ist ihre Schutzfunktion. Diese ist, wie bereits MONTAGNA betont hat, gebunden an die nach Art einer holokrinen Drüsensekretion erfolgenden Produktion der Hornschicht, in deren Verlauf die äquipotenten, undifferenzierten basalen Epidermiszellen zur Hautoberfläche gelangen, absterben und in Horn umgewandelt werden. Dieser normale Differenzierungs- oder Reifungsvorgang beginnt bereits im Str. basale und führt, wie wir durch elektronenmikroskopische Untersuchungen wissen, über eine Reihe von Zwischenstufen in Str. spinosum, Str. granulosum und der Barriere, von denen jede durch eine Zelle mit einer typischen morphologischen Struktur gekennzeichnet ist. Hingewiesen sei nur auf das sich wandelnde Verhalten der Mitochondrien, der Tonofilamente und Tonofibrillen, der Kern-Plasma-Relation sowie das Auftreten der Keratohyalingranula. Es ist klar, daß diese Wandlungen der Feinstruktur der Basalzelle auf ihrem Differenzierungsweg zur Hautoberfläche verbunden sein muß mit Änderungen im funktionellen-metabolischen Verhalten. — In der Tat geht aus zahlreichen histochemischen Untersuchungsergebnissen hervor, daß auch der Zellstoffwechsel der äquipotenten, undifferenzierten basalen Epidermiszellen auf ihrem Differenzierungsweg mit zunehmendem Differenzierungsgrad und damit zunehmender Entfernung von der O_2-Versorgung beachtlichen Umstellungen unterworfen ist. Der Zellstoffwechsel der Epidermiszellen ist also ein anderer im Str. basale als im Str. spinosum oder Str. granulosum. In der folgenden Diskussion sollte der Versuch unternommen werden, einige typische Verhaltensweisen des Stoffwechsels in Abhängigkeit von der Differenzierungsstufe der Epidermiszellen zu erarbeiten, wobei Enzymen des Kohlenhydrat- und Atmungsstoffwechsels, mit dem Verhornungsvorgang verbundenen Enzymen, ferner Kohlenhydraten, Lipoiden und Nukleinsäuren besondere Aufmerksamkeit geschenkt werden wird. Vergleichend soll der Epidermisstoffwechsel unter solchen pathologischen Bedingungen betrachtet werden, wo der normale Differenzierungsvorgang quantitative oder qualitative Störungen aufweist, die sich histologisch als Akanthose, Atrophie, Hyper-, Para- und Dyskeratose manifestieren. Dabei wird auch epidermalen Tumoren (Basaliom, spinozelluläres Karzinom) Aufmerksamkeit geschenkt. Schließlich soll das wichtige Problem induktiver und regulativer Einflüsse des Bindegewebsstromas auf die Vorgänge der Differenzierung diskutiert werden, soweit sich aus der Histochemie der Blutgefäße, mesenchymaler Grundsubstanz, Fibroblasten und Fasersysteme Hinweise ergeben.

Introduction. It is the concern of this symposium to discuss the histochemistry of the differentiation of the epidermis under normal and pathological conditions. One of the most important rôles of the epidermis is its protective function. This protective function is — as already mentioned by MONTAGNA — linked to the formation of the horny layer, which follows the principle of holocrine secretion. During this process the equipotent undifferentiated basal cell of the epidermis rises to the cutaneous surface, dies and is subsequently transformed into horn. This normal differentiation or maturing process starts in the Str. basale and includes — as we know from electron microscopic findings — a number of intermediary stages in the Str. spinosum, Str. granulosum and the barrier. Each stage is characterized by a typical morphological structure of the cell. The changing behaviour of the mitochondria, the tonofilaments and tonofibrils, the nuclear-cytoplasmic relationship, and the occurrence of keratohyalin granules are a few examples. It is quite obvious that these changes, which take place in the structure of the basal cell during the differentiation, are combined with changes in its functional-metabolic behaviour. — In fact, numerous histochemical results show that the metabolism of the equipotent undifferentiated basal cell is subjected to considerable changes as the stage of differentiation advances and the distance from the source of O_2 increases. The metabolic pattern of epidermal cells in the Str. basale is therefore different from the metabolic pattern of epidermal cells in the Str. spinosum or Str. granulosum. During the forthcoming discussion an attempt should be made to outline some of the typical reactions of the cell metabolism and their dependence on the grade of differentiation of the epidermis. Special attention should be given to the enzymes of carbohydrate and respiratory metabolism, to enzymes which are related to the process of cornification, to carbohydrates, lipids and nucleic acids. Comparative studies of the metabolism of the epidermis will be made in cases where the normal differentiation shows disturbances which become manifest as acanthosis, atrophy, and hyper-, para- and dyskeratosis. Attention will also be given to epidermal tumours (basal cell carcinoma and spinocellular carcinoma). Finally, the important problem of the inductive and regulative effects of the stroma of the connective tissue on the process of differentiation will be discussed, making use of all the evidence so far obtained from histochemical determinations of blood vessels, mesenchymic matrix, fibroblasts and fibrous systems.

Universitäts-Hautklinik, 355 Marburg/Lahn (Deutschland).

A. Histotopochemie in Beziehung zur epidermalen Proliferation
Histotopochemistry in Relation to Epidermal Proliferation

Y. NOGUCHI: The Cytochrome System in Normal Epidermal Differentiation.

Histoenzymological study of the cytochrome system in fetal and postnatal epidermal differentiation both of human and mouse skin has been made, whereas, cytochrome oxidase and several other oxidative enzymes including succinic dehydrogenase, nicotinamide-adenine dinucleotide (NAD, or so-called DPN) diaphorase, nicotinamide-adenine dinucleotide phosphate (NADP, or so-called TPN) diaphorase as well as some of dehydrogenases linked with NAD or NADP have been studied as they are related in transportation of electron. — Some current histochemical procedures such as N-phenyl paraphenylen diamine method (after BURSTONE), ADN method (after NACHLAS et al.), G-nadi reaction for cytochrome oxidase and some tetrazolium reactions using nitro-BT or MTT for various dehydrogenases and diaphorases have been employed by the authors for demonstrating activity of the enzymes. — From the present data the author would like to agree with the general opinion that activity of such enzymes as relating biological oxidation or respiration generally predominates in those epidermal cells at the stage of development, nevertheless, heart muscle has shown outstanding positive reaction at all times. — Further investigation on the intramitochondrial localization of some oxidative enzymes has been made by means of electron histochemical technique using nitro-BT, MTT-cobalt or Burstone method.

3—17 Yushimatenjin-cho Bunkyo-ku, Tokyo (Japan).

F. SERRI and D. CERIMELE: Enzymes of the Citric Acid Cycle and Cytochrome System in Pathologic Epidermal Differentiation.

Among the different enzymatic components of the Krebs cycle, succinic, malic and isocitric dehydrogenases are histochemically demonstrables. Succinic dehydrogenase is the most widely studied enzyme in normal and pathologic epidermal differentiation. — Cytochrome-oxidase is the best studied enzymatic component of the cytochrome system in the same conditions. — Succinic dehydrogenase activity is generally increased in metabolically active processes such as in proliferative, hyperkeratotic and inflammatory diseases, especially in the invasive areas of basal cell epitheliomas, in psoriasis, in verruca vulgaris; is diminished in the keratinization areas, as near the pearls in the squamous cell epitheliomas and in senile and seborrhoic keratosis. — The patterns of distribution of cytochrome oxidase and of succinic dehydrogenase are in close correlation and are thought to be identified with the distribution of mitochondria. — Cytochrome oxidase reaction is increased in acanthotic epithelium, particularly in the mitotically active basal cell layers while normal hyperkeratotic and parakeratotic layers are not reactive. In benign tumors and in basal cell epitheliomas the cytochrome oxidase activity is normal, while in squamous cell epitheliomas, particularly in parenchimal areas near the stroma, the activity is sensibly increased. — In pigmented nevi and in melanosis circumscripta the cells proliferating in the epidermis are more oxidase positive than the adjacent ones.

Istituto di Clinica Dermatologica dell'Università di Sassari, Via Mamco 52, Sassari (Italia).

G. ACHTEN: Carbohydrates in Normal and Pathological Epidermal Differentiation.

Amongst the polysaccharides can be distinguished the following: 1. The polysaccharides which are closely concerned with the general metabolism of the cell. Glycogen (glycogen and mitosis-glycogen and keratinization-glycogen and glandular activity) belongs to this group. — 2. The polysaccharides of support. The ground substance, the cement which surrounds the collagen and elastic fibres and keratin, belongs to this second group. In dermatology the importance of this fundamental substance is noted in: the diffusion of the microbic eczema as a result of the hyaluronidase secreted by the infecting organisms; the penetration of the dermatophyte into the keratin by its action on the complex keratin polysaccharide;

the changes in the diseases "said to be of the collagen", which appear, in fact, to be diseases of the ground substance. — 3. The polysaccharides "of induction", that is those which act as biochemical links in the differentiation of the cell. The importance of these mucopolysaccharides is well-known in the early development of the egg, and similarly in the development of the limb buds. — The skin is not exempt from this differentiation: the life of the hair follicle (its appearance in the embryo — the catagenic, anagenic and telogenic stages — its destruction by various means) is bound up with the presence of acidic mucopolysaccharides in the papilla and around the hair follicle. — Cicatrization of wounds is further evidence of this. The "mastocytic attack" of the embryo (increase of the cutaneous mast cells towards the fifth month of pregnancy), the presence of abundant mastocytes in the various skin disorders show how the mucopolysaccharides are concerned in the processes of embryological induction.

Clinique de Dermatologie et de Syphiligraphie, Hôpital Universitaire Saint Pierre, Rue Haute, 322, Bruxelles (Belgium).

H. G. GODLEWSKI: Intracellular Carbohydrates in Epidermal Tumours.

Following substances can be distinguished among the histochemically demonstrable carbohydrates: glycogen, neutral and acid mucopolysaccharides including sialic acid. The character of distribution of the said substances frequently depends on the form of a skin tumour examined. — A typical basocellular carcinoma contains practically no glycogen. On the other hand, the spinous tumour cells are characterized by an abundant deposit of this compound. Glycogen granules can also be found in cells surrounding the cornified epithelial pearls. Glycogen in these cells may be considered as a sign of the earliest phase of the keratohyaline formation. The next phase of this process may be recognized within the tumour owing to the presence of the PAS positive material resistant to amylase digestion. — In some foci of basocellular carcinoma undergoing hyalinic degeneration the PAS positive substances indicate the presence of highly polymerized neutral glycoproteins. The same is true when the cornified material is considered, sometimes encountered in the basal cell tumours. — The presence of acid mucopolysaccharides is to be ascribed chiefly to the connective tissue stroma: the nature of these mucopolysaccharides is mentioned. — Some examples of skin tumours varying in degree of their histologic maturity are discussed in the light of polysaccharide contents and distribution.

Institute of Experimental Pathology, Polish Academy of Sciences, Dworkowa 3, Warszawa 12 (Poland).

R. W. GOLTZ: Enzymes of the Citric Acid Cycle and Glycolysis during Normal Epidermal Differentiation.

The study to be discussed represents an attempt to clarify in some measure the pathways of carbohydrate metabolism in normal epidermis. The method used consisted of examination of the distribution, and the quantities at various levels, of two histochemically demonstrable enzymes concerned with glucose metabolism. The first of these, an anaerobic phase enzyme, phosphorylase, was demonstrated by a modification of Takeuchi's method, using either Lugol's solution or periodic acid Schiff for demonstration of the resulting glycogen. The second enzyme, succinic dehydrogenase, was chosen as representative of the aerobic phase. It was demonstrated by a tetrazolium method. — In both cases, relative amounts of enzyme activity were estimated by visual inspection and by use of photometric mensuration. Results of these measurements will be presented, and an attempt made to integrate the findings into meaningful patterns. Several hypotheses to explain the progression of changes in carbohydrate metabolism as epithelial cells mature and as keratinization occurs will be proposed for discussion by members of the panel and audience.

Division of Dermatology, Medical School, University of Minnesota, Minneapolis 55, Minn. (USA).

J. HEWITT, R. WEGMANN and M. GUIGON: A Comparative Study of Enzymes of Glycolysis, Citric Acid Cycle and Hexose Monophosphate Shunt in Pathologic Epidermal Differentiation*.

The following enzymes were studied in relation to epidermal differentiation. Their activity was compared in some pathologic cases to their activity in normal epidermal differentiation. 1) Anaerobic glycolysis: phosphorylase, glyceraldehyde-phosphate-dehydrogenase and, in connection, lactic-dehydrogenase. 2) Isocitric cycle: cis-aconitase, isocitrate-, succinic- and malic-dehydrogenases. 3) Hexose monophosphate shunt: glucose-6-phosphate- and 6-phosphogluconate-dehydrogenases. — *In parakeratosis.* Among glycolysis enzymes, only phosphorylase has a great activity and develops in parallel with glycogen, while the others are below normal level. Isocitric acid cycle enzymes are less active, but their localization is superficial. Hexose monophosphate shunt enzymes show a great activity, especially in superficial layers. It seems that in parakeratosis and above all in psoriasis, this activity is the main metabolic change. — *In hyperkeratosis.* Enzymatic activities of the 3 pathways follow a pattern close to those of normal epidermis. A slight increase of phosphorylase without modification of glycogen, a hyperactivity of glyceraldehyde-phosphate-dehydrogenase and, unlike parakeratosis, a very slight activity of hexose monophosphate shunt are to be noted. *In dyskeratosis (Darier's disease).* Phosphorylase is almost inactive in pathological sites, without glycogen, but glyceraldehyde-phosphate-dehydrogenase and lactic-dehydrogenase show a great activity. The most striking evidence is the complete lack of succinic-dehydrogenase. A slight decrease of hexose monophosphate shunt enzymes is observed.

* Aided by a grant from Institut National d'Hygiène, Paris.

Hôpital Broca et Institut d'Histochimie Médicale, Paris (France).

B. Histotopochemie in Beziehung zur Verhornung
Histotopochemistry in Relation to Keratinization

O. BRAUN-FALCO: Interzelluläre Kohlenhydrat-Verbindungen bei normaler und pathologischer Differenzierung der Epidermis.

Bereits seit Jahren ist bekannt, daß im Interzellularraum der Epidermis PAS-reaktive neutrale Kohlenhydrat-Verbindungen vorkommen. Vor einigen Jahren konnten wir zeigen, daß auch saure Mucopolysaccharide zwischen den Epidermiszellen nachweisbar sind. Das Studium ihres Verhaltens in den verschiedenen Schichten der Epidermis läßt erkennen, daß diese beim Mechanismus der Verhornung von Bedeutung sind. Sowohl histochemische als auch biochemische Untersuchungen legen die Vermutung nahe, daß bei Störungen der normalen Differenzierung der Epidermis Änderungen der interzellulären Kohlenhydrat-Verbindungen vorkommen, die auch bei Störungen der Verhornung besonders eindrucksvoll sind. Die Funktion dieser Substanzen bei normaler und pathologischer Differenzierung ist noch wenig klar.

Intercellular Carbohydrate Compounds in Normal and Pathological Differentiation of the Epidermis. It has been known for several years that PAS-reactive neutral carbohydrate compounds are present between the cells of the epidermis. Some years ago we were able to show that acid mucopolysaccharides are also demonstrable between these cells. The study of their behaviour in the various layers of the epidermis shows that they are important for the mechanism of cornification. The results of histochemical and biochemical investigations suggest that changes in the intracellular carbohydrate compounds occur if the normal differentiation of the epidermis is disturbed. These changes are especially impressive if the process of cornification is disturbed. So far, little is known about the function of these substances in normal and pathological differentiation.

Universitäts-Hautklinik, 355 Marburg/Lahn (Deutschland).

H. G. GOSLAR: Keratine und Lipoide bei normaler epidermaler Differenzierung.

An Hand von Modellversuchen bei Reptilien und Ergebnissen an der Säugerhaut wird über morphologisch und histochemisch differenzierbare Phasen der Keratinbildung berichtet und der Versuch unternommen, jeder dieser Phasen ein Fermentmuster zuzuordnen. Besonders wird dabei auf die Rolle des Pentosephosphatzyklus eingegangen und die Frage, inwieweit man aus den enzymtopochemischen Gegebenheiten auf einen autokatalytisch ablaufenden Vernetzungsprozeß der Sulfhydrylgruppen schließen kann, diskutiert. Die methodischen Schwierigkeiten einer Lipoiddarstellung in den epidermalen Bereichen werden an Hand polarisationsoptischer Befunde von W. J. SCHMIDT beleuchtet und über die Lipoidausscheidung durch die Epidermis und den Einbau in das Keratingefüge berichtet.

The Role of Keratins and Lipids in Normal Epidermal Differentiation. On the basis of experiments on reptiles and on mammalian skin the author reports on the various stages of keratin formation which can be differentiated morphologically and histochemically. He also tries to associate each stage with a specific enzyme pattern. Special attention is given to the function of the pentose phosphate cycle. The question whether the enzyme topochemical data indicate the occurrence of an autocatalytic process of cross linkage of the sulphydryl groups is discussed. The practical difficulties which arise from the determination of lipids in epidermal zones will be discussed on the basis of polarization optical findings by W. J. SCHMIDT. An account is also given on the secretion of lipids through the epidermis and their incorporation into the keratin structure.

Anatomisches Institut der Universität, Nußallee 10, 53 Bonn (Deutschland).

A. JARRETT: Lipids and Keratins in Pathological Differentiation of the Epidermis.

Lipids: The location and possible origin of lipids in normal and abnormal skin will be mentioned with special reference to phospholipids. The lipid content of keratin will be considered with respect to the degree of development of the granular layer. The work to determine the exact nature of the phospholipids by means of enzymatic digestion will be discussed. — *Keratins:* Differences between the various normal and abnormal keratins as seen by fluorescence microscopy will be reported. The artificial production of keratinization of epidermal cells by pressure is similar to that of the naturally occurring inner root sheath keratin. The production of this keratinization in non-malignant tumours of the skin is thought to limit their extension into the dermis. The speed of epidermal turnover is partly responsible for the type of keratin produced. Conditions such as psoriasis and squamous cell carcinomas have a high rate of epidermal proliferation and tend to produce parakeratotic keratin, whereas in lichenification there is a thickening of all parts of the epidermis, and it is probable that this is related to a decreased rate of epidermal cell production.

Dept. of Dermatology, University College Hospital, Medical School, University Street, London W.C.1 (England).

A. KINT: Nucleic Acids in Normal and Pathological Differentiation of the Epidermis.

Study of the nucleic acids in normal epidermis and in various forms of acanthosis: lichenification, benign tumours (verruca seborrhoica, kerato-acanthoma), precanceroses, malign tumours. The RNA is demonstrated by the Unna-Bracht method and with Gallocyanin. The DNA is coloured by the Feulgen-technic and histophotometrically determined. — The relation between the nucleic acid content of the cells and the forms of acanthosis is studied.

Kliniek voor Huidziekten, Pasteurlaan 2, Gent (Belgium).

R. A. ELLIS: Hydrolytic Enzymes in the Epidermis.

There is now definite evidence that a variety of enzymes serving different functions show distinct and measurable differences in their reactivity in the epidermis covering diverse parts of the body. The hydrolytic enzymes such as acid phosphatase and the esterases are most abundant in the upper level of the Malpighian layer. The fine localization of the acid phosphatase activity has been attempted with the electron microscope in both fetal and newborn skin. Since no lysosome-like particles have been demonstrated in epidermal cells, the correlation of epidermal cell fine structure with hydrolytic enzyme localization is of special interest. The results suggest that the hydrolytic enzymes are important in the protoplasmic differentiation of the epidermal cells and play a significant role in the process of keratinization.

Dept. of Biology, Brown University, Providence 12, Rhode Island (USA).

G. K. STEIGLEDER: Hydrolytische und proteolytische Enzyme bei pathologischer Differenzierung der Epidermis.

Histiocytome und Fibrome induzieren eine Wucherung der Epidermis, welche die Akanthose bei der Psoriasis, die verschiedenartigen Epithelwucherungen bei seborrhoischen Keratosen und sogar die Struktur der oberflächlichen Basaliome nachahmt. Diese Induktion geht offensichtlich von der Cutis aus. Sie ist von charakteristischen Verschiebungen hydrolytischer und proteolytischer Aktivität im Histiocytom selbst, in den Kapillaren des Papillarkörpers und schließlich in der Epidermis begleitet. Den Veränderungen der enzymatischen Aktivität geht das Auftreten metachromatischer Substanz parallel. — Aus diesem Modellfall lassen sich Hinweise für das Zustandekommen mancher Dermatosen, wie der Psoriasis, aber auch bestimmter Hauttumoren gewinnen.

Hydrolytic and Proteolytic Enzymes in Pathological Epidermal Differentiation. Histiocytomata and fibromata induce a proliferation of the epidermis which imitates the picture of acanthosis seen in psoriasis, the various epithelial proliferations in seborrhoeic keratoses, and even the structure of superficial basal cell carcinomas. This induction starts apparently in the cutis. It is accompanied by a characteristic shifting of hydrolytic and proteolytic activity in the histiocytoma itself, in the capillaries of the papillary layer and finally in the epidermis. Parallel to the changes in enzymatic activities goes the occurrence of metachromatic substances. — This natural model gives information not only about the development of certain dermatoses (as for example psoriasis) but also about certain skin tumours.

Universitäts-Hautklinik, Ludwig-Rehn-Str. 14, 6 Frankfurt a.M.-Süd 10 (Deutschland).

C. Induktion und Regulation der epidermalen Differenzierung
Induction and Regulation of Epidermal Differentiation

G. MORETTI and E. RAMPINI: Alkaline Phosphatase and Epidermal Differentiation under Normal and Pathological Conditions.

The epidermis of many mammalians' species contains little or no alkaline phosphatase which might be histochemically revealed. The epidermal appendages, on the contrary may contain, according to the species, variable amounts of this enzyme. — Thus a mere review of the literature on the title's matter does not really enlighten about the precise role of this enzyme in the processes of epidermal differentiation. — A few reliable hints to this regard, however, might perhaps be gathered through the histochemical observation of its behaviour in a variety of species and both physiological and pathological conditions. Out of many skin samples of human fetuses and adult subjects or of other mammalians we have studied, therefore, a number of fresh frozen sections treated with the alkaline phosphatase technique of GOMORI as modified by MONTAGNA. — Within the skin of some of these animals, furthermore, our analysis was completed by observations on the regional behaviour of the histochemical reaction. — Finally the epidermal reactivity was examined in a group of pathological skin conditions ranging from the simple inflammatory to the neoplastic dermatoses. — After having briefly shown their data and their observations, the authors discuss their hypothetical significance.

Clinica Dermatologica della Università di Genova, Viale Benedetto XV, Genova (Italia).

R. E. MANCINI: Mesenchymal Ground Substance and Epidermal Differentiation under Normal and Pathological Conditions.

Fibroblasts, collagen and mucopolysaccharides of dermal tissue were studied during epidermal differentiation in rats at different stages. Biochemical, radioautographic and histochemical methods were applied. From birth to maturity, the following was observed: 1. early fibroblasts changed into typical adult cells, fibrocytes, with gradual diminution of cytoplasmic nucleoproteins, periodic acid Schiff positive material, succinodehydrogenase, B-glucuronidase and leucylaminopeptidase. Electron microscopic obseravations revealed the presence of an abundant reticuloendoplasmic system and labyrinthic cisternal in young fibroblasts which tend to decrease in the mature cells. Methionine-C^{14} and adenine-C^{14} were seen more in the young cells and negligible amount in the intercellular substance. Also, proline-H^3, glycine-C^{14} and S^{35} (sodium sulphate) were trapped more by the young fibroblast and they have been observed first in the fibroblast and then in the intercellular ground substance. 2. This gradual decrease in the cytochemical activity is paralleled in the dermal tissue by a diminution in the total hexosamine, uronic acid and sulphated radicals, as well as in the total soluble collagen (neutral salt and citrate soluble fractions). Also an accumulation of a noncollagenous protein rich in proline was noticed only at post-natal stages. Some differences were observable in the aminoacid composition of soluble collagen from young and adult animals. The incorporation of the above mentioned radioisotopes by the epidermis and hair follicles were higher in the post-natal stages.

Instituto de Anatomia General y Embryologia, Universidad de Buenos Aires, Buenos Aires (Argentina).

Interference Microscopy and Histology

Chairman: R. BARER, Sheffield (England)

Participants: R. D. ALLEN, Princeton (USA), E. J. AMBROSE, London (England), G. GER-ZELI, Pavia (Italy), D. GLICK, Palo Alto (USA), H. ROELS, Gent (Belgium), K. F. A. Ross, Newcastle (England), W. SANDRITTER, Giessen (Germany), H. G. SCHIEMER, Frankfurt/M. (Germany), U. STENRAM, Uppsala (Sweden).

R. BARER: Introduction.

This meeting will take the form of an informal panel discussion of important theoretical and practical problems in interference microscopy. It is now a little over ten years since interference microscopy was introduced as a quantitative cytological method for mass determination. It will be the purpose of the meeting to discuss the general value of the method, its validity, special problems und future developments. In general the validity of the results have been fairly well supported by other measurements such as biochemical estimations, X-ray microanalysis and so on. Methods have been developed for the estimation of specific components by extraction and enzymic digestion procedures. Some enzymes reactions have been followed by measuring the change in mass of the product at active sites. In some cases it has been possible to make measurements on intracellular inclusions and the combination of several measurements made in media of different refractive indices has given information on dry and wet mass, concentration and thickness of cells. One of the most fundamental difficulties is that measurements usually have to be made at a single point so that integration over a large area is difficult. Automatic scanning techniques have been evolved and these, in combination with computors, may enable results to be obtained from large populations of cells. Applications in the fields of general cytology, tissue culture, cell growth and cancer research will be discussed and illustrated by films.

Dept. of Human Biology and Anatomy, University of Sheffield (England).

Neurohistochemistry

Chairman: A. Pope, Boston (USA).

Alfred Pope: Perspectives in Neurohistochemistry.

The critical interdependence of structure and function in neuronal units and assemblies has made neurological applications of the principles and techniques of histochemistry of particular significance. Broadly defined, these comprise all procedures that enable establishment of correlations between biochemical composition and histological and cytological structure. This array includes classical prodedures for *in situ* localization of chemical constituents and sites of enzymatic activities, the use of differential gradient and centrifugation methods for isolation of cellular and sub-cellular particles combined with biochemical analyses upon them, and the techniques of quantitative histochemistry in which tissue samples of microgram or sub-microgram sizes are directly analyzed by microchemical procedures. — Historically, the histochemistry of nervous tissues has been studied for a half-century. However, an explosive development has occurred during the last two decades, and a voluminous literature now exists upon neurohistochemical observations of many kinds. Clearly, it is impossible to review this subject in one afternoon. It is, however, pertinent to consider certain current activities of particular interest and importance. Relevant advances in methodology deserve consideration such as the reduction of *in situ* histochemistry to the level of the electron microscope, the use of fluorescence for cytologic localization of neurochemically important compounds, and the refinements of microchemistry enabling quantitative determinations upon single cells and even parts thereof. Extensive applications of all forms of histochemistry to regional neurocytology and neuropathology also merit discussion. — Accordingly, the program of this Symposion has been structured so as to present topics in (1) technical developments in histochemistry of special importance to neurobiology, (2) recent insights concerning biochemical fine structure in the central nervous system, in peripheral nerve, and in neuro-muscular end organs, and (3) applications of histochemistry to selected problems in neuropathology.

Harvard Medical School, McLean Hospital, Belmont, Mass. (USA).

Russel J. Barrnett: Fine Structural Demonstration of Enzyme Activity Implicated in Transport Phenomena in the Nervous System.

Contrary to the localization of nucleoside phosphatase activity in unfenestrated capillaries of various tissues (e.g. striated muscle) where this activity is related only to the pinocytic invaginations and vesicles of endothelial cells, the endothelial cells of small blood vessels of the brain and retina show no enzymatic activity. However, the endothelial cells of vessels of the choroid, pineal, and area postrema, which are considered outside the blood-brain barrier, contain enzymatically active vesicles. Some cerebral and retinal capillaries show enzymatic activity in the basement membrane region and for all of these small vessels, activity is demonstrable in relation to the plasma membrane covering the glial end feet that abut on the capillary bed. — Distant from the vascular areas, nucleoside phosphatase activity is manifested at the interface of glial-glial elements, glial-neuronal elements but not neuronal-neuronal elements in both the retina and cerebral cortex. Similar localization of this enzyme activity occurs also at the interface between the invaginated membranes of Schwann cells and those of unmyelinated axons in smooth muscle. These observations shall be discussed with reference to transport phenomena in the nervous system.

Yale University School of Medicine, New Haven, Connecticut (USA).

Olavi Eränkö: Histochemical Demonstration of Catecholamines in Nervous Tissues.

Aromatic monoamines such as adrenaline, noradrenaline, dopamine and 5-hydroxytryptamine can be made fluorescent by treatment with dry formaldehyde vapour. This method is sensitive enough to render it possible to study the distribution of these amines in nervous tissues by fluorescence microscopy. Discrimination between different amines is possible by studying the effect of temperature and the length of formaldehyde treatment necessary to produce the fluorescence, and the colour of the fluorescent compound thus produced. By limiting the length of formaldehyde treatment it is possible to convert, e.g., noradrenaline fluorescent without destroying enzymic activity in the frozen dried piece of tissue. Enzymes such as acetylcholinesterase can thereafter be demonstrated with a histochemical reaction in the same section after photographic registration of the distribution of noradrenaline. Comparison of these two reactions in individual neurons shows that a high concentration of noradrenaline may be associated with an intense acetylcholinesterase activity, an observation which suggests that a neuron may at the same time be both adrenergic and cholinergic.

Dept. of Anatomy, University of Helsinki, Siltavuorenpenger, Helsinki (Finland).

D. B. McDougal, jr.: Micro-Assay Methods for Acetyl Choline, Choline Acetylase, and Choline Esterase*.

A specific chemical micromethod for the assay of acetyl choline (ACh) would be useful for the measurement of choline acetylase (ChA) and cholinesterase (ChE) as well. However, chemical methods for ACh and ChA have not kept pace with those for ChE partly because 1) the level of activity of ChE is usually one or two orders of magnitude higher than that of ChA; 2) small amounts of enzymes are generally easier to measure than small amounts of substances which lack a known catalytic function; 3) and there are no known specific, sensitive chemical reactions suitable for measuring ACh. At present chemical assays of ChE activity can be carried out in the Cartesian diver on about 0.5 mμg of average brain (GIACOBINI). Ten thousand times as much brain is needed for the most sensitive chemical assay of ChA (McCAMAN), whereas still another hundred thousand times more brain is needed to assay the ACh content chemically. — The refinement of cycling methods by LOWRY et al. (1961, J. Biol. Chem., *236*, 2746) has placed a premium upon systems in which one of the final products is a pyridine nucleotide. Therefore the following scheme has been devised for ACh:

1. $\text{ACh} + \text{H}_2\text{O} \xrightarrow[\text{or OH}^-]{\text{AChE}} \text{Acetate} + \text{Choline.}$

2. $\text{Acetate} + \text{Coenzyme A} + \text{ATP} \xleftrightarrow{\text{aceto-CoA-kinase}} \text{Acetyl-S-Coenzyme A} + \text{AMP} + \text{pyrophosphate.}$

3. $\text{H}_2\text{O} + \text{Acetyl-CoA} + \text{Oxalacetate} \xleftrightarrow[\text{enzyme}]{\text{condensing}} \text{Citrate} + \text{CoA} + \text{H}^+$

4. $\text{Malate} \xleftrightarrow[\text{dehydrogenase}]{\text{malate}} + \text{DPNH} + \text{H}^+.$

DPN^+

The problems which arise in the development of such a system due to peculiarities of the constituent enzymes, contamination by interfering enzymes, and the presence of interfering substances in the tissues will be discussed. Results of the use of the assay in tissues will also be given.

* Supported by NIH grant 5-K 3-NB-18, 005-02 and by United Cerebral Palsy Grant No. R-96-61 C.

Dept. of Pharmacology, Washington University, St. Louis, Missouri (USA).

M. A. GEREBTZOFF: Histochemical Localization of Cholinesterase in Central and Peripheral Nervous Systems.

Since our First International Congress, histochemistry of cholinesterases has been dominated by the KOELLE concept of "internal or reserve" and "external or functional" enzymes. While differences in activity of acetylcholinesterase at various levels of the central nervous system were confirmed and submitted to a detailed study, a great deal of attention was brought to bear on the exact localization of both specific and non specific cholinesterases in neurons, in nerve fibres and at synaptic sites. The presence of the two types of enzymes was demonstrated in some neurons, while others showed only that of acetylcholinesterase, and again in others histochemical techniques were unable to reveal any cholinesterase activity. Acetylcholinesterase was generally found in neuro-neural synapses, with the exception of some non cholinergic transmission sites. The question of pre- or postsynaptic localization (or major localization) appears to have different answers according to the central or peripheral synapse studied. Nerve fibres seem also to possess both internal and external acetylcholinesterase. This is quite evident for peripheral myelinated fibres where the internal enzyme is present all along the axoplasm, while the external acetylcholinesterase is limited to the Ranvier nodes, where this functional enzyme would be bound to saltatory nerve conduction.

Institut d'Anatomie, 20, rue de Pitteurs, Liège (Belgium).

C. W. M. ADAMS: Lipids, Proteins and Enzymes in Peripheral Nerve Myelin*.

Phospholipid, sphingomyelin, phosphoglycerides, cerebroside and free cholesterol can be localized histochemically in peripheral nerve myelin by the osmium tetroxide-α naphthylamine (OTAN), NaOH-OTAN, gold hydroxamate, modified PAS and perchloric acid naphthoquinone (PAN) methods. At least two types of lipoprotein are present in peripheral nerve myelin; one protein is digestible in trypsin while the other is both resistant to this enzyme and is insoluble in anhydrous chloroform-methanol. This latter protein gives rise to the neurokeratin artefact of peripheral nerve. If peripheral nerve is fixed in buffered OsO_4 and embedded in methacrylate, the neurokeratin network cannot be seen by either light- or electron-microscopy. — Variable amounts of alkaline phosphatase activity are seen in the neurokeratin artefact of peripheral nerve and a neutral proteinase is also present in the myelin. The following enzymes could not be convincingly demonstrated by histochemical means in internodal myelin of rat and man: succinic dehydrogenase (menadione), malic dehydrogenase, glutamic dehydrogenase, α-glycophosphate dehydrogenase (DPN), glucose-6-phosphate dehydrogenase (TPN), 6-phosphogluconate dehydrogenase, lactic dehydrogenase, $NADH_2$-tetrazolium reductase, $NADPH_2$-tetrazolium reductase, cytochrome oxidase, indoxyl and α-naphthyl aliesterase, acetyl and butyryl thiocholinesterase, acid phosphatase, 5′-nucleotidase, glucose-6-phosphatase, ATPase, cytidine triphosphatase (?-nonspecific), L-leucyl-β naphthylamidase (aminopeptidase) and α-glutamyl-β naphthyl-amidase (aminopeptidase). Density gradient centrifugation showed moderate activity of L-leucyl-β naphthylamidase but only trivial amounts of ATPase, acid phosphatase, alkaline phosphatase and glucose-6-phosphate dehydrogenase in the CNS myelin fraction. It is concluded that the myelin sheath is nearly inert in respect of respiratory, phosphoesterase and carboxyesterase enzymes.

* Aided by grants from Messrs. Pfizers, Sandwich, Kent and the National Fund for Research into Poliomyelitis and other Crippling Diseases. — This work was carried out in collaboration with Drs. Y. H. ABDULLA, A. N. DAVISON, G. G. GLENNER, N. A. GREGSON, M. Z. M. IBRAHIM, N. A. TUQAN and Miss O. B. BAYLISS.

London University, Guy's Hospital Medical School, London (England).

GERARD M. LEHRER and MURRAY B. BORNSTEIN: Quantitative Enzyme Histochemistry of the Rat Cerebellum during Development in vitro*.

Six enzymes of carbohydrate metabolism have been studied in cytologically similar areas of fragments of newborn rat cerebellum, propagated on a reconstituted collagen substrate. 0.01 to 0.04 μg samples were dissected from homologous areas of frozen-dried sister cultures at various times after explantation, weighed on quartz fiber balances, and analysed in hexuplicate. Consistent results were obtained within replicate samples and there was good agreement between replicate cultures in four sets of sister cultures, each set coming from one litter. — Detailed data will be presented for hexokinase, lactic dehydrogenase, malic dehydrogenase, glucose-6-phosphate dehydrogenase, isocitric dehydrogenase, and β-D-glucuronidase. Comparisons will be made with similar data on developing granularis of the cerebellum *in vivo* and the significance of the data will be discussed as related to *in vitro* development and myelin formation.

* Supported in part by grants Nr. 293-2 and 246-4 from The National Multiple Sclerosis Society, Nr. CA-05766-03 from The National Cancer Institute and Nr. NB-01913-05 from NINDB, USPHS.

Laboratories of Neurochemistry and Cellular Neurophysiology, The Mount Sinai Hospital, New York, N.Y. (USA).

PAUL BERND DIEZEL: Histochemische Untersuchungen am Nervensystem bei der Ahornsirupkrankheit.

An einem weiblichen Säugling, der am 8. Lebenstag an einer Ahornsirupkrankheit verstorben ist, wurde eine pathologisch-anatomische Untersuchung vorgenommen. In der Familienanamnese beider Eltern ist eine latente familiäre Störung im Bereich des Aminosäurenstoffwechsels bekannt. Eine Schwester des von uns untersuchten Falles war 2 Jahre zuvor am 10. Lebenstag an derselben Erkrankung verstorben. — Klinisch hatte sich neben einer massiven Ausscheidung der Verzweigtkettenaminosäuren Leucin, Isoleucin und Valin hauptsächlich eine cerebrale Störung bemerkbar gemacht, welcher der auch anatomisch im Vordergrund stehende zentralnervöse Befund entspricht. Es findet sich eine Störung in der Markreifung vor allem der stammesgeschichtlich jungen Gehirnanteile, die bereits im Foetalleben begonnen haben muß. Außerdem ist in der weißen Substanz des Gehirns ein feinporiger Status spongiosus ausgebildet. Die schwammige Gewebsauflockerung ist teilweise auch in der grauen Substanz sowie im Nebennierenmark anzutreffen. Die Hohlraumbildungen des Status spongiosus enthalten eine schwach PAS-positive, leicht wasserlösliche, eiweißreiche Flüssigkeit, wie erstmals durch eine histochemische Untersuchung nachgewiesen werden konnte. — Eine Vakuolenbildung wird auch in den Leber- und Harnkanälchenepithelien beobachtet. An frischen Gefrierschnitten lassen sich in diesen Hohlräumen teilweise auskristallisierte und büschelförmig angeordnete Eiweißkörper nachweisen. Nach Alkoholfixierung lassen sich außerdem in mehreren Organen kleinere kristalline Gebilde im Bereich der Zellkerne feststellen. Offenbar besteht ein Zusammenhang zwischen der Menge dieser Kristallstrukturen und der Intensität des Eiweißstoffwechsels in den einzelnen Geweben. — Die Vakuolisierung des Nebennierenmarkes, der Leber- und Harnkanälchenepithelien und der Status spongiosus des Gehirns müssen, ebenso wie ein gestörter zellulärer Aufbau im Windungsbild einzelner Großhirnabschnitte und des ganzen Kleinhirnes, auf die Aminosäurenstoffwechselstörung zurückgeführt werden.

Histochemical Study at the Nervous System in Maple Syrup Disease. There was made a pathological anatomical investigation on a female infant, who died on the eighth day of life. A latent derangement concerning the aminoacid metabolism is known in the family anamnesis of both parents. A sister of the infant studied by us had died of the same disease two years ago on the tenth day after her birth. — Clinically seen chiefly a cerebral derangement had made itself conspicuous beside a solid secretion of the branched aminoacids Leucine, Isoleucine and Valine, a derangement to which the central nervousdiagn osis being anatomically as well striking corresponds. There is disturbance in developing of myelin especially in the phylogenetically young parts of the brain, which must have begun already in the fetal period. Besides a status spongiosus with fine pores was found in the white matter. The spongious structure can partly be found as well in the grey substance as in the marrow of the adrenal gland. — The cavities of the status spongiosus contain a liquid which is faint PAS-positive, easily soluble in water, and highly albuminous, as could be proved for the first time by a histochemical examination. — We observed a formation of vacuoles in the epithelium of liver and in those of the tubuli in the kidney. On fresh frozen sections albumin substance partly crystallized and being together in tufts, can be proved in these vacuoles. Crystals were found also in status spongiosus of brain as well as in some glia cells especially in foot of astrocytes near the capillaries. Besides, after fixation in alcohol small crystalline forms near the nuclei can be found in several organs. Evidently there is an interrelation between the great number of these crystal structures and the intensity of the albumin metabolism in the single tissues. — The formation of vacuoles in the marrow of the adrenal gland, in the epithelium of liver, and tubuli of the kidney and the status spongiosus of the brain as well as a disturbed structural formation in gyri of single cerebrum parts and of the whole cerebellum must be referred to the derangement of the aminoacid metabolism.

Pathologisches Institut der Universität, 69 Heidelberg (Deutschland).

Histochemistry of the Mitochondrial System

Chairman: TIBOR BARKA, New York (USA).
Co-Chairman: GÜNTHER BAHR, Washington (USA).

TIBOR BARKA: Introduction.

It has been customary to introduce any discussion on the mitochondrion by summing up its functions as the energy center of the cell. Because of the extensive reviews and symposia in this field in recent years such an introduction would be repetitious and will therefore be omitted. — The relative ease by which mitochondria can be isolated from various cells and the considerable purity of such fractions permitted an analysis of the overall functions of mitochondria and the extrapolation of these results, on a statistical basis, to a single mitochondrion. Further fragmentation of isolated mitochondria into sub-units that carry segments of the functional activity of the whole organelle significantly contributed to the understanding of this complex system. With the discovery of morphologic sub-units, using high resolution electron microscopic techniques, a stage is approached where the inextricable connection between structure and function becomes a tangible reality rather than a cliché, so widely used by biologists. — Concentration on the basic aspects of mitochondrial organization and the ingenuity and industriousness by which this approach is pursued have somewhat diverted our attention from problems faced at a higher level of organization. I am referring to such questions as the origin and fate of individual mitochondria, structural and biochemical variations within a cell and in various cells in an organ, intracellular interactions of mitochondria with other cell organelles, or in other words, the problems in which mitochondria with their structural and morphologic facets are viewed from cell and organ physiologic points of view. Here the potentialities of *in situ* techniques, including histochemical staining methods, are obvious. — The program of this symposium intends to reflect the two levels of operation at which investigations of mitochondrial systems are pursued. One refers to the basic problem of molecular organization. The other approach, including the use of histochemical techniques, is not refined at present to a molecular level of organization, but it nevertheless offers other parameters for the understanding of the biology of mitochondria. — In the followings Dr. ANDRÉ will discuss the general morphology of mitochondria. Dr. STOECKENIUS will review the evidence for the existence of morphologic sub-units. Dr. SIEBERT will focus on some biochemical aspects of mitochondria structure. Drs. BAHR and ZEITLER will deal with the problems of origin and fate of mitochondria, in general, while Dr. LUCK will present evidence indicating that mitochondrial populations increase by division of existing mitochondria and that the mitochondrial mass is augmented by addition of new units to the existing frame-work. Drs. FARBER and MORRISON's and Dr. BARRNETT's talks should provide us with the critical understanding of the potentialities of the *in situ* staining techniques. Finally, Dr. BALOGH assumed the difficult task of discussing how histochemical technics may contribute to the investigation of organ physiologic problems.

The Mount Sinai Hospital, 100th street and fifth Avenue, New York 29, N.Y. (USA).

JEAN ANDRÉ: The Morphological Basis of Mitochondrial Physiology.

Mitochondrial fine structure varies from organ to organ and from organism to organism. At the present time, a classification based on phylogeny is impossible, whereas a classification based on organic function shows some features in common. — A number of normal, pathologic and experimental processes cause mitochondrial structure to change. Mild changes consist in an alteration of the matrix or a deposition of materials in the matrix; extreme changes constitute a metamorphosis during which the conventional structures disappear and are subsequently replaced by an entirely different pattern; bodies exhibiting three-dimensional periodicity may occur; substances usually absent in mitochondria may appear. A striking example of the latter is the accumulation of PAS-positive material. The fate of the enzymatically active molecules during these changes is not known. It is suggested that they become organized into the structures having three-dimensional periodicity. — The localization of the enzyme molecules in normal mitochondria is in the process of being elucidated; granules similar to those seen after negative staining, can now occasionally be seen in sectioned material. The possibility that these granules represent enzyme molecule assemblies is discussed. — The morphologic data which would account for the presence of nucleic acids are scant; they are discussed in light of genetic data. — The problem of mitochondrial duplication is difficult to solve from a morphological approach. Favourable cases however, tend to show that mitochondria are formed by growth and subsequent division of preexisting mitochondria.

Institut de Recherches sur le Cancer, Villejuif (France).

WALTHER STOECKENIUS: Structure of Mitochondrial Membrane Surfaces.

Mitochondria consist of a matrix enclosed by the mitochondrial envelope. The envelope is made up of an inner and an outer membrane which differ slightly in their morphology, but both show the typical unit membrane structure. The cristae appear as evaginations of the inner membrane of the envelope and have the same morphological characteristics. In negatively stained preparations roughly spherical particles of ~85 Å diameter are observed on the surface of the membrane which faces the matrix. They are connected to the membrane proper by a short stem. The space between the inner and outer membrane varies in width according to the preparation procedures, but is usually narrow or completely obliterated. Its actual width *in vivo* is unknown. Contradictory views have been published on the functional significance and chemical composition of the particles observed on the inner membrane surface. It seems that at present no definite conclusions on this point can be reached.

The Rockefeller Institute, New York, N.Y. (USA).

GÜNTHER SIEBERT: Biochemical Aspects of Mitochondrial Structure.

Biochemistry contributes to a more profound understanding of the relationship between structure and function in mitochondria as will be illustrated by the following three examples: 1) the sub-units of mitochondria, defined primarily as functional units are linked together by such compounds as phospholipids and proteins. A recombination of isolated sub-units to form higher functional complexes is feasible and has been achieved. Some of the consequences of the existence of and interaction among sub-units with regard to the kinetics of metabolic processes catalyzed within mitochondria will be considered. 2) The mitochondrial membrane is apparently impermeable for pyridine nucleotides. On the other hand the nucleus is the only site of NAD synthesis. The nucleomitochondrial interaction will be discussed for pyridine nucleotides. A hydrogen shuttle catalyzed by extramitochondrial glycerophosphate dehydrogenase and intramitochondrial glycerophosphate oxidase may explain how the permeability barrier for pyridine nucleotides is overcome functionally in the cytoplasmic-mitochondrial interaction. 3) The elaborate internal structure of mitochondria is reflected biochemically by the detection of metabolic compartments. The different metabolic pools of ATP within mitochondria will be discussed in order to illustrate this phenomenon. Permeability of the mitochondrial membrane for adenosine phosphates is a requirement for a power plant-like function.

Dept. of Physiological Chemistry, Johannes Gutenberg University, 65 Mainz (Germany).

GÜNTHER F. BAHR and ELMAR H. ZEITLER: Origin and Fate of Mitochondria; Facts and Speculations*.

In reviewing the current concepts of mitochondrial origin, namely: 1) *De novo* generation of mitochondria from particles that are beyond the resolving power of the light microscope; 2) Derivation of mitochondria from the nuclear or cellular membranes or the endoplasmic reticulum; 3) Development of mitochondria via an intermediate step involving so-called microbodies; 4) Omnis mitochondrium e mitochondrio, the authors find the evidence for the latter alternative the most convincing one. — All support for alternative (1) stems from light microscopic observations, but the essential events remain inaccessible to observation. The literature contains no biochemical or electron microscopical evidence in favor of (1). — Situations in support of (2), observed with the electron microscope, most reasonably reflect cell ecology by proximity of energy donor to energy consumer. — Microbodies (3) most probably belong to the group of cytoplasmic granules generally termed lysosomes. It is highly unlikely that these particles would divest themselves of all enzymes in acquiring an entirely new biochemical function and a new macromolecular structure. — There is older light microscopic evidence that mitochondria divide (4), and more recently biochemical, cytochemical, and qualitative as well as quantitative electron microscopic data have accumulated that are unquestionably in favor of the concept that mitochondrial material and structures derive from pre-existing mitochondria only. — The amount of mitochondrial mass per cell is a fair indicator of cellular energy requirements. This reflects itself in increasing mass during cell growth, constant mass in nonproliferating tissues, and decreasing mitochondrial requirements, as in keratinizing cells or erythrocytes. If the mitochondrion is defined as the cell's principal energy producer, cytopathological events—as, for example, edema, rhexis, shedding of the exterior membrane, and uptake or accumulation of material alien to its normal function—may end its existence as a mitochondrion.

* This investigation was supported in part by grants from the American Cancer Society and the United States Atomic Energy Commission and in part by a research contract, Project Number 3A012501A818, from the Medical Research and Development Command, U.S. Army, Washington, D.C. 20310.

Armed Forces Institute of Pathology, Washington, D.C. 20305 (USA).

David J. L. Luck: Formation of Mitochondria in Neurospora crassa*.

Radioautographic studies on mitochondria of logarithmically growing cultures of *Neurospora* (Luck, J. Cell Biol. *16*, 483, 1963) support the hypothesis that in vegetatively growing cells, the mitochondrial mass is increased by addition of new lipid units to the existing framework. These studies also suggest that the number of individuals in the population is increased by division which distributes existing mitochondrial material uniformly, or nearly uniformly, to progeny mitochondria. The insertion of new material apparently takes place at many points; the mitochondrion has no growing tip. — Based on these findings and interpretations, it is possible to ask whether the mitochondrion can be viewed as a growing mosaic of equivalent functional units of fixed chemical composition (as implied by many current views of mitochondrial organization). Experiments with mutant and wild type strains indicate that availability of precursor substances can substantially alter structure and composition of mitochondria, without seriously altering their functional capacity. — In a choline requiring mutant, for example, growth at two levels of supplemental choline (both adequate for maximal growth) produces mitochondria which show a twofold variation in phospholipid to protein ratio and a significant density difference which is easily resolved by equilibrium sedimentation in sucrose gradients. It is possible to take advantage of these density differences to test models of mitochondrial biogenesis. When cultures grown at low choline conditions (in which low density mitochondria are produced) are shifted to high choline conditions (in which high density mitochondria are produced), two possible results can be expected: either a new mitochondrial population with density equivalent to high choline will be formed (*de novo* formation, or formation from intracellular membrane precursors) without change in density of existing mitochondria, or there will be a gradual modification, as a single population, of the density of existing mitochondria toward the lower density (accretion-division). The second result was obtained. — The findings indicate that the cell does not use a rigid chemical formula in the production of whole mitochondria and can adjust the chemical composition of these organs independently of cell division.

* Supported by a grant from the National Science Foundation.

The Rockefeller Institute, New York, N.Y. (USA).

Emmanuel Farber and Margaret Morrison: The Localization of Oxidative Enzymes in Mitochondria by Histochemical Staining Procedures.

Many methods have been reported for the cellular localization of oxidative enzymes with the use of tetrazolium salts. These enzymes fall into two groups (1) flavoproteins which react directly with the appropriate tetrazolium salt or other indicator and (2) NAD (DPN)- or NADP (TPN)-linked dehydrogenases which react only indirectly through an appropriate flavoprotein and must, therefore, be firmly bound to the flavoprotein. Of the former, control studies have clearly shown that the mitochondrial enzymes, succinic dehydrogenase, α-glycerophosphate dehydrogenase (non-NAD linked), and choline dehydrogenase, as well as D-amino acid dehydrogenase (oxidase) and NADH- and NADPH-tetrazolium reductases (diaphorases) can be accurately localized. Each of these show distinctive distribution patterns in many organs. Of the latter, the only enzyme that is accurately and reproducibly localized is the tightly-bound NAD-linked mitochondrial enzyme, β-hydroxybutyric dehydrogenase. Great uncertainty still exists as to the accuracy of localization of all other NAD- and NADP-linked dehydrogenases because (a) the identity and cytochemical distribution of the NADH- and NADPH-tetrazolium reductases with which the non-flavoprotein dehydrogenases must react are not known and (b) many become solubilized and possibly displaced during incubation. These two difficulties will be discussed in detail and an analysis will be presented to show the inherent difficulties in the accurate cytochemical localization of NAD- or NADPH-linked dehydrogenases.

Dept. of Pathology, University of Pittsburgh School of Medicine, Pittsburgh, Pennsylvania, 15213 (USA).

Russell J. Barrnett and Kazuo Ogawa: Cytochemical Demonstration of Oxidative Enzyme Activity in Relation to the Fine Structural Elements of Mitochondria.

This presentation will consider the problems concerned with the fine structural demonstration of oxidative enzyme activity (succinic dehydrogenase and DPNH diaphorase) in mitochondria of tissue cells and of mitochondria isolated by differential centrifugation. This data will be compared and contrasted with data on the fine structural detail of isolated and purified oxidative enzyme systems. Finally, some current experiments to demonstrate oxidative enzyme activity in relation to the elementary particles of heart muscle mitochondria will be presented.

Depts. of Anatomy, Yale University School of Medicine, New Haven, Connecticut (USA), and Kyoto University School of Medicine, Kyoto (Japan).

KÁROLY BALOGH: Functional Implications in the Histochemistry of Mitochondrial Enzymes in Endocrine Systems.

Histochemical methods for the localization of mitochondrial enzyme activity not only provide information on normal distribution patterns, but also can be useful in experimental studies and in disease. This is especially true for tissues where structural complexity allows the histochemical method to provide information not obtained by the use of routine quantitative biochemical techniques. — There is no evidence that mitochondrial enzymes are specifically related to hormone synthesis, but factors affecting the metabolism of endocrine systems or their target organs can produce selective alterations in enzyme activities which might be detected by histochemical methods. For instance, gonadectomy depresses the activity of several enzyme systems in the target organs, and this effect can be reverted by substitutive hormone therapy. It is well known that onkocytes in thyroid and parathyroid glands are filled with mitochondria and show strikingly high levels of mitochondrial enzyme activity; nevertheless, onkocytes are considered not to have any endocrine function. These and other examples illustrate the use of histochemistry in correlating morphology with cell function. It should be emphasized that histochemical observations have to be critically compared with pertinent data obtained by other investigative methods.

Dept. of Pathology, Harvard Medical School, Boston, Massachusetts (USA).

Histochemistry of Carbohydrates in the Intercellular Space

Chairman: J. A. SZIRMAY, Leiden (The Netherlands).
Co-Chairman: E. A. BALAZS, Boston, Mass. (USA).

Participants: S. GARDELL, Lund (Sweden), G. QUINTARELLI, Birmingham (USA), J. E. SCOTT, Taplow (England), S. SPICER, Bethesda (USA).

J. A. SZIRMAI: Introduction.

The intercellular space of various connective tissues is characterized — among others — by the presence of several types of macromolecular carbohydrates. The knowledge of the chemical structure of these substances, of their physico-chemical properties and metabolism forms the basis for the elucidation of their biological function, but this knowledge is still far from complete. Especially concerning the microscopical and submicroscopical organization of these compounds, many essential data are still lacking. What contributions can be expected from the application of histochemical methods in this field? To what extent is it possible, by the histochemical approach in the widest sense, to locate these macromolecules and to distinguish individual carbohydrates or carbohydrate-protein complexes?

Before considering the potential value of histochemical methods of the staining type, it is necessary to pay attention to the histological fixation. In this respect the behaviour of substances in the intercellular space can be fundamentally different from that of cellular structures. How can one fulfill the classical criteria of tissue fixation and preserve at the same time these macromolecular components as required for their histochemical detection and identification? How far can our present knowledge of the structure and physical chemistry of these substances be utilized in this respect?

Since histochemical reactions should be based on chemical or physico-chemical principles, their exact nature and quantitative character must be known. The staining of a number of substances in question with cationic dyes (metachromasia or basophilia) is based on their properties as anionic polyelectrolytes. Under certain conditions, this interaction appears to be quantitative and fulfills the criteria of a specific histochemical test; however, unsuitable fixation and other circumstances of the histological procedure easily disturb the quantitative nature of the staining, whereas in general this type of reaction suffers from lack of specificity in distinguishing individual macromolecules. How far can one overcome these shortcomings and what are the possibilities of distinguishing individual carbohydrate macromolecules by, for example, blocking of the reactive groups or by the use of specific enzymes?

Other staining principles — such as the ones based on periodate oxidation and subsequent demonstration of aldehyde groups — have become increasingly popular in the past decade. In many cases they proved to be of significant value, yet often their uncritical use has resulted in confusing information. The exact mechanism and the quantitative nature of these procedures require critical reappraisal.

For many investigators, the histochemical approach is synonymous with staining procedures at the light microscope level. Nevertheless several other approaches, originating from histochemistry in a wider sense, seem applicable, such as the use of physical-optical procedures or the methods of micro- and ultramicrochemistry. Which principles would have potential value in this respect? Which contributions can be expected from, for example, microchemistry in providing new data or in assessing the validity of the classical histochemical tests?

These problems and questions are essential for the histochemist when evaluating and assessing the validity of his tools; furthermore, they might be of importance in attempts to clarify the structural organisation of the intercellular space and its biological functions. In spite of its modest value at present, histochemistry should be able to contribute to this field, especially when its results will be integrated with those obtained by other disciplines.

Research Laboratories, Dept. of Rheumatology, University Hospital, Leiden (The Netherlands).

Montag, Monday, Lundy, 17. August 1964

Kreislauforgane · Organs of Circulation · Organs de circulation

1.

ALFONSO GIORDANO, ANGELO CANTABONI, CARLO OKELY and FRANCO RILKE: Histo-chemical Investigation on the Mucopolysaccharides of the Aortic Wall in the Fetus and Newborn.

The aortas of feti in the last gestational months, and of newborn babies up to one year of age, were investigated and special attention was paid to mucopolysaccharides of the ground substance. The cases were devided into three groups: normal subjects, cases with congenital malformations of the cardiovascular and other systems, and subjects with infectious diseases.— During the last months of fetal life the aortic wall undergoes continuous changes consisting of increase of the ground substance and of progressive separation of the elastic layers with alterations of their regular parallel disposition. — Maturative processes may be disturbed by infectious diseases which are able to induce changes directly or by means of the endo-crine system (adrenal glands) in the ground substance such as focal accumulations of chromo-trope substance in the intima and in the media, destruction of elastic systems and increase of collagen fibers. — The first change in fetal aortic walls due to pathological factors is re-presented by an alteration in the composition of the ground substance, which appears stained deeper by Alcian-blue on account of an increase of the degree of polymerisation of muco-polysaccharides. — Special importance is attributed to this kind of changes in the aortic wall, before or after birth within the first year of age, and to their possible significance in view of later alterations to which the vessel may undergo in adult life.

Istituto di Anatomia e Istologia Patologica della Università degli Studi di Milano, Via Francesco Sforza 38, Milano (Italy).

2.

EEVA LEVONEN and KIMMO K. MUSTAKALLIO: In Vitro Interaction between Chlortetracycline, Calcium and Phospholipoproteins in Sections of Human Aorta*.

Tetracyclines have affinity for sites where calcium is known to deposit, like on the sur-face of hydroxyapatite crystals in bone (URIST and IBSEN 1963) and at the sites of necro-biosis (MALÉK 1962). The phospholipids are known to be cation exchangers (WOOLLEY and CAMPBELL 1962) and FELS (1961) suggests that the insoluble phospholipoproteins bind calcium in aorta. — We treated acetone fixed cryostat sections of human aortas with 1 mg/ml solutions of different tetracyclines in water. Ultraviolet fluorescence was seen only in deposits of calcium, demonstrable with Morin. When chlortetracycline was dissolved in 0,01 M calcium chloride, the sections treated with it showed yellow tetracycline fluorescence also in the early atherosclerotic plaques in intima where no calcium could be demonstrated. — No affinity of intima for chlortetracycline was present in sections treated with boiling chloro-form-methanol. Obviously the substance binding calcium chelate of chlortetracycline is of lipid nature. The phospholipid stains Luxol Fast Blue and Rhodamine 6 G and also fluorescein labelled anti human β-lipoprotein accumulated at the sites corresponding to the chlortetracycline-calcium fluorescence. — LACKO et al. (1959) have demonstrated in serum precipitation of lipoproteins with chlortetracycline in presence of calcium. Apparently the chlortetracycline-calcium is precipitated also by tissue phospholipoproteins.

* Aided by a grant from the Emil Aaltonen Foundation, Tampere, Finland.

Dept. of Forensic Medicine, University of Helsinki, Snellmaninkatu 10, Helsinki (Finland).

3.

TOICHIRO KUWABARA and DAVID G. COGAN: Histochemistry of the Retinal Capillary*.

The wall of the retinal capillary consists of two kinds of cells: endothelium and mural cells. Selective degeneration of the latter cells was specifically seen prior to the vascular pathology of diabetic retinopathy. These findings have been reported in a series of papers. — Fresh intact retinas were incubated in various media to demonstrate tetrazolium reductases with glycolytic and TCA cycle substrates, glycogen synthesis, phosphatases, and esterase.

Formalin fixation of the incubated tissue was followed by the trypsin digestion technique. — Tetrazolium reduction with the presence of DPN and the substrates of the glycolytic pathway was conspicuously active in the mural cells. Higher glycogen synthesis activity was also demonstrated in the mural cells. Intense staining with the lactic acid dehydrogenase system revealed the whole view of the mural cell with its fine processes extending in the capillary wall. Activity and distribution of other enzymes are about the same in both endothelial and mural cells. Species differences and pathological changes will also be demonstrated.

* Supported by a PHS Research Grant No. NB 03015-04 from the National Institute of Neurological Diseases and Blindness, US Public Health Service.

Howe Laboratory of Ophthalmology, Harvard University Medical School, Boston, Massachusetts (USA).

4.

JÜRGEN SCHLÜNS: Zum Vorkommen von C—N-Bindung spaltenden Hydrolasen in den Hülsenzellen der Schweigger-Seidelschen Kapillarhülsen der Milz.

Die Hülsenzellen der Schweigger-Seidelschen Kapillarhülsen werden von verschiedenen Autoren als besonders aktive Anteile des Retikuloendothelialen Systems angesehen. Die häufig zu beobachtenden phagozytierten Partikel, etwa Erythrozyten, in diesen Zellen, scheinen eine solche Auffassung zu stützen. Über den histochemischen Nachweis C—N-Bindung spaltender Hydrolasen in den Hülsenzellen, die an dem Abbau derartiger phagozytierter Substanzen und an der Resynthese von Peptiden beteiligt sein könnten, liegen bisher in der Literatur keine Angaben vor. Unlängst konnten wir in den Hülsenzellen der Schweinemilz eine γ-Glutamyltranspeptidase-ähnliche Enzymaktivität aufzeigen, die jedoch bei Huhn, Taube und Sperling unter unseren Versuchsbedingungen nicht nachzuweisen war. Im vorliegenden wird über die Darstellung von Aminosäurenaphthylamidaseaktivität mit der Methode von BURSTONE und FOLK (Substrat: L-Leucin-β-naphthylamid, Trispuffer pH 7,1, Inkubationsdauer 2 Std bei 20⁰ C) in den Hülsenzellen von Huhn und Taube berichtet. Nach der angegebenen Inkubationszeit zeigen die Hülsenzellen starke Anfärbung. Daneben lassen noch einzelne Retikulumzellen der roten Pulpa schwache Aktivität erkennen. Die Enzymaktivität wird durch 0,05 m Dinatriumäthylendiamintetraessigsäure oder 0,001 m Kupfersulfat stark, durch 0,005 m Jodessigsäure oder 0,005 m p-Chlormercuribenzoat schwach gehemmt. Mit 0,002 m Natriumfluorid oder 0,001 m Diisopropylfluorphosphat ist keine Inhibition zu erzielen. Da aus der Literatur bekannt ist, daß L-Leucin-β-naphthylamid nicht spezifisch von Leucinaminopeptidase gespalten wird, haben wir Hühnermilzhomogenat mit der Stärkegelelektrophorese nach SMITHIES untersucht. Es lassen sich bis zu drei Aminosäurenaphthylamidasefraktionen trennen, deren Herkunft und Eigenschaften noch weiter differenziert werden.

The Occurrence of C—N Linkage Splitting Hydrolases in the Cells of the Schweigger-Seidel Sheaths of the Spleen. The cells of the Schweigger-Seidel sheaths are regarded by some authors as particularly active parts of the reticuloendothelial system. This idea seems to find support in the fact, that phagocytosed particles, such as erythrocytes, can frequently be observed in these cells. So far there are no data published concerning the histochemical determination of peptidases in the sheath cells which could take part in the decomposition of these phagocytosed particles and in the re-synthesis of peptides. Recently we succeeded in demonstrating a γ-glutamyltranspeptidase-like substance in the sheath cells of pig spleen. Under the same experimental conditions it was, however, not possible to demonstrate a similar enzyme activity in chicken, pigeon, or sparrow. Now we would like to describe the determination of amino-acid naphthylamidase activity in the sheath cells of chicken and pigeons. The method of BURSTONE and FOLK was used; substrate: L-leucyl-β-naphthylamide; tris-buffer pH 7.1; incubation for 2 hours, 20⁰ C. After incubation a strong reaction could be observed in the sheath cells. Some reticulum cells of the red pulp also showed weak activity. Disodium ethylenediaminotetra-acetic acid (0.05 M) or copper sulphate (0.001 M) had a strong inhibitory effect on the enzyme activity whereas 0.005 M iodoacetic acid or 0.005 M p-chloromercuribenzoic acid caused a weak inhibition only. Sodium fluoride (0.002 M) or 0.001 M di-isopropylfluorophosphate had no effect at all. As is well known leucine aminopeptidase does not split L-leucyl-β-naphthylamide specifically. For that reason we examined a homogenate of chicken spleen with the starch gel electrophoresis technique (SMITHIES). It was possible to isolate up to three amino acid naphthylamidase fractions, the origin and properties of which have to be further investigated.

Veterinäranatomisches Institut der Freien Universität Berlin, Histologisch-Embryologische Abteilung, Drosselweg 1—3, 1 Berlin 33 (Deutschland).

5.

A. Tóth* und T. H. Schiebler: Histochemische Untersuchungen am Herzmuskel und am Reizleitungssystem während der Entwicklung.**

Alle Muskelfasern von Rattenherzen enthalten bis zum 10. Tag nach der Geburt reichlich histochemisch nachweisbares Glykogen. Für das Meerschweinchenherz gilt dies nur bis zum 48. Embryonaltag. In der Zeit des reichlichen Glykogenvorkommens sind in den Herzmuskelfasern beider Tierarten nur sehr wenig Cytochromoxidase, Bernsteinsäuredehydrogenase und β-Hydroxibuttersäuredehydrogenase vorhanden. Die DPNH-Reaktion ist kräftig, wenn auch schwächer als beim erwachsenen Tier. Die Herzkapillaren enthalten keine alkalische Phosphatase. In der Folgezeit — bei der Ratte beginnend am 11.—12. Lebenstag, beim Meerschweinchen etwa ab 50. Embryonaltag — nimmt der histochemisch erfaßbare Glykogenbestand von der Herzoberfläche her zum Endokard hin fortschreitend ab. Bei der Ratte ist am Ende der 3. Lebenswoche, beim Meerschweinchen schon kurz vor der Geburt, Glykogen in der Arbeitsmuskulatur der Kammerwand kaum noch nachweisbar. Reichlich Glykogen behalten während des ganzen Lebens nur die Reizleitungsfasern. Die Vorhofsmuskulatur ist viel länger glykogenreich als die Kammermuskulatur, beim Meerschweinchen auch im ausgewachsenen Zustand. Der Glykogenabnahme folgt in den glykogenfreien Zonen der Arbeitsmuskulatur eine fortschreitende Zunahme von Cytochromoxidase, Bernsteinsäuredehydrogenase und β-Hydroxibuttersäuredehydrogenase, sowie in den Herzkapillaren von alkalischer Phosphatase. Das Reizleitungssystem bleibt stets arm an den genannten Fermenten. — Vorliegende Befunde lassen offen, ob die Änderung des histochemischen Glykogenmusters des Herzens auf eine Änderung der Glykogenzusammensetzung oder der Glykogenmenge zurückzuführen ist.

A Histochemical Investigation of Cardiac Muscle and the Impulse Conducting System during the Course of Development. Up to the 10th day after birth all muscle fibres of rat hearts contain abundant glycogen, which can be demonstrated histochemically. In Guinea pigs, however, this applies only up to the 48th embryonic day. At the time when there is abundant glycogen present, only small amounts of cytochrome oxidase, succinate dehydrogenase and β-hydroxybutyrate dehydrogenase are present in the cardiac muscle fibres of both species. The DPNH-reaction gives positive results which are, however, weaker than those obtained in adult animals. No alkaline phosphatase is found in the cardiac capillaries. During the following period — in rats beginning with the 11th/12th day of life, in Guinea pigs round about the 50th embryonic day — the histochemically demonstrable amount of glycogen will decrease; this decrease starts at the surface of the heart and moves progressively towards the endocardium. At the end of the third week of life (rats) or just before birth (Guinea pigs), hardly any glycogen will be found in the endocardium of the ventricle. Only the fibres of the impulse conducting system retain ample glycogen for the complete lifetime. The musculature of the atrium will contain glycogen for a longer period than does the musculature of the heart chamber. In Guinea pigs this applies also to adult animals. The decrease in glycogen is followed by a progressive increase in cytochrome oxidase, succinate dehydrogenase and β-hydroxybutyrate dehydrogenase in the glycogen-free regions of the endocardium. At the same time an increase in alkaline phosphatase will be observed in the cardiac capillaries. The impulse conducting system continues to contain very little of the enzymes mentioned. — The present findings do not show whether the changes in the histochemical glycogen pattern of the heart are due to an alteration in the structure of the glycogen or to an alteration in its quantity.

* Stipendiatin der A. v. Humboldt-Stiftung.
** Mit Unterstützung der Deutschen Forschungsgemeinschaft.

Anatomisches Institut der Universität, Koellikerstraße 6, 87 Würzburg (Deutschland).

6.

P. P. Rumyantsev: An Autoradiographic Study of the Problem of Myocardial Growth and Regeneration Using H³-Thymidine.

Myocardial histogenesis, unlike that of skeletal muscles, is characterized by a prolonged persistence of the high index of labelled nuclei. The "saturation" with H³-thymidine (repeated injections) results in the labelling of 40—50% of the myocardial nuclei of newborn rats, which is in agreement with a much slower accumulation of the myocardial actomyosin blocking DNA synthesis. — 1—32 days after injury followed by single or repeated H³-thymidine injections DNA synthesis in the bulk of the myocardial nuclei of adult mice is tightly blocked despite the increase in the size of nuclei and nucleoli, RNA accumulation etc. In contrast to this nuclei label of injured skeletal muscles in the mice "saturated" with H³-thymidine is extremely high. — In less differentiated adult frog myocardium reactive DNA synthesis occurs in numerous muscle nuclei (5—40 days after the trauma) and is followed by their mitotical division and slow redifferentiation of newly-formed portions of muscle fibers. — The labelling of frog M. sartorius regenerating fibers nuclei with H³-thymidine is limited within a shorter period of time; it begins after the 5th day and practically ends on the 16th—20th day. — Thus, both myocardial histogenesis and regeneration are characterized by slow differentiation of fibers and prolonged period of nuclei proliferation (DNA synthesis). It seems to be adaptation providing for the growth in the system with a constant loss of energy for contractions. — Differences in the regeneration capacity of the myocardial and skeletal muscles most strongly pronounced in mammals seems to be related to a lesser degree of its dedifferentiation after injury which can be due to higher resistance of myocardial actomyosins. — Autoradiographical data support the idea that the blocking of DNA synthesis in myocardial nuclei is related not to the onset of actomyosin synthesis but to accumulation of some "limiting mass" of this protein.

Laboratory of Cell Morphology, Institute of Cytology of the Academy of Sciences of the USSR, Leningrad, F 121, Prospect Maclina, 32 (USSR).

7.

FERENC JOÓ: Histochemical and Ultrastructural Features of the Intercalated Discs in Mammalian Cardiac Muscle.

Recent electronmicroscopical investigations prove the non-syncytial character of the cardiac muscle. It has been shown that the intercalated discs are specialized junctions between cellular units of the myocardium. Electrophysiological investigations published by SPERELAKIS suggest that there are no low resistance pathways for spreading of excitation between adjacent cardiac cells, thus the heart muscle can not be regarded as a functionals syncytium either. In our previous works it has been shown that the intercalated discs display a well-defined acetylcholinesterase activity, probably associated with the conduction of excitation in a similar way as in the myoneural junction. In the present study acetylcholinesterase activity, PAS-positivity and a "lead-reactive substance", all characterizing peripheral synapses operating with a neurohumoral transmission mechanism, could be demonstrated histochemically in the intercalated discs both of auricle and ventricle muscles. The ultrastructure of the intercalated discs has been investigated under the polarization microscope by means of ROMHÁNYI'S precipitation anisotropic staining. It was found that the arrangement of protein and lipid parts of the intercalated discs show characteristic differences in the auricle as contrasted with the intercalated discs of the ventricle. Further investigations revealed that the more complicated molecular organization of the intercalated discs in the ventricle develop dynamically during postnatal life by mechanical effects exerted by the myofibrils. During this period oxidative enzymes appear in the intercalated discs.

Dept. of Anatomy, University Medical School, Szeged (Hungary).

8.

D. ONICESCO, I. DICULESCO and R. WEGMANN: Variation of Dehydrogenase Activity in Cardiac Muscle Tissue.

The activity of several dehydrogenases in the heart of fish, frog, chicken, mouse, dog, horse and man were investigated. The dehydrogenase activity variations were followed up in the ontogenetic development of the heart of chicken and human embryo, as also in the development of the experimental infarctus (from 3 hours to 7 days). — Characteristic for the adult heart is the intense activity of the dehydrogenases of the Krebs cycle, DPNH-diaphorase and dihydrothiocticdehydrogenase as also the very low activity of dehydrogenases linked to the anaerobic utilization of pyruvic acid. — These characteristics can be very well observed both in the ontogenetic as also in the phylogenetic development of the heart. — During the development of the experimental myocardial infarct the enzymes of the anaerobic metabolism of pyruvic acid are activated.

Facultatea de Medicină, Catedra de Histologie, Bd. Dr. Petru Groza 8, Bucharest (Rumania).

9.

LUCILLE BITENSKY, N. R. NILES and J. CHAYEN: The Histochemical Detection of Changes in Human Myocardium during Operation.

There has been no explanation of the post-operative morbidity frequently encountered with prolonged open-heart surgery. Thus it seemed likely to be due to a sub-cellular biochemical lesion of the myocardium. Since only minute biopsies could be obtained during the operation, the investigation had to depend on rigorous histochemical procedures performed on specially prepared sections rather than on grosser biochemical techniques. — It has been shown that, in relatively healthy human myocardium, there is a system by which hydrogen, removed by succinic dehydrogenase, is transported directly to the myofibril. The effect of prolonged open-heart surgery is to reduce this flow of energy from the sarcosomes. Concomitantly, the localization of the activity of adenosine triphosphatase, which previously had been restricted to the bands of the fibres, appeared to become distorted and clumped. Another effect was the progressive unmasking of phospholipid throughout the muscle. The possible relationship between these changes, with particular respect to oxidative phosphorylation and the supply of energy for contraction, will be discussed.

Royal College of Surgeons of England, London, W.C.2 (Great Britain).

10.

A. I. STRUKOV, Moscow/USSR: Dynamics of Metabolic Derangements in Myocardial Infarction.

Tagesrhythmus · Diurnal Rhythm · Rythme quotidien
Lunge · Lung · Poumon
Verschiedenes · Divers · Divers

1.

G. HORVÁTH: Quantitativer und qualitativer Nucleinsäurebestand normaler Rattenorgane.

Leber, Milz, Skelettmuskel und Hirn von Ratten wurden auf den RNS- und DNS-Gehalt zu verschiedenen Tageszeiten biochemisch und histochemisch untersucht. Diese Organe normaler Ratten stammten aus männlichen und weiblichen, genetisch homogenen Tiergruppen gleichen Lebensalters (4 Monate $\pm$ 4 Tage), welche unter stabilisierten Bedingungen (standardisiertes Rattenfutter ad libitum, gleiche Temperatur $\pm$ 0,5° und gleiche Luftfeuchtigkeit) gehalten wurden. Zur Untersuchung wurden die Tiere während 24 Std in 2 Std-Intervallen durch Genickschlag getötet und entblutet. — Für die beiden untersuchten Substanzen besteht in Leber und Milz eine ausgesprochene Tagesrhythmik, wobei sowohl die RNS als auch die DNS Quantitätsschwankungen bis zu 300% aufweisen. Bei Muskel und Hirn ist der Nucleinsäuregehalt ebenfalls nicht stabil. — In der Tagesrhythmik herrscht ein bedeutender Sexualunterschied, wobei die sexualspezifischen Kurven von der Jahreszeit abhängen. Es kommen ferner Qualitätsunterschiede der DNS vor. Die biochemisch objektivierten Quantitätsunterschiede erklären die histochemischen Erscheinungsbilder, welche mit den biochemisch gewonnenen Werten verglichen wurden.

The Quantitative and Qualitative Content of Nucleic Acids in Normal Rat Organs. Using biochemical and histochemical methods liver, spleen, skeletal muscle and brain were tested for their RNA and DNA content at various times of the day. These normal rat organs were obtained from genetically homogeneous male and female animals of the same age (4 months $\pm$ 4 days) which were kept under stabilized conditions (standardized food *ad libitum*, constant temperature $\pm$ 0.5°, and constant atmospheric humidity). At 2 hour intervals the animals were killed (e.g. within 24 hours) by breaking the animal's neck and draining out the blood. It appeared that there was a marked diurnal variation of the two substances in question in liver and spleen; both RNA and DNA varied in quantity up to 300%. The nucleic acid content of brain and muscle was also unstable. — There was, however, a pronounced difference in the diurnal variation between the two sexes; the sex-specific pattern furthermore was seasonally dependent. There were also variations in the quality of the DNA. — The variations in quantity, as found by biochemical means, accounted for the histochemical findings which were compared with the biochemical results.

Laboratorium voor Cytologie en Histologie, Universiteit, Nijmegen (The Netherlands).

2.

CHR. JERUSALEM: Histometrische Untersuchungen zum Nucleinsäure-Stoffwechsel.

Es ist bekannt, daß einesteils Desoxyribonucleinsäuren quantitativ fast ausschließlich im Zellkern angetroffen werden und daß andererseits eine bestimmte Relation zwischen Kerngröße und DNS-Gehalt bestehen kann. Befunde über tageszeitliche Schwankungen von Total-DNS-Werten waren der Anlaß, gleiche Organe (Leber), an denen biochemische DNS-Bestimmungen durchgeführt wurden, auch karyometrisch zu untersuchen. Hierzu wurden insgesamt Lebern von 104 Ratten (im Alter von 4 Monaten $\pm$ 4 Tage) verwendet, von denen je 8 Tiere (4 ♀, 4 ♂) alle 2 Std (10^{00}—10^{00} h des folgenden Tagels) gleichzeitig getötet worden waren. Ferner wurden 20 zu verschiedenen Jahreszeiten gewonnene Leberpräparate von Ratten in die Untersuchung einbezogen. Es zeigten sich Ergebnisse, die besonders für cytophotometrische Untersuchungen von Interesse sind: tagesrhythmische Kerngrößenschwankungen sind nachweisbar. Sie erreichen jedoch nicht ein derartiges Ausmaß, daß allein hiermit die Veränderungen des Gesamt-DNS-Gehaltes erklärbar wären. Es treten zudem deutliche jahreszeitliche Schwankungen auf, die uns bei langfristigen Experimenten zur Gewinnung von neuem Kontrollmaterial nötigen. Ferner können derartige histotopographische Kerngrößenveränderungen beobachtet werden, die es bei vergleichenden Untersuchungen an der Leber notwendig erscheinen lassen, nur homologe Stellen (periportal, intermediär, perizentral) gegeneinander auszuwerten.

Histometrical Investigations on the Metabolism of Nucleic Acids. DNA is found quantitatively almost exclusively in the cell nucleus and there may be a definite relationship between the size of the nucleus and its DNA content. Biochemical investigations demonstrated diurnal variations in the total DNA content of the nucleus. The same material (liver) as used for the biochemical investigations was therefore tested kariometrically. A total of 104 rat livers was tested. The animals (aged 4 months ± 4 days) were killed in groups of 8 (4 ♀, 4 ♂) at intervals of 2 hours (10 a.m. to 10 a.m. of the next day). Twenty rat liver preparations, obtained at different times of the year, were also tested. The results were of special cytophotometrical interest: there were diurnal variations in the size of the nucleus. They were, however, not so marked as to explain the variation of the total DNA content. Furthermore there were pronounced seasonal fluctuations, which make it necessary to provide more control animals in long-term experiments. — Because of these variations in the size of the nucleus it is advisable to compare homologous parts only (periportal, intermediary, pericentral), when doing comparative studies of the liver.

Laboratorium voor Cytologie en Histologie, Universiteit, Nijmegen (The Netherlands).

3.

CH. PILGRIM, K. WEGENER, K. J. LENNARTZ und S. HOLLWEG: **Autoradiographische Untersuchungen über tageszeitliche Schwankungen der Zahl der DNS-synthetisierenden und in Mitose befindlichen Zellen in normalen Geweben der Maus, verglichen mit fetalen Zellen und Carcinom-Zellen.**

H-3-Thymidin ermöglicht eine selektive Markierung derjenigen Zellen, die sich im Stadium der DNS-Synthese befinden. Mit Hilfe dieses DNS-Vorläufers wurde untersucht, ob auch die Zahl der DNS-synthetisierenden Zellen tageszeitlichen Schwankungen unterliegt, wie sie für die Mitoserate bei Tieren und Pflanzen schon lange bekannt sind. Hierzu wurden Mäuse zu verschiedenen Tageszeiten mit H-3-Thymidin injiziert und kurze Zeit nach der Injektion getötet. Auf Autoradiogrammen verschiedener Gewebe wurde die Zahl der H-3-markierten Zellen und die Zahl der Mitosen bezogen auf die Gesamtzellzahl (H-3-Index bzw. Mitose-Index) als Funktion der Tageszeit bestimmt. Dabei ergaben sich in einigen Geweben beträchtliche Schwankungen des H-3- und Mitose-Index. Interessanterweise liegt das Maximum des Mitose-Index mehrere Stunden später als das des H-3-Index. Der Abstand der beiden Maxima entspricht etwa der zeitlichen Distanz zwischen der Mitte der S-Phase und der Mitte der Mitose. — Daraus wird geschlossen, daß die bereits bekannten tageszeitlichen Schwankungen des Mitose-Index mindestens zum Teil die Folge eines Synchronisationsprozesses sind, der eine Gruppe von Zellen bevorzugt aus der G_1-Phase in die S-Phase eintreten läßt. Es erscheint prinzipiell möglich, die Schwankungen des H-3-Index und möglicherweise aud die des Mitose-Index auf Schwankungen in der Länge der G_1-Phase zurückzuführen. — In diesem Zusammenhange wurde auch das Verhalten einiger fetaler Zellarten der Ratte und zweier Ascites-Tumoren der Maus untersucht. Im Gegensatz zu den normalen Zellen des erwachsenen Tieres waren in beiden Fällen der H-3-Index und der Mitose-Index zu den verschiedenen Tageszeiten konstant. Die tagesperiodischen Synchronisationen scheinen sich also, wie das auch für andere tagesperiodische Funktionen des Organismus bekannt ist, erst nach der Geburt zu entwickeln. Bei den Tumoren wurde das Fehlen dieser Schwankungen als Folge der mit der Cancerisierung verbundenen Verkürzung der G_1-Phase gedeutet.

Autoradiographic Investigations on Diurnal Fluctuations of the Number of DNA Synthesizing Cells and Mitoses in Normal Mouse Tissues, as Compared with Fetal Cells and Carcinoma Cells. H-3-thymidine permits selective labeling of cells engaged in DNA synthesis. It was investigated by means of this DNA precursor whether the number of DNA synthesizing cells is too subject to diurnal fluctuations, as has been known for the mitotic rate in animals and plants for many years. — Mice were injected with H-3-thymidine at intervals throughout the day and killed shortly after the injection. In autoradiographs of various tissues the percentage of labeled cells (H-3-index) and the percentage of cells in mitosis (mitotic index) were estimated as a function of the time of day. In some tissues considerable fluctuations of the H-3-index and the mitotic index were found. It is of interest that the maximum of the mitotic index follows that of the H-3-index by several hours. The interval between the two maxima corresponds approximately to the distance between the middle of the S-phase and the middle of the mitosis. — From that it is concluded that the well-known diurnal fluctuations of the mitotic index are at least partially a consequence of a synchronizing process which causes a cohort of cells to enter prematurely into S-phase from G_1-phase. On principle, it seems to be possible to refer the fluctuations of the H-3-index and perhaps also those of the mitotic index to fluctuations of the G_1-phase duration. — In connection with these investigations the behaviour of some fetal cell types of the rat and of two ascites tumours of the mouse was examined. In contrast to the normal cells of the adult animal, the H-3-index and the mitotic index at the different times of day were constant. Thus, the diurnal synchronisations seem to develop only after birth, as is known for other diurnal functions of the organism. As for the tumours, the lack of fluctuations could be due to the shortening of the G_1-phase which is connected with the process of cancerisation.

Institut für medizinische Isotopenforschung der Universität Köln, 5 Köln-Lindenthal, Kerpener Str. 15 und Pathologisches Institut der Universität Köln, 5 Köln-Lindenthal, Lindenburg (Deutschland).

4.

R. LESKE: Qualitative und quantitative Unterschiede des Leberglycogens.

Im Anschluß an die grundlegenden Untersuchungen von HOLMGREEN wurde der Glycogengehalt von Rattenlebern innerhalb zweier Jahre histochemisch und biochemisch untersucht. Es ergab sich, daß der Glycogentagesrhythmus wesentlich von der Jahreszeit und vom Geschlecht der Tiere beeinflußt wird. Die biochemisch-quantitative Bestimmung liefert demnach in jeder Jahreszeit und für jedes Geschlecht Kurven bestimmter Form, wobei auch die durchschnittliche Glycogentagesmenge innerhalb des Jahres schwankt. Unterschiedliche Glycogenpräparationen ergeben zu den einzelnen Jahreszeiten durchaus nicht immer gleiche Glycogenmengen. Dies verursacht Formunterschiede der Quantitätskurven nach verschiedener biochemischer Aufarbeitung (KOH- oder TCA-Extraktion). Daraus kann geschlossen werden, daß das Glycogen zu den einzelnen Tageszeiten Qualitätsunterschiede aufweist. — Die Stärke der histochemischen Reaktion steht meist in Einklang mit einem der mit biochemischen Methoden gewonnenen Werte. Das histochemische Bild und der daraus resultierende optische Eindruck wird darüber hinaus von der Fixierungsart beeinflußt. Auch hier läßt sich feststellen, daß einige der bekannten Glycogenfixantien zu den verschiedenen Tageszeiten unterschiedlich wirken. Offenbar beeinflussen die Qualitätsunterschiede des Glycogens wesentlich das histochemische Bild. Die topographische Verteilung des Glycogens in der Leber wird vom Biorhythmus bestimmt.

Qualitative and Quantitative Differences in Liver Glycogen. Following HOLMGREEN's fundamental studies, the glycogen content of rat livers has been investigated by histochemical and biochemical methods over a period of 2 years. The tests employed demonstrated that the diurnal variation of glycogen depended essentially on the season and the sex of the animal. Quantitative biochemical determinations therefore showed definite sex- and season-specific patterns. They also showed that the average daily amount of glycogen varied with the seasons. Different methods of processing used at different seasons did not always yield the same amount of glycogen. The results of quantitative determinations therefore vary with the biochemical method chosen (KOH or TCA extraction). It can be concluded, that the quality of the glycogen varies diurnally. — The intensity of the histochemical reaction was in most cases in agreement with the biochemical findings. In addition the histochemical picture was dependent on the fixative employed. Some of the well-known glycogen fixatives seemed to work differently at different times of the day. The histochemical result was evidently influenced by the quality of the glycogen. The topographic distribution of the glycogen in the liver was subject to diurnal variations.

Laboratorium voor Cytologie en Histologie, Universiteit, Nijmegen (The Netherlands).

5.

H. von MAYERSBACH, K. PHILIPPENS und YAP PAUL HONG-KIE: Die Einflüsse biologischer Tagesschwankungen auf fermenthistochemische Untersuchungen.

Der Einfluß von Biorhythmen auf den Ausfall histochemischer Reaktionen ist mit Ausnahme von Glycogenveränderungen völlig unbekannt. Systematische Untersuchungen an einem homogenen und unter standardisierten Bedingungen gehaltenen Tiermaterial (Ratten beiderlei Geschlechts) ergaben bedeutende tageszeitliche Unterschiede des histochemischen Reaktionsausfalles bei Anwendung der Nitro-Reaktion für Succino-Dehydrogenase sowie des 5-Brom-Indoxylazetates für Esterasen. Die Veränderungen betreffen sowohl die histo-topographische Verteilung der Enzyme in der Leber als auch die Art der zellulären Ferment-ablagerung sowie den quantitativen Gehalt. Die Fermentaktivität wurde weiters durch biochemische Analysen quantitativ bestimmt. — Qualitativ-histochemische und quantitativ-biochemische Veränderungen dieser beiden Enzyme liefern eine „Cirkadia"-(Tag-Nacht-) Rhythmik, wobei bedeutende Geschlechtsunterschiede zu beobachten sind. Die Aktivitäts-muster der einzelnen Geschlechter unterliegen weiter starken Saisoneinflüssen, obwohl die Tiere unter standardisierten Umweltsbedingungen (Futter, Temperatur, Luftfeuchtigkeit und Licht) gehalten wurden. — Die Form- und Größenänderungen sind so bedeutend, daß sie bei experimentellen Untersuchungen unbedingt berücksichtigt werden müssen; sie er-klären ferner die zahlreichen und in der Literatur oft divergierend angegebenen Beschrei-bungen von histochemischen Reaktionsausfällen.

The Influence of Biological Diurnal Variations on Histochemical Investigations of Enzymes. With the exception of glycogen and its metabolism, the influence of diurnal variations on the outcome of histochemical reactions is largely unknown. Systematic investigations, using a homogeneous strain of animals (rats of both sexes raised under standardized conditions), have shown considerable differences in the results of histochemical tests (e.g. Nitro-BT reaction for the determination of succinate dehydrogenase and acetyl-5-bromoindoxyl for esterases) if carried out at different times of the day. These changes involved not only the histotopographic distribution of the enzymes in the liver, but also the amount and the manner in which the enzyme is deposited in the cells. The activity of the enzymes was checked by quantitative biochemical analysis. — Qualitative histochemical and quantitative biochemical variations of these two enzymes resulted in a "Circadian" (night/day) rhythm. They showed, furthermore, an important difference between the sexes. The "activity pattern" in both male and female rats was largely influenced by the seasons, though the animals were kept under standardized environmental conditions (food, temperature, atmospheric humidity and light). — The changes in form and size were so considerable that they will have to be taken into consideration when doing similar experiments. They may also account for the various and often divergent descriptions of histochemical results which can be found in the literature.

Laboratorium voor Cytologie en Histologie, Universiteit, Nijmegen (The Netherlands).

6.

Marie De Groodt-Lasseel and Marcel Sebruyns: **Morphological and Histochemical Aspects of the Lipid Globules in the Alveolar Lining of the Axolotl Lung*.**

Lipid inclusions have been demonstrated in the alveolar lining of the *Axolotl's* lung. There is neither agreement about the chemical constitution, physical state and appearance of the alveolar lipid inclusions, nor about their significance. — Reinforced pulmonary venti-lation brings about an increase of the osmiophilic inclusions. They are regular outlined and stain positively as well with Sudan Black as by means of the Baker technique for demon-stration of phospholipids. Study with the electron microscope shows that those phospho-lipid inclusions consist of concentrically arranged layers, the periodicity of which, on a transversal section, is about 45 Å. Frequently the phospholipid globule begins to decompose into separate units suggesting the possibility that cyto-membranes may be derived from the globule material. Prolonged hyperventilation results into dissociation, fragmentation and desintegration of those layers. In the first stages of hyperventilation numerous mitochondria are found in a close association with the lipid granules. When the latter increase lysosomes appear besides them. — These results suggest that the appearance of lipid globules in the alveolar wall of the lung visualizes an initial step of a progressive reaction. The enzyme-rich organelles, which accumulate at the sites of phospholipid granules, probably participate in the process of building up and decomposing the lipid bodies.

* Supported by grants from the "Fonds voor Wetenschappelijk Geneeskundig Onderzoek" Brussels.

University of Ghent, Institute for Histology, Godshuizenlaan 4 (Belgium).

7.

Erich Schiller: **Mikrointerferometrische Beiträge zur Lungenreinigung*.**

Der Bau eines Mikrointerferometers nach dem „Round-the-square"-System von Mach und Zehnder durch die Firma Leitz gibt erstmalig die Möglichkeit, Objekte beliebiger Größe und Strukturverteilung ohne Beeinträchtigung der Interferenzverhältnisse zu untersuchen. Rattenlungen wurden nach Quarzstaub-Inhalation (Teilchengröße $< 3\ \mu$m) und Anwendung verschiedener Vitamine, Hormone und anderer die Lungenreinigung beeinflussender Pharmaka mit Formol, Zenker-Formol, Susa oder Carnoyscher Flüssigkeit durchspült. Ungefärbte Paraffinschnitte wurden in Wasser oder Glycerin eingeschlossen und mit der Ölimmersion 100×, A 1.36 im Streifenfeld und im homogenen Feld (Interferenzkontrast) photographiert. Neben den Trockenmassen von Zellstrukturen und Interzellularsubstanzen wird den inhalierten Quarzteilchen besondere Beachtung geschenkt. Der Vorteil der Mikrointerferometrie gegenüber der von BREITENECKER in die Silikoseforschung eingeführten Historadiographie wird herausgestellt.

A Contribution to the Problem of the Removal of Dust Particles by the Lung. The construction of a microinterferometer by Leitz (according to Mach and Zehnder's "Round-the-square" system) makes it possible to analyse objects of any size and structure without any lessening of the interference conditions. Lungs of rats, which previously have been exposed to quartz dust (size of the quartz particles: < 3 μm) and treated with various vitamins, hormones and drugs which have an influence on the clearing function of the lung, were perfused with formalin, Zenker-formalin, Susa, or Carnoy fixative. Unstained paraffin sections were mounted in water or glycerol and photographed (interference contrast, oil immersion, 100×, A 1.36). Apart from the dry substance of cell structures and intercellular substances special attention was given to the inhaled quartz particles. Special stress is laid upon the advantages of microinterferometry over historadiography which has been introduced by BREITENECKER for use in silicosis research.

* Mit finanzieller Unterstützung der Europäischen Gemeinschaft für Kohle und Stahl, Luxemburg.

Med. Institut für Lufthygiene und Silikoseforschung, Medizinische Akademie, 4 Düsseldorf (Deutschland).

8.

IMRE SCHNEIDER und FERENC SZARVAS: **Enzymchemische Untersuchungen nach Serotonin-Verabreichung.**

Aus der Literatur ist bekannt, daß nach Serotonin-Dauermedikation in der Leber eine nachweisbare Verminderung der Aktivität gewisser Enzyme eintritt. Bekannt ist auch die hodenschädigende Wirkung des Serotonins. In vorangegangenen Untersuchungen konnten wir nach der Serotonindosierung einen Anstieg des Transaminasegehaltes im Serum der Ratten beobachten. Ziel unserer Versuche war das Studium der serotoninbedingten Enzym-Aktivitätsverminderung in der Leber, den Hoden und der Haut von Ratten. Versuchstiere waren rund 200 g schwere männliche Albinoratten. Die Tiere der ersten Gruppe erhielten 5 mg/200 g Serotonin-Kreatininsulfat subkutan, die Tiere der zweiten Gruppe außerdem noch 300 Gamma/200 g Deseryl und die der dritten Gruppe nur physiologische Kochsalzlösung. Nach 8 Std wurden die Ratten ohne Narkose getötet, anschließend ihre Lebern, Hoden und Stückchen der durch vorhergehende Enthaarung in den Zustand des anagenen Haarwuchses gebrachten Rückenhaut histologisch (Hämatoxylin-Eosinfärbung) und mit histochemischen Methoden (Succinodehydrogenase-Aktivität, alkalische Phosphatase, Sudan-Schwarz und PAS) untersucht. Auffallend an den Ergebnissen war, daß in den Leber-, Hoden- und Hautpräparaten der serotoninbehandelten Ratten im Verhältnis zu den übrigen Tiergruppen eine ausgesprochene Herabsetzung der Succinodehydrogenase-Aktivität festzustellen war. Die mit anderen Methoden durchgeführten vergleichenden Untersuchungen ließen — im Verhältnis zur Succinodehydrogenase-Aktivität — weniger ausgesprochene Abweichungen zwischen den einzelnen Gruppen erkennen.

Enzyme Histochemical Examinations Following Serotonin Administration. According to the literature following long-term administration of serotonin a decrease in the activity of certain enzymes in the liver ensues. It is also known that serotonin impairs the testes. In previous experiments we observed that after serotonin administration the transaminase level in the serum of rats rises. The aim of the experiments was to study the decrease of the enzyme activity induced by serotonin in the liver, testis and skin of rats. The experiments were performed on albino male rats weighing about 200 g; one group was administered subcutaneously 5 mg/200 g serotonin creatinine-sulphate. The second group was given in addition to serotonin 300 gamma/200 g deseryl. Animals of another group were only administered physiological saline. The animals were killed after 8 hours without anaesthesia. After decapitation the liver, the testes and a piece of the dorsal skin which was in the anagen stage of the development of the hairs initiated by previously performed epilation were examined by histological (haematoxylin eosin staining) and histochemical (succino-dehydrogenase activity, alkaline phosphatase, Sudan black and PAS) methods. The results of the examinations demonstrated strikingly that in the liver, testis and skin of the rats treated with serotonin as compared to the same tissues of the other groups a pronounced decrease in the succino-dehydrogenase activity had ensued. The comparative studies carried out with the other methods showed as compared to the succino-dehydrogenase method less marked differences between the various groups.

Universitäts-Hautklinik Szeged (Ungarn) und Medizinische Universitäts-Klinik, 2 Hamburg-Eppendorf (Deutschland).

9.

SIGURD RAUCH: Mikrogasometrische Befunde zur segmentären Organisation der Cochlea*.

Die Stria vascularis des Innenohres gehört zu den Geweben, die einen außerordentlich hohen Sauerstoffverbrauch haben (CHOU u. RODGERS 1961). Wenn man mit Hilfe des Kartesischen Tauchers den O_2-Verbrauch der Stria vascularis aus den vier Windungen der Meerschweinchenschnecke getrennt bestimmt, findet man eine deutliche Abnahme des Sauerstoffverbrauchs von basal nach apikal. Andere biochemische und histochemische Befunde an der Meerschweinchenschnecke (z. B. Gehalt an Carboanhydrase, Melanin, Tyrosin, Fetttropfen in den Hensenschen Zellen, DC-Potential usw.) lassen auf eine segmentäre Organisation in den einzelnen Windungen der Cochlea schließen. Die mikrogasometrischen Ergebnisse decken sich weitgehend mit diesen Befunden. Sie geben damit eine neue Stütze für die Annahme einer segmentären Regulation der Stoffwechselvorgänge innerhalb des Hörorgans.

Microgasometric Investigations on the Segmental Organization of the Cochlea*. The stria vascularis of the inner ear belongs to the tissues having a very high rate of oxygen consumption (CHOU and RODGERS 1961). The stria vascularis, gathered from the four whorls of the cochlea of the guinea pig, shows a clear diminution of oxygen consumption from the basal to the apical whorl (measured with the Cartesian diver method). Other biochemical and histochemical results of the inner ear (e.g. carboanhydrase, melanin and tyrosine content, droplets of fat in HENSEN's cells, DC-potential etc.) suggest a segmental organisation in the different whorls of the cochlea. The microrespirometric results also give a new indication that the hearing organ has a segmentally regulated metabolism.

* Mit Unterstützung der Deutschen Forschungsgemeinschaft.

Biochemisches Laboratorium der Hals-Nasen-Ohrenklinik der Medizinischen Akademie, 4 Düsseldorf (Deutschland).

10.

L. VOLLRATH: Das Enzymmuster der Meerschweinchenplazenta während verschiedener Schwangerschaftsstadien*.

Es wird versucht, die Kenntnisse über das Enzymmuster der Meerschweinchenplazenta zu vervollständigen, um Einblick in die Funktion verschiedener Plazentarbezirke zu bekommen und den Zeitpunkt ihrer Funktionsaufnahme zu erfassen. — Im Interlobär- und Randsyncytium von 45 Tage alten Plazenten geben DPN-, DPNH-Diaphorase, α-Glycerophosphat-, Laktat-, Glutamat-, Bernsteinsäuredehydrogenase und Cytochromoxydase stärker positive Reaktionen als im Labyrinth. Geringe Aktivität von Bernsteinsäuredehydrogenase und Cytochromoxydase und starke Aktivität von α-Glycerophosphat- und Laktatdehydrogenase lassen daran denken, daß glykolytische Prozesse dominieren. Vermehrtes Vorkommen von Glukose-6-Phosphatdehydrogenase zeigt möglicherweise den Pentosezyklus als zusätzliche Energiequelle an. Da alkalische Phosphatase, Glucose-6-Phosphatase und Leucinaminopeptidase verstärkt im Labyrinth nachweisbar sind und sich mütterliches und fetales Blut hier in engem Kontakt befinden, scheint dieser Bezirk eine besondere Rolle im Intermediärstoffwechsel und bei den Austauschvorgängen zwischen Mutter und Fetus zu spielen. — Im Subplazentasyncytium sind Dehydrogenasen und Diaphorasen, nicht jedoch Phosphatasen und Leucinaminopeptidase nachzuweisen. — Bei 16—20 Tage alten Plazenten lassen sich das neugebildete Syncytium und die die Decidua durchdringenden Syncytiumsprossen durch positive Reaktionen für Dehydrogenasen (außer Bernsteinsäuredehydrogenase und Cytochromoxydase) und Diaphorasen darstellen. Die Fermentaktivität ist jedoch geringer als bei älteren Stadien. Sobald das Plazentalabyrinth vom 20.—23. Tag durch Einsprossen von Allantoisgefäßen entstanden ist, sind alkalische Phosphatase und Glucose-6-Phosphatase verstärkt, Leucinaminopeptidaseaktivität ist dagegen erst am 27. Tag vorhanden. Offensichtlich beginnen im Labyrinth Kohlenhydrat- bzw. Fettstoffwechsel einige Tage vor dem Eiweißstoffwechsel.

The Enzymatic Pattern of the Guinea Pig Placenta during Various Stages of Pregnancy.

This work has been undertaken in order to complete the knowledge of the enzymatic pattern of the Guinea pig placenta and thus get some information of the function of various placental zones and the moment they start functioning. In the spongy zone of 45 day-old placentas

DPN-, DPNH-diaphorase, α-glycerophosphate-, lactate-, glutamate-, and succinate dehydrogenase, and cytochrome oxidase, give stronger positive reactions than they do in the labyrinth. A low activity of succinate dehydrogenase and cytochrome oxidase and a high activity of α-glycerophosphate- and lactate dehydrogenase seem to indicate a preponderance of glycolytic processes. The increase in glucose-6-phosphate dehydrogenase shows that the pentose cycle acts probably as an additional source of energy. As an increased activity of alkaline phosphatase, glucose-6-phosphatase and leucine aminopeptidase is found in the labyrinth, where there is a close contact between maternal and foetal blood, this region seems to be important for the intermediary metabolism and the exchange processes between mother and foetus. In the subplacental syncytium dehydrogenases and diaphorases, but not phosphatases and leucine aminopeptidase, are present. In 16—20 days old placentas the newly formed syncytium as well as syncytial roots, which spread out into the decidua, can be demonstrated by means of positive dehydrogenase and diaphorase reactions (exceptions: succinate dehydrogenase and cytochrome oxidase). The enzymatic activity, however, is lower than it would be in later stages. As soon as the placental labyrinth is formed, by the spreading of allantoic vessels (20th to 23rd day), an increased activity of alkaline phosphatase and glucose-6-phosphatase can be observed, whereas leucine aminopeptidase activity occurs only on the 27th day. These findings indicate that in the labyrinth carbohydrate and fat metabolism starts several days prior to protein metabolism.

* Mit Unterstützung durch die Deutsche Forschungsgemeinschaft.

Anatomisches Institut der Universität, Koellikerstraße 6, 87 Würzburg (Deutschland).

11.

J. TURCHINI: Transferts enzymatiques et induction au cours du développement embryonnaire.

Par divers exemples, concernant en particulier le développement du foie et celui de l'appareil urinaire, l'auteur se basant sur les recherches de son élève CATAYEE et les siennes propres montre l'importance considérable des transferts enzymatiques au cours du développement embryonnaire agissant comme de véritables inducteurs.

Enzymatic Transfers and Induction during the Embryonic Development: The author who takes as a basis the researches of CATAYEE, his pupil, and his own researches, shows up the considerable importance of the enzymatic transfers which act like real inductors, through various examples concerning especially the development of the liver and the development of the urinary apparatus.

Laboratoire d'Histologie, Falcuté de Médicine, Montpellier (France).

Endokrine Organe · Endocrine Organs · Glandes endocrines

1.

SALLY B. FAND: Histochemical Evidence for a Neuroendocrine Connection in the Anterior Pituitary of Man: The Role of the Pars Tuberalis*.

We have previously reported the utility of the histochemical method for the oxidative enzyme α-glycerophosphate dehydrogenase in depicting certain cells of the anterior pituitary in the rat. Extension of these studies to more than 50 human pituitary glands reveals similar selective anterior lobe enzymic staining in cells which may be thyrotrophs. In addition, cells running along pituitary stalk in the human have been demonstrated to have high complements of this enzyme. These are the cells of the pars tuberalis. Of particular interest is the clear cut finding that these positively stained cells enter into and merge with the basophils of the coronal portion of anterior lobe at their junction with the pars intermedia. They do not enter posterior lobe as does the stalk. Since these cells accompany stalk along much of its course from hypothalamus to hypophysis, it is suggested that they may serve as an anatomical link between brain, on the one hand, and the non-nervous portion of the hypophysis, on the other.

* Aided by U.S. Public Health Service Grant AM 07430-02 END.

Veterans Administration Hospital and Dept. of Medicine, State University of New York at Buffalo, Buffalo, New York (U.S.A.).

2.

T. P. J. VANHA-PERTTULA: Studies on the Adenohypophyseal Esterases.

Esterase activity of the rat anterior pituitary was studied using crude homogenate, high speed sedimentation fractionation, DEAE-cellulose chromatography, starch gel electrophoresis and paper electrophoresis. Histochemical localisation of the esterases was demonstrated in fresh and formol-chloralhydrate-sucrose fixed cryostat sections with 5-bromo-indoxyl acetate as substrate. Specific and non-specific acetylcholinesterases were demonstrated with GOMORI's thiocholine method. PAS-staining was used to aid in localisation after the enzyme histochemical reaction. — Four separate esterase activities were noted and characterized using various inhibitors and activators. 1. C-esterase, resistent to E 600, was identified in starch gel electrophoresis and DEAE-cellulose chromatography. In the fresh tissue sections this esterase was poorly demonstrable but in fixed material it was localised mostly in large peripheral mucoid cells around sinusoids. After ultracentrifugation it was found in the supernatant fraction. In crude homogenate and in ultracentrifuge fractions it is activated by p-chloromercuribenzoate 10^{-4} M. The pH optimum of this esterase is 8.3. 2. Non-specific cholinesterase was separated apart in starch gel electrophoresis using α-naphthyl acetate and butyrate, acetylthiocholine iodide and butyrylthiocholine iodide as substrates. This enzyme was inhibited by lysivane 5×10^{-5} M, but not with 62 C 47 10^{-5} M and 284 C 51 10^{-5} M. In histochemical section it was found in basophil cells. 3. Two also E 600 sensitive B-esterases were noted in starch gel electrophoresis differing profoundly in mobility. These B-esterases seemed to be more diffusely divided in various cell types. The maximum activity was near pH 7. These esterase activities were clearly separated from peptidases and no demonstrable protease activities were found to be connected with them.

Dept. of Anatomy, University of Turku (Finland).

3.

HAROLD JOHN SOBEL and JACK GELLER: **Comparison of the Cytochemistry of Thyroid Glands of I¹³¹ Irradiated and Experimental Thyroiditis Animals*.**

Thyroid glands of irradiated rat, experimental thyroiditis guinea pig and control animals were studied by means of enzyme-visualization techniques for plasma membranes and Golgi apparatus (nucleoside phosphatases), lysosomes (Ac Pase) and mitochondria and endoplasmic reticulum (DPNH Nitro-BT reductase). Animals were killed 3 days after intraperitoneal administration of 2 microcuries of I^{131} per gm. body weight (SOBEL, BIEMPICA and NOVIKOFF 1962) and 6—12 weeks after subcutaneous or intramuscular injection of an homologous thyroid/Freund's adjuvant mixture (4 mg. thyroid). — These animals revealed nucleoside phosphatase activity of the luminal surface of many acini, increased size and complexity of the Golgi apparatus with many Golgi lamellae curled into spherical bodies and increased number and size of lysosomes. The DPNH Nitro-BT reductase technique revealed large formazan deposits. In the irradiated animals these deposits were basal, not granular and mitochondria appeared unremarkable. They were thought to indicate an alteration of the endoplasmic reticulum. The thyroiditis glands had many swollen mitochondria and the deposits were both apical and basal and granular; these deposits are probably mitochondrial. — These studies illustrate the similarity of the reaction of the thyroid acinar cell to various injuries and the usefulness of cytochemical staining methods for the better understanding of pathological processes.

* Aided by research grant AM-05516 from the National Institute of Arthritis and Metabolic Diseases.

Albert Einstein College of Medicine, Bronx, New York 10461 (USA).

4.

SYDNEY S. LAZARUS: **Histochemical Specificity and Ultramicroscopy of Rabbit Pancreatic B Cell Glucose-6-Phosphatase*.**

Glucose-6-phosphatase (G-6-Pase) was demonstrated previously in pancreatic B cells of various species. Because of the responsivity of B cells to blood glucose concentration it is believed that this enzyme must play a role in regulating insulinogenesis. However, that it would function as in the hepatic parenchymal cell to release glucose to the cell exterior seemed questionable. In the present study the specificity of this enzyme was tested against various hexose, pentose, triose and nucleosidephosphates as well as other phosphorylated substrates which are of metabolic importance. Hexoses and pentoses with phosphorylated terminal carbon and unbound 1-hydroxyl were hydrolyzed. These findings indicate a similarity of the B cell enzyme to the G-6-Pase of liver and epididymis. Ultramicroscopically, as in liver, the enzyme was localized to the inner surfaces of the lamellae of the endoplasmic reticulum (E.R.) and nuclear membranes. Connections were readily demonstrable between E.R. and nuclear membrane but not between E.R. and the plasma membrane. With the exception of granules no continuities with other cell organelles were noted. The possible significance for the B cell of these findings and of an enzyme which sequesters free glucose to the cisternae of the E.R. will be discussed.

* Supported by Grants NSF G1888 and NIH A2203.

Isaac Albert Research Institute, Brooklyn 3, New York (USA).

5.

JOZEF MESTDAGH: DNA-Content of the Nuclei in the Adrenal Cortex of the Mouse in Different Experimental Conditions.

The Feulgen-DNA-content of the interphase nuclei in the different layers of the adrenal cortex of the mouse has been assessed by the cytophotometric technique of LISON. Prepuberal male mice, 12 days old, were used. — A decreasing DNA gradient from the fasciculata externa to the X-zone has been observed in the control animals. The mean DNA-content of the nuclei in the zona glomerulosa and in the zona fasciculata externa was equal. — Prepuberal castration does not influence the nuclear DNA-content in the X-zone. It induces however a decrease in the zona glomerulosa. The same observations were made in LH treated prepuberal castrated mice. — The stimulation of the glucocorticoid function by administration of exogenous ACTH results in an increase of the nuclear DNA-content in the inner layers. — These observations do not show a stimulating effect of LH neither on the X-zone, nor on the other layers. The increase of the nuclear DNA-content in the two inner layers after ACTH-administration suggests that these cells may be stimulated by this hormone.

State University of Ghent, Dept. of Human Anatomy, Bijlokekaai 1, Ghent (Belgium).

6.

MARIAPIA VIOLA: A Decrease of DNA in Adrenal Medulla Cells Following Exposure to Low Temperature.

The possibility of quantitative changes of DNA in nuclei, in relation to cell metabolism irrespective of phenomena of polyploidy and polyteny, cannot be considered as demonstrated. The largest amount of known observations have been performed by means of microphotometric methods, against which objection of general as well technical character have been raised. However the reported changes were small. — Researches have been performed on adrenal medulla cells of rats exposed to cold. A considerable decrease of DNA content per nucleus (32—42%) has been observed using microphotometric, microinterferometric and chemical methods; this decrease was followed by a restoration to normal or supranormal values upon withdrawal of the stimulus. In our experiments two processes can therefore be distinguished: 1) a decrease of DNA content induced by the action of the stimulus, 2) a recovery upon cessation of this. The decrease cannot be attributed to phenomena of cellular degeneration, since no degenerating cells have ever been observed. The recovery cannot be due to mitotic processes, since the adrenal medulla cells, in adult rats, are known to be in intermitotic irreversible state. — The agreement of the results obtained by three different approaches, and the amount of DNA variations strongly suggest a real participation of DNA to the metabolism of the adrenal medulla cells.

Istituto di Patologia Generale dell'Università, Via Roma 55, Pisa (Italia).

7.

J. H. TRAMEZZANI, S. R. CHIOCHIO, and G. F. WASSERMANN: A New Technique for Localization of Noradrenaline with the Light and Electron Microscope.

Fixation of adrenal gland with glutaraldehyde 6.5% in Millonig buffer at pH 7.2 during 1 hour develops a yellow colour in noradrenaline secreting cells while adrenaline containing cells remain uncoloured. When frozen sections of adrenal gland fixed in glutaraldehyde are treated with an ammoniacal silver solution, silver is reduced by noradrenaline cells in less than 30 seconds. — A deep brown or black colour according to the silver solution used, was observed in noradrenaline containing cells while the adrenaline ones remained uncoloured. — Under a dissecting microscope small pieces of chromaffin tissue containing noradrenaline cells and others containing adrenaline cells were removed and embedded in Epon. The electron microscopical study of these two types of cells confirmed the optical observations and showed that the silver appeared only in the cathechol containing granules of the noradrenaline secreting cells, whereas no silver was present in the granules of the adrenaline containing cells.

Instituto de Biologia y Medicina Experimental, Obligado 2490, Buenos Aires (Argentina).

8.

GIULIO MURATORI: Histochemical Characteristics of the Aortico-Pulmonary Paraganglia in the Adult Pig.

In the adult pig numerous aortico-pulmonary paraganglionic bodies are present under the aortic-arch on the superior surface of the distal extremity of the pulmonary artery at its bifurcation; they are placed in the adventitia or near to it or in the periadventitial fat.

Their transversal maximal dimensions vary from 210 μ to 180 μ; the smaller ones of about 140 μ to 70 μ are included in nervous bundles or ganglia. After fixation in Helly fluid and subsequent prolonged chromization the cytoplasm and the granules of the paraganglionic cells appear yellow. After treatment with potassium jodate at pH 2,5 or pH 10, both the cytoplasm and the granules of the specific cells of the paraganglionic lobules are distinctly stained in brown-yellow. With the Bodian activated protargol method, after fixation with acetic-formol the granules of all the paraganglionic cells are stained in black. In frozen sections of pieces fixed in 10% formalin examined in Wood's light a part of the paraganglionic bodies showed a brilliant yellow green fluorescence. Chromaffin reaction, jodaffin reaction at various pH, and Bodian argentophilic reaction all being positive, yellow fluorescence after formalin fixation also being present, it is concluded that these aortico-pulmonary bodies of the pig contain catecholamines.

Istituto Anatomico, Università di Ferrara, Via Fossato di Mortara 66, Ferrara (Italy).

9.

RICHARD B. COHEN, KAROLY BALOGH, Jr., and HUBERT WOLFE: Histochemical Observations on Enzymes of the Pentose Phosphate Pathway in Steroid Hormone Producing Tissues.

The pentose phosphate pathway is important in the metabolism of steroid hormones. This importance may be derived largely from the fact that in the course of metabolism of glucose via the pathway, reduced triphosphopyridine nucleotide (TPNH) is generated. TPNH is essential for hydroxylation of the steroid molecule at several sites and is probably essential for splitting the side chain of the cholesterol molecule. Our histochemical studies encompassing the adrenal, testes, and ovary indicate high activity of glucose-6-phosphate dehydrogenase (G-6-PD) and 6-phosphogluconic dehydrogenase (6-PGD) in the fascicular and reticular zones of the adrenal cortex, Leydig cells of the testes and theca interna of the ovarian follicle and corpus luteum. The glomerulosa of the rat adrenal shows scant activity unless the animal is placed on a sodium restricted diet at which time the broadened glomerulosa zone shows marked increase of enzyme activity. The significance of the histochemical observations in relation to the role played by the pentose phosphate enzymes in steroid hormone metabolism will be discussed. Morphological observations on aspects of the developement of Leydig cells as determined by G-6-PD activity will be considered.

The Massachusetts General Hospital, Boston, Massachusetts (USA).

10.

IRA GHOSH and ASOK GHOSH: Influence of Age on Cytochemically Demonstrable Avian Catechol Hormones.

The present investigation was pursued in an attempt to ascertain whether the methylation mechanism for conversion of noradrenaline to adrenaline in juvenile birds remains equally effective as in adults. Adrenal glands of preadolescent and mature birds belonging to eight different species were used in this study. Chromate-dichromate and iodate reactions were followed to demonstrate total catechol hormone and noradrenaline-containing cell groups respectively. Histometric studies of the noradrenaline and adrenaline-containing cellular areas in both age groups indicate that the values for cytochemically detectable catechol hormones in juvenile birds are almost similar to that of the adults. It may be concluded that the enzyme system responsible for methylation of catechols possess the same degree of efficacy in both predolescent and adults of the avian species studied, barring only *Corvus splendens* which has a definite higher noradrenaline content in its early days of life.

Histophysiology Laboratory, Dept. of Zoology, University of Calcutta, Calcutta 19 (India.)

11.

HELMUT H. WOLFF: Elektive Darstellung der thyreotropinbildenden Zellen des Hypophysenvorderlappens mit Dichlorpseudoisocyanin*.

Die Pseudoisocyanine sind in Gegenwart eng benachbarter stark elektronegativer Gruppen zu reversibler Polymerisation fähig, wobei sich ihr Absorptionsspektrum in charakteristischer Weise ändert; diese Reaktion zeigt sich im histologischen Schnitt als Metachromasie und ist z.B. zum histochemischen Insulinnachweis benutzt worden (SCHIEBLER und SCHIESSLER, Histochemie 1, 445, 1959). — Die Eiweißkomponente des Glykoproteids Thyreotropin zeichnet sich, wie auch aus biochemischen Untersuchungen hervorgeht, durch hohen Gehalt an Cystin aus; durch Perameisensäure werden dessen Disulfidgruppen zu Sulfogruppen oxydiert und dem Nachweis mit der erwähnten metachromatischen Reaktion zugänglich gemacht. — Als beste Fixierung für die Hypophyse bewährte sich Formol-Sublimat. Die entparaffinierten

Schnitte werden 5 min in Perameisensäure oxydiert, danach ca. 4 min in 2×10^{-4} M N,N'-Di-äthyl-6,6'dichlorpseudoisocyaninchlorid** gefärbt und unter Vermeidung von Alkohol aus wäßrigem Milieu oder nach Entwässerung mit Aceton-Xylol eingedeckt. Bei der Ratten-hypophyse ergibt sich eine elektive metachromatische Darstellung der S-Zellen, einer Unter-gruppe der klassichen ,,Basophilen", die nach geltender Anschauung als Thyreotropinbildner angesehen werden. Die Identität mit diesen durch Aldehydfuchsin und Perameisensäure-Alcianblau darstellbaren Zellen wird durch Umfärbung sichergestellt. Blockierung der Sulfo-gruppen durch Methylierung hebt die Färbbarkeit auf, anschließende Verseifung stellt sie wieder her. Die Färbung ist auch bei anderen Spezies — (unter anderem Mensch, Kaninchen, Hund, Maus, Schwein, Rind) mit Erfolg durchführbar. Die metachromatische Färbung wird bei Betrachtung im monochromatischen Licht der Wellenlänge 578 nm sowie im Fluor-eszenzmikroskop besonders deutlich erkennbar.

Selective Staining of the Thyrotropin Producing Cells of the Anterior Pituitary by Dichlor-pseudoisocyanin*. The pseudoisocyanins are able to form reversible polymers in presence of closely neighbouring electron negative groups; this reaction leads to a shift of their absorption maxima, which histologically can be shown as metachromasia. It has been used for the histo-chemical detection of insulin (SCHIEBLER and SCHIESSLER, Histochemie 1, 445, 1959). — The protein component of the glycoproteid thyrotropin is characterized by its high amount of cystine, as results from biochemical investigations. Performic acid oxidizes its disulphide groups to sulpho groups and thus makes it demonstrable by means of the above-mentioned reaction. — The most suitable fixative for the pituitary seems to be formol-sublimate. The deparaffinized sections are oxidized for 5 min by performic acid, stained for about 4 min in 2×10^{-4} M N,N'Diethyl-6,6'dichlorpseudoisocyaninchloride** and mounted from the aqueous medium by avoiding alcohol or by dehydrating in aceton-xylene. In the rat pituitary we find a selective staining of the S-cells, a subgroup of the classic "basophiles" which are re-cognized at the present to be the producers of thyrotropin. These cells are identical with the aldehyde-fuchsin and performic acid-Alcian blue stainable cells, which could by shown by restaining. Blocking of the sulpho groups by methylation removes the stainability, sub-sequent saponification restores it. The staining is successful also in the pituitary of several other species (e.g. man, rabbit, dog, mouse, pig, cow). The metachromatic reaction can be demonstrated quite distinctly in monochromatic light of 578 nm wavelength or by the fluorescence microscope.

* Mit Unterstützung durch die Deutsche Forschungsgemeinschaft
** Bezug durch Fa. Dr. H. Harms, 509 Leverkusen 3, Alte Landstr. 3

Anatomisches Institut der Universität, 87 Würzburg (Deutschland).

Lipide · Lipids · Lipides

Lipopigmente · Lipopigments · Lipopigments

Leber · Liver · Foie

1.

ILONA DE SIBRIK: Oxidative Enzymes in Intact and Degenerating Nerve and Their Relation to Lipids*.

Experimental degeneration was produced in a number of rats. The technique has been described previously by JOHNSON, MCNABB, and ROSSITER. It seemed of interest to investigate histochemically the changes in the lipids in a degenerating nerve compared and connected with the changes in two oxidative enzymes: a carbohydrate dehydrogenase (succinic) and a fatty acid (β-hydroxybutyric) dehydrogenase, in normal controls and in Wallerian degen-eration. For the identification of the lipid substances 10% buffered formalin fixed frozen sections were taken from the sciatic nerve of rats, and were stained for total lipids with Sudan IV, Sudan Black, and Oil Red 0, for neutral lipids with Nile Blue sulfate and for plasmalogens with the Schiff reagent. For the localization of the dehydrogenases, the tissues were rapidly frozen by dry ice and after cutting transferred directly into the substrate solution containing nitro-blue tetrazolium according to the method described by NACHLAS and col-laborators. In normal nervous tissue the two different dehydrogenases exhibited similar patterns. The distribution of both enzymes followed the course of the nerve fibers. The intensity of hydroxybutyric dehydrogenase was somewhat weaker. In the degenerated nerve the change occurring in the enzyme system was striking. The fatty acid dehydro-genase still exhibited a faint reaction, showing only a little elevation in activity at the first stage compared with the intact nerve. Succinic dehydrogenase activity increases after 14 days of operation when the first invasion of fat cells appears between the degenerated fibers. Both of these enzymes show activity in adipose tissue. Succinic dehydrogenase exhibited an increasing activity during the process of demyelination. The enzyme was represented as extremely fine and uniform granules arranged along the nerve fibers. The changes occurring in the degenerated myelin lipids will be described and the significance of these findings will be discussed in relation to the enzyme changes in Wallerian degeneration.

* Supported by the Multiple Sclerosis Association of Greater Washington.

Georgetown University, Dept. of Neurology, Washington, D.C. (USA).

2.

J. CASPER and M. WOLMAN: Study of Myelin Figures with Ultraviolet Light as a Histochemical Test for Polar Lipids.

Myelin figures are formed whenever lamellar structure of polar lipids are destroyed due to an increase in hydrophilia. Lipids whose molecules contain hydrophobic fatty acid chains at one end and hydrophilic groups at the other end — like phospholipids and gangliosids — are apt to form myelin figures which are easy to recognize by induced fluorescence. Fresh frozen unfixed tissues obtained from autopsies and biopsy-specimens are cut in the cryostat, stained by Thioflavine TFS (Ed. Gurr, London) and Hematoxylin as suggested by VASSAR and CULLING, mounted in Apathy's medium and examined under ultra-violet light. Unfixed smears, e.g. of fresh spleen tissue, stained by Thioflavine and mounted in Apathy's medium may be examined in the same manner. After a short time, myelin figures are beginning to appear in tissues and cells containing polar lipids and are fully developed after several hours. This method can be used as a histo-chemical test for the presence of polar lipids in tissues and cells, particularly, of phospholipids. Gangliosides may be identified by a positive PAS-reaction. This technique can be tested easily in normal brain tissue and in peripheral nerves. — The results in frozen sections from formalin-fixed material are unreliable. As examples are demonstrated cases of Niemann-Pick's disease and Familial Amaurotic Idiocy, Phospholipid-containing macrophages in the spleen of case of Thrombocytopenic Purpura and other diseases and lipid-containing cells in a Granular-Cell Myoblastoma.

Pathological Institute, Beilinson-Hospital, Petah-Tiqva (Israel).

3.

SERGIO B. CURRI and M. RASO: The Application of Histochromatography. A Contribution to the Histochemistry of Phospholipids.

The direct thin-layer chromatography of mitochondria and tissue sections, with exclusion of an extraction step, resolves the lipid fractions into their components as well as does thin-layer chromatography of their organic solvent extracts. It seems that the solvents used for the chromatographic run are sufficiently polar to rupture the lipid-protein bonds and to permit the lipids to migrate as separate spots. — As was shown by the correspondence of the results with those obtained by different methods, direct thin-layer chromatography can be used not only for qualitative but also for quantitative analysis. Compared with the other procedures direct thin-layer chromatography is easier, much less expensive, and is useful for the rapid screening of a large number of samples. The time required for quantitative analysis has been considerable reduced both by the elimination of the extraction step with organic solvents and by simplification of the determination of phospholipid phosphorus. — For the tissue sections, direct chromatography (Histochromatography) permits to exact location of the chromatographed lipids when the tissue sections are submitted to a proper histological analysis after chromatography. The direct chromatographic procedure, unlike the method based on a preliminary chemical extraction of lipids, does not destroy the structure of the tissues but only releases the lipid fractions which are under investigation. — The results of chromatographic technique in various pathological conditions are reported.

Dept. of Pathology, University of Padova, Via Gabelli 35, Padova (Italy).

4.

J. CHAYEN, LUCILLE BITENSKY and C. LONG: The Acid Haematein Method and the Detection of Masked Lipid.

The specificity of this method has been examined by the response of samples of purified lipids and lipid components on cigarette paper. It has been shown that the positive reaction depends on the presence of unsaturated groups, in the fatty acid; it may also require that these groups are present in the same molecule as the phosphate moiety. In consequence, the reaction in sections of tissue can be controlled by bromination rather than by the use of fat-solvents which may enhance the reaction by unmasking such structural lipids as are resistant to extraction by virtue of their peculiar binding to other tissue elements.

Royal College of Surgeons of England, London W.C.2 (Great Britain).

5.

CARLO RIZZOLI, PAOLO CARINCI and FRANCESCO ANTONIO MANZOLI: Histochemical Studies of Lipid Metabolism in Ethionine Treated Chick Embryos.

A massive, "physiological" fat accumulation, consisting of neutral fats, cholesterol and cholesterol esters and phospholipids, in the liver cells of the chick embryo has been described in our previous research projects. Presently the effects of ethionine on these "fatty livers" is under investigation in this laboratory. — The administration of ethionine into the yolk sac during the 12th day of incubation of the Leghorn chick was found to give a marked decrease of cholesterol and cholesterol esters by using the Schultz and digitonine tests for these substances. In these same test animals the phospholipids and the total fat content was found to be unchanged. Sex dependent differences were not noted. — The microscopic appearance of this ethionine induced fat accumulation is typical and diagnostic. Instead of the numerous and small droplets seen in the above mentioned "physiological" fat accumulation, we found under ethionine induction large lipid globules engulfed within liver parenchymal cells. This morphological appearance is accompanied by a change in the nature of the fatty acids. By gas chromatography, studies on the 15th day of incubation and at hatching showed an increase in insaturated fatty acids. — In conclusion, ethionine was found to cause profound alteration in lipid metabolism made manifest by decreased cholesterol accumulation and increase of insaturated fatty acids.

Istituto di Istologia ed Fisiologia Generale, Via Belmeloro 8, Bologna (Italy).

6.

I. DICULESCO and D. ONICESCO: Histoenzymological Researches on Lipogenesis.

Histoenzymological researches reveal a highly varied and very intense enzymic activity in cells synthesizing lipids. In the lipogenetic process, *in situ* as also *in vitro*, are active: DPNH-diaphorase; TPNH-diaphorase and DPN- and TPN-linked enzymes of the Krebs Cycle; enzymes of the amino acid metabolism; enzymes of the anaerobe glycolysis as also ATP-ase + Ca, ATP-ase + Mg and 5-nucleotidase. — The histochemically detectable enzymes of the lipid metabolism show characteristic variations according to the tissue type and in the different normal and pathological conditions. Betahydroxyacyl CoA shows maximum activity in adipose tissue and in the mammary gland and is reduced or absent in the rest of tissues and organs. Cellular membranes and particularly the endothelium and sheaths of some nervous filaments reveal intense choline metabolism. Reactions for the 20-beta-hydroxysteroid dehydrogenase are to be found in a limited number of cells and organs. Their activity varies under different conditions, for instance in the cortico-suprarenal and especially in the liver — in state of pregnancy — in the more advanced stages of lipid accumulation *in vitro* (in fibroblast cultures) and also in experimental atherosclerosis. The significance for tweenesterase reaction is being discussed. The relation between enzyme activity of some infrastructures and lipogenesis is shown.

Facultatea de Medicină, Catedra de Histologie, Bd. Dr. Petru Groza 8, Bucharest (Rumania).

7.

P. GEDIGK: Über die Entstehung von Lipopigmenten.

Das Ceroid- und das Lipofuscinpigment erfahren — ebenso wie das Vitamin E-Mangelpigment — im Laufe ihrer Alterung eine kennzeichnende Änderung ihrer färberischen und histochemischen Eigenschaften. Diese Entwicklung folgt einer im Tierexperiment und im Reagenzglas reproduzierten Gesetzmäßigkeit und ist an den oxydativen Umbau der in diesen Pigmenten vorhandenen ungesättigten Fettstoffe gebunden. — Die Farbkomponente lipogener Pigmente ist ein Oxydations- und Polymerisationsprodukt hochungesättigter Fettsäuren. Die Lipopigmente stimmen in ihren Fettbausteinen weitgehend überein. Infolge ihrer verschiedenen kausalen und formalen Genese bestehen jedoch Unterschiede in der Quantität und wahrscheinlich auch Qualität ihrer Eiweißkomponenten. — Elektronenmikroskopisch entsteht das Vitamin E-Mangelpigment, das den Lipofuscinen nahesteht, in der Nähe des Golgi-Feldes von Parenchymzellen unter zunehmender Ablagerung von feinkörnigem und gröberem lipidhaltigen Material. Eine Umwandlung von Cytoplasmaorganellen in Pigmentgranula erfolgt nicht. Der Ceroid tritt in Makrophagen auf und entwickelt sich bei der Phagozytose zugrunde gegangener fetthaltiger Gewebsanteile.

The Development of Lipopigments. The Ceroid-, the Lipofuscin- and the Vitamin-E-deficiency-pigment are altered in a very characteristic manner during the course of their aging. This process is based on the oxidation and polymerisation of unsaturated fats contained in these pigments. It can be observed and reproduced with great regularity in animal experiments as well as in the test tube. — The coloured component of lipogenous pigments is an oxidation and polymerisation product of highly unsaturated fatty acids. The lipid components of these pigments are almost identical. However, because of their different genesis there are differences in the quantity and probably quality of their protein-components. — In the electronmicroscope the vitamin-E-deficiency-pigment, which is only a variant of the lipofuscins, develops near the Golgi-field of parenchymal cells with increasing deposition of finely granular and coarser lipid-containing material. A transformation of cytoplasmic organelles into pigment granules does not occur. The ceroid appears in macrophages after the phagocytosis of fat-containing tissue particles.

Pathologisches Institut der Universität, Robert-Koch-Straße 5, 355 Marburg/Lahn (Deutschland).

8.

J. HEINRICH HOLZNER: Anwendung histochemischer Methoden bei der Untersuchung von Leberbiopsien*.

In der Leberdiagnostik werden seit langem histochemische Methoden verwendet: Lipidnachweise bei Steatose und dystrophischen Fettablagerungen, spezielle Lipoidfärbungen zur Identifizierung von Lipopigmenten und bei Speicherkrankheiten, PAS-Reaktion zum Glykogennachweis und zur Darstellung diastase-resistenter PAS-positiver Substanzen, Schwermetallnachweise, z. B. für „Fe“ (Hämochromatose) und „Cu“ (Wilsonsche Erkrankung) und in neuerer Zeit Methoden der Immunohistochemie zur Darstellung spezieller Antigene und Antikörper. Neue Möglichkeiten eröffnen die Methoden der Enzymhistochemie. Es wird über Erfahrungen an über 500 Leberbiopsien berichtet. — Die Darstellung der ATPase- und 5-Nucleotidase-Aktivität ermöglicht die morphologische Beurteilung der intralobulären Gallenkanälchen, die Erkennung beginnender Umbauprozesse durch atypische Verästelungen; bei Fettleber konnten klinische Zeichen von Cholestase durch mechanische Verdrängung der Canaliculi erklärt werden. Verminderung der APTase-Aktivität in der Wand der Gallenkanälchen findet sich bei Parenchymschäden, Vermehrung in den Endothelzellen bei entzündlichen Prozessen (Hepatitis). Fehlen oder Verminderung von Glukose-6-phosphatase-Aktivität in den Leberzellen kann zur Diagnose einer Glykogenspeicherkrankheit beitragen. Saure Phosphatase-Aktivität findet sich in großer Menge in gewissen Speicherzellen (Gauchersche Krankheit). Ihre Verteilung in der Leberzelle ist bei bestimmten Erkrankungen verändert. Verminderung oder Schwund der Aktivität oxydierender Enzyme kann als Zeichen einer Parenchymschädigung gewertet werden.

The Use of Histochemical Methods for the Examination of Liver Biopsies. For a long time histochemical methods have been in use for the diagnosis of liver diseases: lipid determinations in cases of steatosis and dystrophic fat deposition, specific lipid staining methods for the identification of lipopigments and in cases of storage diseases, the PAS-reaction for the determination of glycogen and diastase-resistant PAS-positive substances, methods for the determination of heavy metals, as for example "Fe" (haemachromatosis) and "Cu" (Wilson's disease) and lately immuno-histochemical methods for the demonstration of special antigens and antibodies. Enzyme-histochemical methods can provide much new information. Experiences acquired by processing over 500 liver biopsies are to be reported. The demonstration of ATPase and 5-nucleotidase activity allows a morphological assessment of the interlobular bile ducts and the recognition of acute transformation processes, which may show non-specific anastomoses. In cases of fatty liver it is possible to demonstrate that the clinical signs of cholestasis are due to a mechanical displacement of the small ducts. A reduction of ATPase activity in the walls of the bile ducts can be found in association with parenchymal lesions, whereas an activity increase in endothelium cells can be observed during inflammatory processes (hepatitis). The absence or reduction of glucose-6-phosphatase activity in liver cells can contribute to the diagnosis of a glycogen storage disease. Acid phosphatase activity can be found in considerable amounts in certain storage cells (Gaucher's disease). The distribution of macrophages in the liver changes in certain diseases. The reduction or disappearance of oxidizing enzyme activity can be regarded as a sign of a parenchymal lesion.

* Unterstützt durch ein Stipendium des U.S. Public Health Service, National Institutes of Health, Bethesda, Md.

Pathologisch-anatomisches Institut der Universität, Spitalgasse 4, Wien IX (Österreich).

9.

YUKIO YANAGIMOTO: Morphological and Cytochemical Studies on the Liver Cell in Case of the Rat Liver Lobe Ligation.

The main theme of this study is the relationship between microcirculation and function of the liver, i.e. the rat liver lobe was ligatured at its root, and both the ligatured and non-ligatured liver cells studied morphologically and cytochemically. The results are to be reported as follows: 1. On the mitosis of the liver cells. The mitosis of the liver cells of non-ligatured side was clearly delayed in its time as compared with that of the liver cells on the affected side. 2. On the liver cell structure. The mitochondria as time goes by show swelling and rough granulation on the ligatured side. After 6 hrs. their stainability shows a marked fall and the mitochondria gradually decrease. Besides the above results of mitochondria the specific findings were observed with the glycogen and fat. The above stated results are to be reported in comparison with the results of the non-ligatured side. 3. The changes of some chemical components. Notable changes were observed by means of quantitative cytochemical methods and biochemical determination of lipase, phosphorylase and phosphatase. Meanwhile, the results of paperchromatogram analysis showed changes in the reducing substance.

Dept. of Anatomy, Osaka City University Medical School, Asahi-Machi, Abeno-Ku, Osaka (Japan).

10.

ÉVA HORVÁTH und K. KOVÁCS: Vergleichende histochemische Untersuchung regenerativer Prozesse.

Verfasser haben die Regenerationsvorgänge der Leber im Anschluß an eine subtotale Hepat-ektomie, bzw. an eine Behandlung mit dem Serum subtotal hepatektomierter Ratten, sowie die Regeneration der Niere nach einer hormonell (Hypophysenhinterlappenextraktverabreichung nach vorhergehender Oestrogenbehandlung) ausgelösten bilateralen Nierenrindenenkrose bei Ratten in akuten und chronischen Versuchen studiert. Es werden die Aktivitätsabweichungen einiger oxydativer Enzyme (Succinodehydrogenase, Cytochromoxydase) und hydrolytischer Fermente (alkalische Phosphatase, ATP-ase), sowie die quantitativen Veränderungen der Nukleinsäuren in den einzelnen Phasen der Regeneration verfolgt. — Die Regeneration nach Parenchymverlust beginnt und endigt in der Leber früher als in der Niere; dement-sprechend verändert sich der Nukleinsäuregehalt beider Organe. Bei der Verabreichung des Serums ist eine langsame anhaltende Steigerung des Nukleinsäuregehaltes der Leber zu beobachten. — Die Regenerationserscheinungen sind in allen drei Experimenten von einer Aktivitätsverminderung der oxydativen Enzyme begleitet; bis zur Beendigung der Regeneration werden keine normalen Werte erreicht. Die Regenerationsvorgänge der Leber werden von einer Aktivitätssteigerung der alkalischen Phosphatase und/oder der ATP-ase begleitet. Die morphologische Restitution geht in beiden Organen der histochemischen voran.

Comparative Histochemical Studies of Regenerative Processes. The authors studied the activity of some oxidative and hydrolytic enzymes (succinate dehydrogenase/cytochrome oxidase and alkaline phosphatase/ATP-ase) as well as quantitative changes in nucleic acids, during the regeneration of the liver and the kidney in the following groups of animals: 1) sub-totally hepatectomised rats; 2) rats injected with the serum of sub-totally hepatectomised rats; 3) rats with bilateral necrosis of the renal cortex (caused by oestrogen treatment which was followed by an application of extracts from the posterior part of the hypophysis). — Tests were carried out at various stages of the regeneration. The regeneration of liver parenchyma commenced and ended earlier than the regeneration of kidney parenchyma. The content of nucleic acid changed accordingly in both organs. A slow but lasting increase of nucleic acids was observed in the liver of the animals of group 2. During the regeneration a decrease in the activity of oxidative enzymes was observed in all three groups of animals. This decrease in activity would last even after the regeneration had been completed. The regenerative process of the liver was accompanied by an increased activity of alkaline phosphatase and/or ATP-ase. In both organs the morphological restitution preceded the histochemical refunctioning.

Pathologisch-anatomisches Institut und I. Medizinische Klinik der Universität, Szeged (Ungarn).

11.

JÁNOS KOVÁCS and BORBÁLA HAFIEK: Histochemical Study of the Effect of Neutral Red on the Liver Cells of the Mouse.

After the injection of neutral red, in the cytoplasm of the liver cells a great number of dye containing granules there appear peribiliary in distribution. In the second hour after the administration of the neutral red a strongly basophil material appears on the surface of the granules (Chlopin's crinom). The crinom substance contains RNA and a PAS positive material. This latter, however, may be demonstrated only after formol or Helly fixation. The crinom substance does not contain proteins, or if only a very small amount. Since the dyestuff accumulates mainly in the centrilobular cells, these are the cells in which the crinom formation is the most intensive. Parallel with the accumulation of crinom substance, the basophilia of the cytoplasm decreases. Likewise, the activity of acid phosphatase decreases in the dye-laden cells. — Carbon tetrachloride causes a decrease in the capacity of centrilobular cells to accumulate neutral red granules. In this case, deposits of dye appear in the perilobular cells and only here may the crinom formation be observed. — According to our data we may suppose that there is close relationship between the neutral red granules and the lysosomes and so between the crinom formation and the lysosomes.

Dept. of General Zoology, Eötvös Lorand University, Budapest (Hungary).

Blut · Blood · Sang
Knochen · Bone · Os

1.

GABRIEL KELENYI and ELISABETH ZOMBAI: Electronhistochemical Investigations on the Granules of the Eosinophil Leukocytes.

The granules of the eosinophil leukocytes of the rat give an intense benzidine-peroxidase reaction. Pretreatment with phosphotungstic or phosphomolybdic acid partly inhibits the reaction: instead of the end product of the reaction, i.e., "benzidine-brown", a quinone-diimine-derivative, the granules appear in bluish-grey colour, suggesting that in intermediary product, "benzidin-blue" is formed. — Electronmicroscopical observations show that both benzidine-peroxidase activity and phosphotungstic acid staining are localized to the outer part of the granules, the externum. The enzyme myeloperoxidase with its high isoelectric point (pH 10—11) is responsible for the peroxidase reaction and may be the cause of the phosphotungstic acid staining of the granules. The partial inhibition of the reaction in the pretreated granules is probably connected with local acidic medium produced by the free radicals of phosphotungstic acid. These stabilize the benzidine-peroxidase reaction in the stage of "benzidine-blue", a phenomenon which may be observed light microscopically during benzidine-peroxidase reaction, when the medium is acidified.

University of Pécs, Dept. of Pathology, Pécs (Hungary).

2.

BOHDAN J. KOSZEWSKI: Cytochemical Studies on the Iron Content of Leukocytes*.

Hemosiderin inclusions in hemic cells are known to appear under physiological and pathological conditions in man and animals. Hemosiderin-loaded phagocytes occur under normal conditions in bone marrow and are markedly increased in cases of iron-overload. They can be easily identified by the Prussian blue stain. We have been working for many years with a modified Turnbull blue method which permits a better identification of the involved cells, as slides already stained with panoptic methods can be decolorized and re-checked for their hemosiderin content. With this method we were able to demonstrate hemosiderin inclusions in bone marrow plasma cells in cases of hemochromatosis with megaloblastic anemia. Similar particles have been seen in lymphocytes and monocytes of peripheral blood in spontaneous hemochromatosis. Artificial iron overload by parenteral injections led to hemosiderin inclusions in lymphocytes and monocytes by the process of phagocytosis. Recently, we noticed that the granulocytes of normal rabbits and squirrels showed a positive iron stain. A few iron-positive polymorphonuclear cells were also seen in birds, cats and pigs. The findings indicate that the polymorphonuclear leukocytes contain an appreciable amount of iron which may reach high values in some species like rabbits and squirrels. Only the Turnbull blue stain demonstrated the granules indicating that the hemosiderin in leukocytes contains bivalent iron. — In humans the normal granulocytes, as well as those from patients with chronic granulocytic leukemia, polycythemia vera, hemochromatosis, various anemias and Alder's anomaly were hemosiderin negative. Toxic granulation proved to be iron-positive in 14 out of 20 cases. Bone marrow elements of individuals with toxic granulation were hemosiderin-negative. The toxic changes do not consist of retained early azurophilic granulation as was previously assumed, but represent a phenomenon acquired in peripheral circulation.

* Supported in part by a grant from the U.S. Public Health Service.

Hematology Research Laboratory, St. Joseph's Hospital and Creighton University School of Medicine, Omaha, Nebr. (USA).

3.

G. L. Castoldi und **H. Merker**: Zytochemischer Nachweis von Cholindehydrogenase in menschlichen Blut- und Knochenmarkszellen*.

Im Cholinoxydasesystem katalysiert die Cholindehydrogenase den ersten Stoffwechselschritt und führt zur Bildung von Betainaldehyd. In einer weiteren Reaktion wird Betainaldehyd sodann zu Betain oxydiert, einem für den Stoffwechsel wichtigen Methyldonator. Wie Tierversuche zeigten, führt Cholinmangelernährung zu erheblichen Störungen im Fettstoffwechsel und zu Leberverfettung. Weiter haben Untersuchungen der letzten Jahre ergeben, daß auch die Zellen hämatopoetischer Systeme sowohl im strömenden Blut als auch in den Blutbildungsstätten in unterschiedlichem Maße am Fettstoffwechsel beteiligt sind. In diesem Zusammenhang erscheint es wichtig, nähere Kenntnisse über die Lokalisation und die Verteilung der Cholindehydrogenase in den verschiedenen Blutzellen zu erlangen. Von hämatologischem Interesse ist ferner die Feststellung aus biochemischen Untersuchungen, daß die Cholinoxydase ausschließlich an die Mitochondrien gebunden ist. Die hier vorgelegten Untersuchungen wurden an frisch gewonnenen Blut- und Knochenmarkszellen sowohl unter aeroben wie auch unter anaeroben Bedingungen ausgeführt. Besonders geprüft wurde der Einfluß verschiedener Fixationsmittel, Tetrazoliumsalze, intermediärer Wasserstoffakzeptoren und Enzyminhibitoren auf den Ablauf der zytochemischen Reaktion. Optimale Darstellungsbedingungen konnten auf diesem Wege ermittelt werden. Die Untersuchungen an den Blutzellen ergaben eine deutliche Enzymaktivität in den Zellen der Granulopoese. Demgegenüber wiesen die kernhaltigen Zellen der Erythropoese wie auch Megakariozyten und Plättchen nur schwache Reaktionen auf. Die erhobenen Befunde werden abschließend in bezug auf den zellulären Lipidstoffwechsel und die Zellfunktionen diskutiert.

Cytochemical Demonstration of the Choline-Dehydrogenase in Human Blood and Bone-Marrow Cells. In the cholineoxidase-system, the choline-dehydrogenase catalyses the first metabolic step and leads to the formation of betaine-aldehyde. In a further reaction, betaine-aldehyde is oxidized to betaine, one of the important methyl-donors in metabolism. As has been proven in animal experiments, a choline-deficient diet brings about considerable disturbances in fat metabolism and causes fatty liver. Moreover, observations in the last few years have shown that the cells of the hematopoietic system participate in varying degrees in fat metabolism. In this connection it appears important to gather more accurate information about the localisation and the distribution of the choline-dehydrogenase in the various blood-cells. Of hematological interest is also the fact that the cholineoxidase is linked exclusively to the mitochondria. The experiments presented were performed on fresh blood and bone-marrow cells under aerobic and anaerobic conditions. During testing, special consideration was given to the influence of various fixing agents, tetrazolium salts, electron acceptors and enzyme-inhibitors upon the course of cytochemical reactions. Optimal testing conditions could be achieved in this way. The reactions on the blood cells demonstrated a distinct enzyme-activity in the granulopoietic cells. In contrast, the erythroblasts as well as the megakariocytes and the platelets show only a weak positive reaction. These results will be discussed upon closing in connection with cellular lipid metabolism and cell functions.

* Gefördert durch Mittel der Deutschen Forschungsgemeinschaft.

Medizinische Universitätsklinik, 78 Freiburg i. Br. (Deutschland).

4.

M. Wachstein, J. Ortiz and **C. Fernandez**: Light and Electron Microscopic Histochemistry of "Thiolacetic Acid Esterase" in Human Blood Cells*.

Hydrogen sulfide liberated enzymatically from thiolacetic acid can be demonstrated histochemically as dark lead sulfide. A striking reaction occurs in certain cells of human blood and bone marrow, when freshly prepared smears were fixed by formalin vapors for 5 minutes with or without preceding freeze-drying and then incubated in the thiolacetic acid substrate as previously described (J. Histochem. **9**, 325, 1961). Ultra thin sections for observation in the E.M. were also prepared from reacted slides. All eosinophilic granules reacted intensely after 15—30 minutes incubation in light and E.M. sections, as did the granules of basophiles. Most but not all granules in polymorphonuclear leucocytes revealed a positive reaction after 30—60 minutes incubation also. In bone marrow smears, myelocytes and metamyelocytes showed varying amounts of granular activity. — Incorporation of 10^{-3} M eserine, 10^{-2} M E 600 or 10^{-4} M DFP in the substrate or pre-incubation for fifteen minutes in 10^{-4} M parachloromercuribenzoic acid, 10^{-2} M beta phenyl-propionic acid, 8×10^{-2} M sodium fluoride and 10^{-2} M KCN did not influence the reaction. Pre-incubation for fifteen minutes in 10^{-2} M iodoacetic acid, 10^{-3} M $CuSO_4$, 10^{-2} M $AgNO_3$ or 5×10^{-2} M sodium taurocholate inhibited the reaction. Pre-incubation in 5×10^{-2} M sodium cacodylate activated the eosinophilic staining. On the basis of these findings the hydrolases which act on thiolacetic acid in human leucocytes cannot be classified as cholinesterases or as typical A, B, or C esterases.

* Aided by grants H 5590 and AM-7191 from the National Institute of Health.

Dept. of Pathology, St. Catherine's Hospital, Brooklyn, New York, and Beth Israel Hospital, Passaic, New Jersey (USA).

5.

CHR. GOULIS and A. KOVATSIS: **The Alterations of Alkaline Phosphatase of Polymorpho-nuclear White Blood Cells during Pregnancy and the Influence of Progesterone, Oestradiol and Chorionic Gonadotropine on it.***

This paper is dealing with the alterations of the alkaline phosphatase of the polymorpho-nuclear white blood cells during pregnancy and also its alterations under the influence of progesterone, oestradiol and chorionic gonadotropine. Our material consisted of two series of pregnant and non-pregnant healthy women, the later used as controls, and also of one group of experimental animals. The alkaline phosphatase was estimated by the histo-chemical method of KAPLOW. — Our results are presented graphically by curves and tables. These are as follows: I. The alkaline phosphatase of the polymorphonuclear white blood cells shows a gradual increase during the whole pregnancy starting out from the first days. II. The alkaline phosphatase of the polymorphonuclear white blood cells increases to extremely high levels during labor and puerperium. III. The benzoic oestradiol causes an increase of the alkaline phosphatase of the polymorphonuclear white blood cells in rabbits. The progesterone decreases the alkaline phosphatase of the polymorphonuclear white blood cells, while the chorionic gonadotropine has no effect. — We have discussed the possible mech-anisms of the alterations of the alkaline phosphatase in the above groups.

* This investigation was supported by Royal Hellenic Research Foundation.

Aristoteles Université de Thessaloniki, Clinique d'Obstétrique et de Gynécologie, Saloniki (Greece).

6.

ERHARD AMBS, GUNTER F. BAHR and ELMAR ZEITLER: **Cytochemistry with the Electron Microscope: Study of Thrombocytes.**

Thrombocytes are difficult to assess with light microscopy, having a size close to the limit of the resolving power of the microscope. The reliability of size measurements is thus limited and suffer further from the fact that the actual shape of the thrombocyte is rarely as spherical as it appears under phase microscopy or stained preparations. The small size of the platelet also precludes any quantitative work with interference microscopy or micro-radiography. Thrombocytes also exhibit, when dried down, bizarre shapes and considerable inhomogeneities. In whole mounted preparations for electron microscopy the shape of the platelet is equally altered and no sensible size or volume determinations are possible. — Quantitative electron microscopy, being independent of object shape and object inhomo-geneities is ideally suited for quantitative cytochemistry on thrombocytes. Series of measure-ments on normal and thrombocytopathic material are presented and population features discussed. The results imply that the release of thrombocytes from megakaryocytes is not a random process but follows a distinct pattern. — The results of attempts to determine the amount of sulfonic groups in individual platelets using cobalt and copper containing electron stains is reported.

University of Würzburg, 87 Würzburg (Germany) and Armed Forces Institute of Pathology, Washington, D.C. (USA).

7.

HANS E. MÜLLER: **Die Zusammensetzung des Plasmaproteinfilmes an der Oberfläche von Erythrocyten*.**

Plasmaproteine (Reinproteine der Behringwerke, Marburg, und Fraktionen von Serum-elektrophoresen) wurden mit radioaktiven Isotopen (^{131}Jod, ^{75}Selen und ^{35}Schwefel) *in vitro* und *in vivo* markiert. Die so markierten Proteine setzten wir ungerinnbarem Vollblut zu und unterwarfen das Blut nach einer Inkubationszeit von einigen Stunden einer Waschprozedur. Im Verlauf der einzelnen Waschvorgänge nimmt die Menge der in der Lösung frei gelösten Proteine schneller ab als die Menge der an die Erythrocytenoberfläche adsorptiv gebundenen Proteine. Diese Adsorption ist für verschiedene Proteine verschieden groß. Daraus ergibt sich, daß sich der Plasmaproteinfilm an der Oberfläche von Erythrocyten in seiner Zusammen-setzung wesentlich von der Zusammensetzung dieser Proteine im Plasma unterscheidet. Wir errechneten die prozentuale Zusammensetzung dieses Proteinfilmes an der Erythrocytenoberfläche und fanden eine Anreicherung von Präalbumin, Lipoprotein und Gamma-Globulin, eine geringe Anreicherung von Fibrinogen und Alpha$_2$-Makroglobulin und schließlich eine Verarmung an Albumin und Transferrin. — Die Untersuchungen wurden bisher an Menschen- und Kaninchenerythrocyten durchgeführt. In beiden Fällen fand sich das gleiche Ergebnis. — Unter pathologischen Bedingungen zeigten sich Verschiebungen dieser Relationen. Die Anreicherung von Gamma-Globulin sinkt, Albumin wird dagegen vermehrt adsorbiert. Die Veränderungen der übrigen Plasmaproteine waren nicht signifikant.

The Composition of the Plasma Protein Film at the Surface of Erythrocytes. Plasma proteins (pure proteins from Behringwerke/Marburg and fractions from plasma electrophoreses) were marked with radioactive isotopes (Iodine-131, Selenium-75 and Sulphur-35) *in vitro* and *in vivo*. Solutions of the marked proteins were mixed with oxalated whole blood, which was incubated for some hours and then washed. In the course of the various washing procedures the marked protein in solution decreased more rapidly than the marked protein adsorbed at the surface of the red blood cells. This affinity for the red blood cells varies for different proteins. It follows that the composition of the plasma protein film at the surface of the erythrocytes differs considerably from that of the freely soluble plasma proteins. We have calculated the percentage composition of this plasma protein film. There is an increase in pre-albumin, lipoprotein and gamma-globulin, a slight increase in fibrinogen and alpha$_2$-macro-globulin and finally a decrease in albumin and transferrin. — Until now the investigations were carried out on human erythrocytes and rabbit erythrocytes. In both cases the same results are obtained. — Under pathological conditions there is a shift of these relationships. The adsorption of gamma-globulin is decreased while that of albumin on the other hand is increased. There is no significant change in the remaining plasma proteins.

* Mit Unterstützung der Deutschen Forschungsgemeinschaft.

I. Med. Klinik und Poliklinik der Universität, 65 Mainz (Deutschland).

8.

K. Hübner und H. G. Schiemer: Quantitativ-histochemische Untersuchungen zur Funktion von Plasmocytomzellen*.

1290 Plasmazellen 11 verschiedener Plasmocytome wurden interferenzmikroskopisch (Jamin-Lebedeff-System, Zeiss) untersucht. Die Meßwerte wurden zu den jeweiligen Serumeiweißwerten im Blut in Beziehung gesetzt. Es zeigt sich, daß die Funktionsaktivität der Plasmazellen, d.h. ihre Proteinsynthese von der Dichte des Cytoplasmas abhängig ist, während Beziehungen zur Qualität des synthetisierten Paraproteins nicht bestehen. Darüber hinaus lassen sich mit der Regressionsanalyse der Meßdaten drei charakteristische Plasmazellfunktionstypen abgrenzen: 1. Zellen mit gesteigerter Proteinsynthese, 2. Zellen mit geringer oder nahezu fehlender Proteinbildung und 3. Zellen mit gesteigerter Proteinsynthese, aber gestörter Eiweißabgabe aus der Zelle. Die letzteren Zellen zeigen histologisch eine Häufung Russelscher Körperchen im Cytoplasma. Dieser Befund läßt erkennen, daß die Russelschen Körperchen nicht Ausdruck der optimalen Syntheseleistung der Plasmazellen sein können, sondern Ausdruck der gestörten Eiweißabgabe sind. Cytophotometrische Paralleluntersuchungen mit dem UMSP I der Firma Zeiss an Feulgen-gefärbten Kernen zeigen, daß die Eiweißsyntheseleistung der Plasmazellen nicht vom DNA-Gehalt der Zellkerne abhängig ist. Vielmehr geht die Proteinsynthese dem Gehalt an Nicht-DNA-Substanzen der Kerne parallel.

Quantitative-Histochemical Investigations of the Functions of Plasma Cells. 1290 plasma cells, originating from 11 different plasmacytomas, have been investigated by interference microscopy (Jamin-Lebedeff-system, Zeiss). The measurements have been related to the actual protein content of the serum. It appears that the functional activity of plasma cells (i.e. the synthesis of protein) depends on the density of the protoplasm. There is no relationship to the quality of the synthesized para-protein. A regressive analysis of the measurements reveals three characteristic functional types of plasma cells: 1. cells with an increased rate of protein synthesis, 2. cells which synthesize small or almost negligible amounts of protein, 3. cells with an increased rate of protein synthesis but with malfunctioning of protein output. The latter cells show histologically an increase in Russel bodies in the cytoplasm. These findings prove that Russel bodies do not indicate maximum efficiency of the cell but a malfunction of the prctein output. Parallel cytophotometric tests on Feulgen-stained nuclei (UMSP I/Zeiss) show that the protein synthesis of plasma cells does not depend on the DNA-content of the nucleus. The protein synthesis runs parallel to the content of non-DNA-substances of the nuclei.

* Mit Unterstützung der Deutschen Forschungsgemeinschaft.

Pathologisches Institut der Universität, 6 Frankfurt a. M. (Deutschland).

9.

Leonard F. Bélanger: Histochemical Evidence of Proteolytic Activity in Bone*.

Proteolytic activity has been revealed at the level of the large osteocytes of young chick and rat bone, by digestion of gelatin in a photographic emulsion substrate. — The reaction was enhanced by pretreatment of the animals with parathyroid extract. — The enzyme was active in incubation media of pH 3.5 to 7.6 but inhibited in media of pH 8 and above. — The reaction was inhibited by addition to the incubating medium of sodium fluoride or potassium ferrocyanide. It was also inhibited by fixation in ethanol-formaldehyde-acetic acid. — The present results seem to confirm and explain further the resorptive potential of the mature osteocytes.

* With financial support from the Medical Research Council of Canada, the Canada Dept. of Agriculture and the United States Atomic Energy Commission (Contract No. AT. 30-1-2779).

Dept. of Histology and Embryology, Faculty of Medicine, University of Ottawa, Ottawa (Canada).

10.

Harold M. Fullmer and Clifton Link, jr.: Studies on Demineralization with Enzymatic Preservation.

Unfixed femurs from 48 rats aged 21 days were constantly stirred in a demineralizing solution consisting of 10% EDTA in 0.1 N tris (hydroxymethyl) aminomethane buffer pH 6.98 (modification of Balogh, J. Histochem. Cytochem. *10*, 232, 1963) for periods of 0, 1, 2, 3, 4, 5, 6, and 7 days. Thereafter femurs were frozen on a block of CO_2 ice, center sagittal sections were cut at 10 μ in a cryostat, and incubated for 30 min at 37⁰ C for the determination of succinic dehydrogenase according to the method of Glick and Nayyar (J. Histochem. Cytochem. *4*, 389, 1956). Representative sections were mounted for microscopic examination. Others were quantitatively extracted with a 1:1 ethanol-tetrachloro-ethylene mixture and read in a spectrophotometer at 505. Each extracted section was quantitatively analyzed for protein according to the method of Lowry et al. (J. Biol. Chem. *193*, 265, 1951). Fourty-eight determinations for protein and 48 analyses of extracted formazan were made. Data indicate no significant loss of enzymatic activity in specimens demineralized for two days. Approximately 50% enzymatic activity was manifest in specimens demineralized from 3—5 days, and 30—40% in those incubated up to 7 days.

National Institute of Dental Research, National Institutes of Health, U.S. Public Health Service, Dept. of Health, Education and Welfare, Bethesda 14, Maryland (USA).

Dienstag, Tuesday, Mardi, 18. August 1964

Haut · Skin · Peau

1.

J. RAEKALLIO: Enzymatic Response to Injury in Skin Wounds.

The methods of enzyme histochemistry seem suitable for detection of the very first reactions in the wounded tissue (cf. RAEKALLIO, Nature *188*, 234, 1960). — Circular excision wounds, 5 mm in diameter, were cut in the dorsal skin of several hundreds of albino rats and guinea pigs. The animals were decapitated at intervals of $^1/_2$, 1, 2, 4, 8, 16, 24, 48, 72, 96, and 120 hours after the injury. Skin flaps containing the wounds were removed immediately and frozen fresh or fixed. The activity of the following enzymes was visualized: nonspecific phosphatases, adenosine triphosphatase, nonspecific esterases, β-glucuronidase, transglucosylases, aminopeptidase, cytochrome oxidase, monoamine oxidase, and succinic dehydrogenase. — In the immediate vicinity of the wound edge, a central or superficial zone, 200 to 500 μ in depth, showed decreased enzyme activity in the connective tissue cells. This decrease, commencing one to eight hours after injury, was considered as an early sign of imminent necrosis. Surrounding the central area, a 100 to 300 μ deep peripheral zone exhibited an increase in enzyme activity, appearing one to eight hours after wounding. This initial increase seemed to represent an adaptive defence mechanism by the local connective tissue cells, as a response to injury.

Dept. of Forensic Medicine, University of Turku, Turku 3 (Finland).

2.

KLAUS WOLFF: Enzyme Histochemical Studies on the Healing-Process of Split Skin Grafts.

The transplantation of different tissues has been the subject of numerous investigations during the past decades. Recently enzyme-histochemical methods have been applied to studies on vital and post mortem skin wounds and important informations have been obtained regarding the distribution of different enzymes in the traumatized tissue (RAEKALLIO). In this study Minnesota pigs were used for transplantation experiments and the healing process of auto- and homeografts was followed enzyme-histochemically up to 8 weeks after the operation. The activities of the following enzymes were demonstrated: Nonspecific alkaline phosphatase, nonspecific acid phosphatase, 5-nucleotidase, ATPase, indoxylesterase, aminopeptidase, DPNH-diaphorase and succinodehydrogenase. The comparison of healing normal wounds, auto- and homeografts reveals characteristic enzymatical changes both within the transplant- and wound tissues. Especially the activities of DPNH-diaphorase and succinic dehydrogenase disappear relatively soon from the graft and return only after the vascular connection between graft and wound has been completed. On the other hand there is strong evidence that prior to this event the survival of the transplant is supported by diffusion mechanisms. The healing process of autotransplants corresponds with normal wound healing to a certain extent while the loss of the homeograft is accompanied with typical enzyme-histochemical changes.

I. Universitäts-Hautklinik, Alserstr. 4, Vienna IX (Austria).

3.

KURT MACH: Über das Vorkommen von Tweenesterase in der menschlichen Haut.

Abweichend von den bisher bekannten Ergebnissen über die Tweenesterasen in der menschlichen Haut konnte mit Polyoxyäthylensorbitan-monooleat (Tween 80) als Substrat sowohl in der Gefäßwand der mittleren Gefäße und Haarfollikel als auch in den Talg- und Schweißdrüsen sowie deren Ausführungsgängen bis zum Str. corneum ein spaltendes Ferment nachgewiesen werden. Es ist anzunehmen, daß dieses Ferment für den Fettgehalt und für den pH-Wert der Hautoberfläche von besonderer Bedeutung ist.

Tween-Esterase Activity in the Human Skin. An enzyme hydrolysing Polyoxyethylen-sorbitan-monooleate (Tween 80) was demonstrated in the walls of medium-sized vessels, hair follicles, in the sebaceous and eccrine sweat glands as well as in the sweat gland ducts up to the level of the epidermal horny layer. This does not correspond entirely with the results of hitherto published investigations. It is assumed that this enzyme may be of some importance for the presence of lipids and for the pH of the epidermal surface.

I. Universitäts-Hautklinik, Alserstr. 4, Vienna (Austria).

4.

J. Hasegawa: Naphthylamidase, a Peptide Hydrolase, in Wheat Germ and the Human Skin*.

Accumulating evidence indicates that a peptide hydrolase, not leucine aminopeptidase nor any other isolated enzyme, is responsible for the tissue hydrolysis of L-leucyl-β-naphthylamide. A study of a wheat germ preparation containing a large concentration of this naphthylamidase reveals that this enzyme prefers as its substrate a structure resembling the β-naphthylamine structure in the residue containing the peptide N, 2- a peptide linkage in the β position of the naphthylamine ring and 3- a free α amino group in the residue containing the peptide C. Two isoenzymes of the wheat germ naphthylamidase hydrolyze the naphthylamides of L-arginine and L-methionine at different rates although each fraction hydrolyzes both as well as other naphthylamidase. — A study of the human epidermis and dermis separately indicates that naphthylamidase is present in greater concentration in crude enzyme extracts than peptidases which hydrolyze L-leucine amide, glycylglycine, DL-alanylglycine and L-leucyl-glycylglycine. The enzyme is present in greater amounts in the epidermis than in the dermis and the naphthylamidase of the skin hydrolyzes the naphthylamides of L-leucine, L-methionine and L-arginine more rapidly than DL-alanine. The function of this enzyme in cellular metabolism is still unknown.

* Aided by USPH grant No CA-09791 from the National Cancer Institute and AM-07275 from the Nationa Institute of Arthritis and Metabolic Diseases, United States Public Health Service.

Northwestern University Medical School (Dept. of Dermatology) and the Veterans Administration Research Hospital, Chicago, Illinois (USA).

5.

Maire Bradshaw and Max Wachstein: Adenosine Triphosphatase Activity in Neoplasms of the Human Skin*.

In previous studies on human skin adenosine triphosphatase activity was demonstrated in melanocytes and in epidermal cells (Bradshaw, M., Wachstein, M., Spence, J. and Elias, J. M., J. Histochem. and Cytochem. *11*, 465—473, 1963). In continuation of these studies enzyme activity is being demonstrated in glands and in neoplasms of the skin. Mitochondrial adenosine triphosphatase is more active in the eccrine glands than in any other glands of the skin and the staining pattern is like that described previously for the salivary glands (Wachstein, M., Bradshaw, M. and Ortiz, J., J. Histochem. and Cytochem. *10*, 65—74, 1962). In neoplastic conditions, nevus glands, both pigmented and non-pigmented, contain numerous mitochondria demonstrating enzyme activity. Melanocarcinomas, also, show much mitochondrial adenosine triphosphatase activity as well as non-mitochondrial activity and non-specific staining. In sharp contrast is the basal cell carcinoma as well as the squamous cell carcinoma where the pattern of mitochondrial adenosine triphosphatase activity is strikingly decreased. This is in contrast to the behavior of other mitochondrial enzymes, such as succinic dehydrogenase, which are very active at the growing edges of these tumors. The activity of acid phosphatase, 5-Nucleotidase, alkaline phosphatase, diphosphopyridine nucleotide diaphorase and of triphosphopyridine nucleotide diaphorase in neoplasms of the human skin will be discussed also.

* Aided by Grant No. AM-7191 from the National Institutes of Health.
St. Catherine's Hospital, Brooklyn 3, New York (USA).

6.

Kimmo K. Mustakallio: Phospholipids of Human Epidermis*.

Sphingomyelin, lecithin, phosphatidylserine, phosphatidylethanolamine and two as yet unidentified phospholipid fractions were detected by means of a thin layer chromatographic method in epidermal sheaths separated *in vivo* from intact human skin with the suction blister method of Kiistala and Mustakallio. — The deepest layer of stratum corneum was quite sharply delineated from other epidermal layers by its affinity for stains and fluorochromes interacting with phospholipids and lipoproteins. This emerged from a study of cryostat sections freeze-substituted in acetone and treated by using the following methods: (1) Luxol Fast Blue MBS for phospholipids (Klüver and Barrera), (2) Rhodamine 6 G in chloroform for acidic phospholipids (Harris and Gambal), (3) Calcium-chelate of chlortetracycline for phospholipoproteins (Levonen and Mustakallio), (4) Toluidine blue staining of heparin, precipitated by phospholipoproteins in the presence of calcium (Mustakallio and Levonen), and (5) Fluorescein labelled antibody for human beta-lipoprotein. — Apparently the phospholipoproteins concentrated in the deepest layer of stratum corneum play an important role in the barrier functions of human epidermis.

* Aided by a grant from Emil Aaltonen Foundation.
Dept. of Dermatology, University Central Hospital, Helsinki (Finland).

7.

JOHN FABIANEK, ANTHONY HERP and WARD PIGMAN: Acid Mucopolysaccharides in the Skin of Some Mammals*.

Hyaluronic acid (HA), chondroitin sulfate (CS) and heparin (H) were determined in the dry defatted skin of guinea pigs, cats, rats, rabbits and men. The mucopolysaccharides extracted from the skin were precipitated with cetylpyridinium chloride. From the precipitate, HA, CS and H were fractionally extracted by successive treatment with 0.4, 1.2 and 2.1 M sodium chloride, respectively, to give fractions indicated by SCHILLER et al. (J. Biol. Chem. *236*, 983, 1961) to consist principally of the three types of mucopolysaccharides. The technique was previously described (Nature *200*, 784, 1963). The amounts of apparent HA, CS and H were determined as uronic acid by the carbazole method. For these species, the total mucopolysaccharide content expressed as uronic acid generally varied between 800 and 1000 μg per gram of dry defatted skin. The proportion of apparent HA generally was in the range 55 to 65% of the total uronic acid. The proportion of the heparin fraction was about 15% for rat and about 10% for cat skin. Guinea pig skin was free of heparin, and the amounts in rabbit and human skin were of borderline significance.

* Supported by grants from the USPHS (AM-04619) and J. Polachek Foundation.

New York Medical College, Biochemistry Dept., New York 29, N.Y. (USA).

8.

M. D. GEDEVANISHVILI: The Seasonal Variation of Mucopolysaccharide Constituents in the Ground Substance of Frog Dermis.

It was demonstrated previously that the distribution pattern of the glycoprotein in the dermal ground substance of *Rana ridibunda* may be changed under certain conditions of physiological experiment (GEDEVANISHVILI, 1960). The present work deals with the results of a histochemical study carried out to determine whether the constitution of the ground substance could be influenced by the change of seasons. Thus the acid mucopolysaccharide located in the upper edge of the compact layer of dermis can be revealed readily during May and June by Alcian blue and has an appearance of a relatively wide furrow. At the same time it shows strong basophilia while staining metachromatically at pH 1.7 with toluidine blue solutions. The basophilia of this mucopolysaccharide in the dermis of winter frogs appears to be diminished significantly according to its binding of toluidine blue at the pH value of 4.0 whereas the Alcian blue reveals only traces if any at all. — The acid mucopolysaccharide of the mucous cutaneous glands differs from that of the dermis in that it can be revealed by the toluidine blue solutions at pH 3.0 and is stainable constantly with alcian blue. These properties of the glandular acid mucopolysaccharide are not altered by the changes of season. — The secretion granules of the fusiform connective tissue cells in the spongy layer of dermis somewhat resemble histochemically the granules of the mammalian mast cells. Thus they show a marked chromotropism towards toluidine blue solutions at the lower values of pH and do not stain by Alcian blue. The histochemical properties of these granules do not change according to seasons but it is noteworthy that the number of the fusiform cells containing these granules appears to be markedly diminished in winter. — The distribution pattern of the glycoprotein of the dermal ground substance did not show any variations according to the changes of season and neither did the glycoprotein of the cutaneous mucous glands. Therefore histochemical methods enabled us to demonstrate regular changes of the properties of the acid mucopolysaccharide in the dermal ground substance according to the season of the year. The number of the fusiform cells containing acid mucopolysaccharide varies too. This fact itself indicates the existence of regulating mechanisms, in the metabolism of the non-epithelial mucous substances, which are responsible for the correlation of general biological status in these animals.

Institute of Pharmacochemistry, Georgian Academy of Sciences, Perovskaya street 22, Tbilisi (USSR).

9.

TIBOR DONÁTH: Fluoreszenzanalyse der PAS-positiven Substanzen im Scheidenepithel.

Im menschlichen Scheidenepithel wurde in den verschiedenen Phasen des Zyklus die Histotopochemie der Polysaccharid-Derivate mittels eines Fluoreszenzfarbstoffes vom Typ des Schiffschen Reagens, mit der Perjodsäure-Choriphosphin 0-Schiff-Reaktion nach KASTEN untersucht. Dieses Verfahren bewirkt eine spezifische intensive Fluoreszenz der Gewebspolyaldehyde. — Im Verlauf der Untersuchungen wurde ferner in der Scheidenschleimhaut ovarektomierter hormonbehandelter Ratten eine für die einzelnen Phasen charakteristische Fluoreszenz beobachtet, die den vorherrschenden Hormonwirkungen entsprechend mit dem

qualitativen und quantitativen Erscheinen der verschiedenen Polysaccharid-Derivate im
Epithel in Zusammenhang gebracht wird. — Die Fluoreszenzintensität der Perjodsäure-
Choriphosphin 0-Schiff-positiven Substanzen wird in den einzelnen Epithelschichten mit
semiquantitativer Methode bestimmt. — Auf Grund des fluoreszenzmikroskopischen Bildes
werden die oberflächlichen und die intercellulären Amylase-resistenten Substanzen im
Vaginalepithel analysiert.

Fluorescence-Analysis of the PAS-Positive Substances in the Vaginal Epithelium. In the
different cyclic phases of the human vaginal epithelium the histotopochemistry of polysaccharide derivatives was examined with a fluorescent Schiff type reagent, the periodic
acid-coriphosphine 0-Schiff reaction of KASTEN, as a result of which a specific intensive
fluorescence is brought about in the tissue-polyaldehydes. — Further examinations of the
vaginal epithelium of ovarectomized and hormone-treated rats revealed a characteristic
fluorescence in the several phases that is supposed to be due to the qualitative and quantitative
appearance—according to the prevailing hormone effects—of the different polysaccharide
derivatives in the epithelium. — In the different epithelial layers the fluorescence-intensity
of the periodic acid-coriphosphine 0-Schiff positive substances has been determined by semi-
quantitative method. — Based on the fluorescent-microscopical picture the superficial and
intercellular amylase-resistant substances are analyzed in the vaginal epithelium.

Anatomisches Institut der Medizinischen Universität Budapest, IX Tüzoltó u. 58 (Ungarn).

10.

**V. PRETO PARVIS and G. ROSSI: Histochemical Findings on Secretion of Human Uterine Tube
Epithelium.**

The authors have carried out researches on secretions of the ampullar segment of surgically
removed salpinges, during the lutein phase. The results with Alcain blue, Hale reaction,
aldehydefuchsin and PAS reaction indicate its aminopolysaccharide character. The staining
with toluidine blue is low and bluish. However if the dye is used in buffers at graded pH, the
secretion appears metachromatic even at pH 0.4, bluish at pH 3.4. Also Alcian positivity starts
at pH 0.5, then is more intense at pH 2.5. With Alcian yellow at pH 2.5, after staining
with Alcian blue at pH 0.5, the secretion appears green. These tests suggest the presence
of acid sulfated and carboxyl mucins; this evidence is proved by loss of dye-binding capacity
after methylation, and by partial restoration after demethylation. The Bracco-Curti (with
benzidine) and Geyer (with dianisidine) reactions confirm the evidence of sulphate groups.
The use of neuraminidase, altering the results to basic blues, suggests the coexistence of
sialic acid. Salivary amylase, reducing PAS reaction of secretion and cytoplasm, shows
the presence, along with PAS-positive glycoproteides, also of glycogen. It is possible that
the secretion, chemically complex and of apocrine type, beside constituting the surface
mucous layer, will supply utilizable materials for germinal cell metabolism.

*Istituto di Istologia ed Embriologia Generale dell'Università di Milano, Via Ponzio 7,
Milano (Italia).*

**Zelle und Zellbestandteile · The Cell and its Components
La cellule et ses éléments**

1.

**G. P. STUDZINSKI and R. LOVE: A Simple Method for Visualization of the Internal Structure
of the Nucleolus*.**

Nucleolar detail can be visualized for light microscopy by utilizing the selective affinity
of the nucleolus for certain heavy metals. Zinc is particularly suitable for this purpose;
when fixed or unfixed HeLa cells grown on coverslips are incubated for 30 min in zinc acetate
at pH 6.5, zinc is deposited within the nucleolus and can be demonstrated by subsequent
treatment with diphenylthiocarbazone (dithizone). The red zinc-dithizone complex is seen
only within the nucleolus, and its deposition in the nucleoli of logarithmically growing cells
presents a reproducible pattern. The characteristic feature is the visualization of multiple
(1—20) granules within the nucleolar substance, which resemble in some respects the "nucleo-
lini" demonstrable by toluidine blue-molybdate (LOVE, R., J. Histochem. Cytochem. *10*,
227, 1962). In addition, the zinc-dithizone reaction stains diffusely, but less intensely, the
bulk of the nucleolar material, with the exception of the nucleolar "vacuoles". One nucleolar
granule, located at the periphery of a vacuole, is usually more prominent than the other
granules. — Treatment of the cells with nucleases, hot TCA and alkali at 37⁰ C shows that

the retention of zinc within the nucleolus is due to RNA. It is uncertain, however, whether the deposition of the colored complex occurs at the sites of its formation, or if the granules which are visualized merely offer conditions favorable for deposition of a diffusible product.

* This work was supported by USPHS Grant CA-05402-03 VR.

Pathology Dept., Jefferson Medical College, Philadelphia, Pa. (USA).

2.

I. B. ZBARSKY, N. G. KHRUSHCHOV and L. P. YERMOLAEVA: On the Composition and Biological Role of the Nucleolus-Associated Heterochromatin.

The nucleolus-associated heterochromatin is cytologically differentiated in its tinctorial properties. Nevertheless, isolated nucleoli obtained by different workers contain large amounts of DNA which appears to originate from euchromatin. — Subsequent extraction of isolated nuclei or tissue slices with 0.14 M NaCl and 1.5 M NaCl removes the bulk of the deoxyribonucleoprotein while nucleoli and nuclear envelopes remain. Cytological analysis shows that even after continuous extraction with solutions of high ionic strength nucleoli preserve the nucleolus-associated chromatin. The ratio of molar sum of adenine plus thymine to that of guanine plus cytosine $\left(\dfrac{A+T}{G+C}\right)$ in the DNA remaining in the nucleolus-associated chromatin was found to be 0.8—1.1. This coefficient, approaching the corresponding value for the ribosomal RNA, makes it probable that the DNA of the nucleolus-associated chromatin may function as a template for the ribosomal RNA synthesis.

Institute of Animal Morphology, Academy of Sciences of USSR, Vavilov street 12/2, Moscow B-133 (USSR).

3.

J. A. VINNIKOV, L. K. TITOVA, I. V. OSIPOVA and A. A. BRONSHTEIN: Cytochemical and Electron Microscopical Investigation of Nucleolar RNA Extruding into Cytoplasm.

After single or repeated exposure of guinea pigs to rotation for 3 minutes at 10 g radial acceleration, cytochemic and electron microscopic methods have revealed extrusion of nucleolar RNA from utricular hair cell nuclei toward basal poles of these cells. After one exposure, the bright red colouring of nucleolar RNA, stained after Brachet, is located close under the nuclear membrane, within the cytoplasm, where original cytoplasmatic RNA has usually disappeared. Following two or three rotations, the amount of extruded RNA proves to have frown to form a sphere, having an apex tapering toward the cell base, and tending to separate from the nucleus. With the electron microscope, granulated nucleolar substance may be seen, as if sifts through pores of the double nuclear membrane, with particular intensity at the cell base. Judging by their histochemical characteristics, the extruded granules represent ribosomes, presumably containing messenger RNA. They induce endoplasmatic membranes to accumulate around them, forming a distinctive spiral complex about the basal pole of the cell.

Setschenov Institute of Evolutionary Physiology, Academy of Sciences of USSR, Leningrad (USSR).

4.

ANDRZEJ VORBRODT and ZENON STEPLEWSKI: Cytochemical Investigations of the Relationship between Nucleolus and Lysosomes.

It is well known that carcinogen, thioacetamide (TAA), when given intraperitoneally at nontoxic levels (50 mg/kg), causes great enlargement of the rat liver cell nucleoli. This enlargement results from the block in the release of nucleolar RNA (some kind of the "enucleolation"). According to that, TAA seems to be very useful in the *in vitro* studies on the metabolic role of nucleolus at least in the rat liver cells. — In our studies the main attention was directed towards the behaviour of lysosomes. As a markers of these cytoplasmic constituents the staining reactions for acid phosphatase, acid desoxyribonuclease and E-600 resistant esterase were applied. The examination of specimens in light and electron microscope reveals that the nucleolar blockade leads to the suppression of the cytochemical reactions for above mentioned enzymes. Lysosomes become smaller and dispersed throughout whole cytoplasm, eventually they are localized close to the nuclear membrane. — These results may indicate that the normal polarity of the liver cell as well as normal elaboration of lysosomes or their enzymes are under nuclear (nucleolar) control.

Dept. of Tumour Biology, Institute of Oncology, Gliwice (Poland).

5.

M. WOLMAN: On the Nature of Lysosomes.

Studies of artificial breakdown of the myelin sheath have shown that in the presence of excess calcium, or in other conditions which enhance hydrophobia of lipid membranes these disintegrate into "water in oil" emulsions. In the presence of excess sodium or in other conditions which enhance hydrophilia of lipid membranes an "oil in water" type of emulsion is formed. These findings suggested the possibility that lysosomes, like microsomes, are products of breakdown of the endoplasmic reticulum (and of the cell membrane ?) and differ from the latter in representing a "water in oil" rather than "oil in water" pattern. Experiments done with GOMORI's lead procedure and the standard azo procedure for acid phosphatase adduced confirmatory evidence for this notion. It has been found that demonstration of acid phosphatase activity in fresh material can be inhibited by excess calcium and that this inhibition is reversible. The inhibition of acid phosphatase activity by calcium is minimal or absent in fixed tissue or in sections extracted with lipid solvents. The findings indicate that in the native state acid phosphatase molecules are bound to a polar lipid. Artificially induced changes in the milieu, but possibly also similar changes *in vivo*, may cause two types of breakdown of the lipid membranes to which these enzyme molecules are attached: in the one type the hydrophilic enzyme bathes in the surrounding medium, whereas in the second type the enzyme molecules are enclosed within a lipid-bounded bag.

Dept. of Pathology, Government Hospital Tel-Hashomer (Israel).

6.

W. STRAUSS: Observations on the Relationship between Lysosomes and Phagosomes in Various Tissues of the Rat by Successive Staining for Acid Phosphatase and Intravenously Injected Horseradish Peroxidase in the Same Sections*.

In previous work, lysosomes were stained red, phagosomes were stained blue, and phago-lysosomes were stained purple in cells of the kidney and liver, by staining formaldehyde-fixed, frozen sections for acid phosphatase by an azo dye procedure and for intravenously injected horseradish peroxidase by benzidine. The double staining procedure has been applied to make a survey, in peroxidase-treated rats, of the occurrence of phagosomes, lysosomes, and phago-lysosomes in various tissues such as spleen, lung, thymus, prostate, pancreas, intestine, and heart. Observations on the size and number of acid phosphatase-positive granules and on the intensity of the acid phosphatase reaction, before and after injection of horseradish peroxidase, and observations on the gradual disappearance of the injected peroxidase in various tissues will be reported. Technical problems, especially concerning the cytochemical reaction of peroxidase with benzidine, will be discussed. Modifications of the conventional procedures will be described which prevent the fading of the blue reaction product of peroxidase with benzidine, crystallization artifacts of "benzidine blue", and artifacts of diffusion and adsorption of injected peroxidase in the tissue.

* Aided by an established investigatorship and a research grant from the American Heart Association.

Institute for Medical Research, The Chicago Medical School, Chicago 12, Illinois (USA).

7.

JAN L. E. ERICSSON and BENJAMIN F. TRUMP: The Significance and Mode of Formation of Acid Phosphatase Containing Cytoplasmic Inclusion Bodies (Cytosomes and Cytosegresomes) in Normal Kidney and Liver Cells*.

By electron microscopy the cytoplasm of renal proximal tubular cells and hepatic parenchymal cells contains a well defined group of single membrane-limited inclusion bodies termed *cytosomes;* these bodies showed acid phosphatase activity as revealed by electron microscopic histochemistry (incubation of $\sim 50\,\mu$ thick sections of glutaraldehyde- or formaldehyde-fixed tissues in the Gomori substrate, post-fixation in osmium tetroxide and embedding in Epon). Acid phosphatase activity was also demonstrated in bodies limited by a single or a double membrane in which various well recognizable cytoplasmic organelles were segregated. Bodies with a similar appearance have been reported in the literature to occur during various pathologic and physiologic conditions. They have often been termed "lysosomes" and "cytolysosomes". Due to the partly controversial and conflicting use of these terms in morphologic and biochemical investigations the purely descriptive and universally applicable name *cytosegresomes* is preferred for such organelles in morphologic studies. The cytosegresomes appear to be formed through the wrapping of membranes, possibly derived from the Golgi apparatus or the endoplasmic reticulum, around diverse cytoplasmic structures. All transitions between cytosegresomes and cytosomes may be encountered; this indicates that at least some of the cytosomes are derivatives of cyto-segresomes. The evidence suggests that the cytosegresomes are involved in the intracellular digestion of cytoplasmic organelles. This process apparently occurs under normal conditions and probably reflects a rapid physiologic turnover of cytoplasmic structures in the metabolically active liver and kidney cells. The number, size and pleomorphism of cytosegresomes appear to increase markedly following several types of cell injury in both humans and experimental animals. Examples of cytosegresomes in both normal and abnormal cells will be shown.

* Aided by a grant from the Swedish Life Insurance Companies and by United States Public Health Science Grant No. AM 07919.

Depts. of Pathology at Sabbatsberg Hospital, Karolinska Institutet, Stockholm (Sweden) and University of Washington, Seattle, Washington (USA).

8.

JØRGEN ROSTGAARD* and RUSSELL J. BARRNETT: Nucleoside Phosphatase Activity in Pinocytic Vesicles of Smooth Muscle and in Other Structures of Tunica Muscularis of Small Intestine.

During a study of the relationship of enzymatic activity to absorption in the small intestine, discrete localization of certain phosphatases was found in the little studied tunica muscularis of the small intestine. These findings were further investigated by a study of frozen sections (60 μ) of glutaraldehyde-fixed small intestine of rat which were incubated in substrate-free media and in media containing ATP, ADP, AMP, ITP, IDP, IMP, CTP, GTP, apha-naphthylphosphate and beta-glycerophosphate at pH 7.2 with lead nitrate as the capture reagent. In addition, alpha-naphthylphosphate at pH 8.2 and beta-glycerophosphate at pH 5.0 were used as substrates in other incubations. Thin sections of this material embedded in Epon were examined with an electron microscope. — Enzyme(s) hydrolyzing nucleoside di- and tri-phosphates were found associated with the membranes of the pinocytic vesicles of smooth muscle cells, and in the narrow space between adjacent cells. Other localizations were: di-phosphatase activity in the Golgi apparatus of the muscle cells; and tri- and di-phosphatase activity in pinocytic vesicles of capillary endothelium, and in the 200 Å space between the plasma membrane of Schwann cells and unmyelinated axons. No enzymatic reaction occurred with monophosphate substrates, regardless of pH, and no non-specific staining occurred in the substrate-free, lead-containing media. The possible relationship of enzyme activity to permeability will be discussed.

* Anna Fuller Fellow. Present address: Dept. of Anatomy, The Royal Dental College, Copenhagen, Ø, Denmark.

Dept. of Anatomy, Yale University School of Medicine, New Haven, Connecticut (USA).

9.

HERBERT VOSS: Zum histochemischen Nachweis der Phagozytose von Polyelektrolytkomplexen.

Polyelektrolyte werden wegen ihrer hohen Oberflächenladung von Zellen schlecht oder gar nicht phagozytiert. So konnten LIEBHABER und TAKEMOTO bei dem Angebot von Polyglucosesulfat-S^{35} an L-Fibroblastenkulturen keine Radioaktivität in den Zellen nachweisen. In der gleichen Weise werden nach eigenen Versuchen weder Heparin noch Dextransulfat von FL-Zellen in Flaschenkulturen phagozytiert, wenn man diese Polyelektrolyte in einem serumfreien synthetischen Medium löst. Mit Hilfe der Astrablaufärbung nach PIOCH läßt sich zeigen, daß die beiden Polyanionen zwar an die Oberfläche der Zellen adsorbieren, aber nicht in das Zytoplasma aufgenommen werden. Bildet man durch Zugabe von positiv geladenem DEAE-Dextran elektroneutrale Polysaccharid-Komplexe, dann verlieren diese ihre Löslichkeit, und das Präzipitat wird leicht in die Zellen eingeschleust. Der Mucopolysaccharidnachweis mit Astrablau zeigt dann beim Heparin-Komplex feine blaue Kugeln, welche die Zellen umgeben und z.T. in ihrem Plasma liegen. Das stärker negativ geladene Dextransulfat bildet feinere Präzipitate mit DEAE-Dextran, die in ähnlicher Weise in die Zellen gelangen. Es wird dadurch wahrscheinlich gemacht, daß Mucopolysaccharid-Antigen-Komplexe *in vivo* in ähnlicher Weise phagozytiert werden. Der nachgewiesene Vorgang scheint von Bedeutung zu sein für die Vorstellungen, welche für die Beseitigung von Antigenen und die Einleitung der Antikörperbildung bei Viruskrankheiten bestehen.

Histochemical Demonstration of the Phagocytosis of Polyelectrolyte Complexes. Because of their high surface tension polyelectrolytes are scarcely or not at all phagocytosed. For that reason LIEBHABER and TAKEMOTO did not succeed in demonstrating radioactivity inside cells after having added polyglucosesulphate-S^{35} to L-fibroblast cultures. I have now established that neither heparin nor dextran sulphate is phagocytosed by FL-cells in bottle cultures provided that these two polyelectrolytes are dissolved in a serum-free synthetic medium. With the help of PIOCH's Astra blue staining it is possible to demonstrate that both the polyanions are absorbed on the cell surface, but that they are not taken up by the cytoplasm. If positively charged DEAE-dextran is added, electro-neutral polysaccharide complexes are formed, which lose their solubility. The precipitate thus formed is easily taken up by the cells. In the case of heparin complexes Astra blue staining reveals fine blue granules, which are grouped around the cells. To some extent they can be found inside the cells. The more negatively charged dextran sulphate forms finer precipitates with DEAE-dextran than does heparin. The dextran sulphate complexes get into the cells in a similar way to the heparin complexes. It is therefore likely that *in vivo* mucopolysaccharide antigen complexes are absorbed in a similar manner. These observations seem to be of importance with respect to theories concerning the removal of antigens and the induction of antibody formation in virus diseases.

Bundesgesundheitsamt Robert-Koch-Institut, Pathologisch-histologisches Labor der Virusabteilung, 1 Berlin 65 (Deutschland).

10.

D. Wittekind und G. Rentsch: Zur Bedeutung der Histochemie und der Vitalfärbung für die Darstellung von Pinocytose-Vakuolen in Phagozyten.

Pinocytose (P.) wurde an Meerschweinchen-Phagozyten teils durch intraperitoneale Injektion von Casein, teils *in vitro* durch Immersion solcher Zellen in hochprozentiger isotonischer Gamma-Globulin-Lösung induziert. Die Untersuchungen wurden mit folgenden Farbstoffen durchgeführt: Acridin-Orange, Neutral-Rot und Toluidinblau, Pyronin und Alcian-Blau. Ferner wurde die PAS-Färbung angewendet sowie enzymhistochemische Verfahren (saure und alkalische Phosphatase u. a.). Die Untersuchungen ergaben, daß P.-Vakuolen, die bereits durch intravitalen Wasserentzug stabilisiert worden waren, zwar die Fixation mit wasserarmen bzw. wasserfreien Fixantien überdauern, allen wäßrigen Medien gegenüber jedoch sehr empfindlich bleiben. Zahlreiche histochemische Verfahren erweisen sich wegen der Wasserlöslichkeit der Einschlüsse als nicht anwendbar. Es werden einige Färbeverfahren so modifiziert, daß bis zur Untersuchung nur wasserfreie Medien zur Anwendung kommen. Bei Vitalfärbung gelingt die Darstellung der P.-Vakuolen nur mit Neutral-Rot — wie bereits von W. H. Lewis (1931) beobachtet — und mit Acridin-Orange. Die anderen basischen Farbstoffe färben diese Einschlüsse nicht. Bemerkenswert ist, daß mit Acridin-Orange fluorochromierte P.-Vakuolen intensive Rot-Fluoreszenz zeigen. Es wird angenommen, daß als Substrate dieser Rot-Fluoreszenz basophile Trägersubstanzen — wahrscheinlich saure Mucopolysaccharide (MPS) im Sinne von Gedigk u. Mitarb. (1956—1960) und Schmidt-Mathiesen (1955) vorliegen. Es ist ungeklärt, worauf die besondere Affinität von Neutral-Rot und Acridin-Orange zu diesen so veränderten P.-Vakuolen beruht. Die Färbung dieser Einschlüsse mit den üblichen Verfahren zur Darstellung der sauren MPS mißlingt wahrscheinlich ebenfalls wegen ihrer veränderten Wasserlöslichkeit. Die Darstellung der alkalischen Phosphatase gelingt in pinocytierenden Zellen unverändert gut und läßt keine morphologischen Unterschiede gegenüber nicht pinocytierenden Zellen erkennen. Die Befunde werden diskutiert.

The Importance of Histochemistry and Vital Staining for the Demonstration of Pinocytosis Vacuoles in Phagocytes. Pinocytosis (P) was induced in Guinea pig phagocytes by either i.p. injection of casein or else *in vitro* by immersing the cells in high-percentage isotonic γ-globulin solution. The following dyes were used for the tests: acridine orange, neutral red and toluidine blue, pyronin and Alcian blue. Furthermore, the PAS-reaction and enzyme-histochemical methods (i.e. acid and alkaline phosphatase) were employed. The tests showed that P.-vacuoles, which had been stabilized by intravital removal of water, survived fixation in fixatives containing little or no water; they remained, however, very sensitive to aqueous solutions. Numerous histochemical methods proved inapplicable because of the water-solubility of the inclusions. For that reason some staining methods were modified in such a way, that water-free media were used throughout. The vital staining of P.-vacuoles was possible with neutral red — a fact pointed out in 1931 already by W. H. Lewis — and acridine orange. All the other basic dyes did not stain the inclusions. A noteworthy observation was that P.-vacuoles, stained with acridine orange, showed an intensive red-fluorescence. It was assumed that a basophil carrier substance — probably acid mucopolysaccharide (MPS) served as substrate for this red-fluorescence (Gedigk et al. 1956—1960; Schmidt-Mathiesen, 1955). The reasons for this exceptional affinity of neutral red and acridine orange for these changed P.-vacuoles are unknown. The conventional methods for the demonstration of MPS proved unsuccessful when employed for the staining of these inclusions. The reason for this may probably be found in their changed water solubility. Alkaline phosphatase could readily be demonstrated in pinocytizing cells and no morphological differences have been found in comparison with non-pinocytizing cells. The results are considered.

Fa. Hoffmann-La Roche AG., Basel (Schweiz).

11.

Hideo Takada: The Regeneration of Cell Fragment of Yeast in Hypertonic Medium with Strontium Chloride.

When the cells of *Saccharomyces ellipsoideus* were incubated in liquid medium containing strontium chloride, the cells ruptured and small particles of protoplasm were derived from the cell to form bubbles. Among particles some subcellular ones were isolated by the peculiar method and were incubated further. Such particles multiplicated, they being designated as K-fragment. After prolonged incubation, K-fragment enlarged in size, in diameter and became K-body. The size of K-body was threefold as large as that of K-fragment. Some interpretation of granules in K-body visible after cytochemical stain was discussed and determined that either some K-fragment of K-body included the essential cell-organelles. After further incubation K-body enlarged in size, calling it K-cell. The protoplasm of K-cell was less well organized, but the early stage of the formation of daughter K-cell as revealed by microscopical observation beared resemblance to the budding found in mother yeast cell.

Laboratory of Plant Physiology, Osaka City University, Osaka (Japan).

Enzyme · Enzymes · Enzymes

1.

K. C. Tsou: Application of Histochemical Enzyme Substrates for Electron Microscopy*.

Even though many chromogenic substrates that form color dyes have been found useful with the light microscope, the extension of their use with the electron microscope has had only limited success. The difficulty lies in the more stringent requirements for E.M. localization than those for histochemical purposes. The major requirements are electron scattering ability, low aggregation, and microcrystallinity. This paper reports exploratory chemical studies related to overcome such difficulties. — Polymeric diazonium salts have been synthesized to reduce crystallinity of the azo-dye method. The slow coupling rate and low penetration during incubation have made these substrates unsuccessful for such purpose. The visualization of succinic dehydrogenase activity at E.M. level in sperm cells several years ago (L. Nelson and K. C. Tsou, J. Histochem. Cytochem. *6*, 94, 1958) encouraged us to synthesize several polymeric tetrazolium salts for dehydrogenase. Aggregation remains the major problem in extending these substrates for E.M. level for this type of cellular localization. Comparison of the formazans of Nitro-BT, TNBT, and the related polytetrazolium salt at E.M. level provides striking demonstration of the problems involved in such development. Attempts have also been made to demonstrate enzyme sites by the indigo dyes at E.M. level. While chemically hydrolyzed 5-bromo-indoxyl does lead to a more amorphous 5,5'-dibromo indigo, the enzymatically hydrolyzed product tends to crystallize into minute oblong rod crystals which obscure the localization site. However, it was noted with interest that spot electron diffraction patterns on such sites give distinct Laue pattern which can be useful for demonstration of enzyme sites at E.M. level. These preliminary observations provide further impetus in application of histochemical substrates to electron microscopy.

* Supported by N.I.H. Grants CY-2350, CY-4361 and CA-0762-01.

Harrison Dept. of Surgical Research, Schools of Medicine, University of Pennsylvania, and the Hospital of the University of Pennsylvania, Philadelphia 4, Pennsylvania (USA).

2.

Enrico Reale und Liliana Luciano: Licht- und elektronenmikroskopische Enzym-Histochemie einiger Zellfraktionen.

Für elektronenmikroskopische Untersuchungen werden enzymhistochemische Methoden immer mehr an Gefriermikrotom- und Kryostatschnitten angewandt. Die Schnittmethoden erlauben die Lokalisierung einiger Enzymaktivitäten in Zellbestandteilen, die oft sehr gut erhalten sind. Sie können auch für histochemische Untersuchungen von Zellfraktionen im Licht- und Elektronenmikroskop (Reale und Luciano, Histochemie *3*, 422—427, 1964) Verwendung finden. Die Zellhomogenate werden in einem Zentrifugenröhrchen aus Zellulose fraktioniert, schnell gefroren, im Kryostaten geschnitten (20—40 μ) und auf einen Kollodiumfilm gebracht. Danach können die Schnitte für lichtmikroskopische Untersuchungen nach einer üblichen Technik gefärbt und mit einer histochemischen Reaktion nachbehandelt werden. Für elektronenmikroskopische Untersuchungen werden sie sofort in einer gepufferten Osmiumtetroxyd-Lösung fixiert und in Kunststoff eingebettet; oder die Einbettung erfolgt nach der doppelten Fixierung (NOVIKOFF, BURNETT and GLICKMAN, J. Histochem. Cytochem. *4*, 416, 1956), in die eine histochemische Reaktion eingeschlossen wurde. Die verschiedenen Behandlungsarten des Materials scheinen keine schweren Zerstörungen der Zellorganellen hervorzurufen. Die Methode des Gefrierschneidens von Homogenaten erlaubte uns, einige Enzymaktivitäten (alkalische Phosphatase, saure Phosphatase, Succinodehydrogenase) in Zellfraktionen zu untersuchen.

Light and Electron Microscopical Investigations of the Enzyme-Histochemistry of Some Cell Fractions. Enzyme-histochemical tests — followed by electron microscopic examinations — are carried out more and more often on frozen or cryostat sections. These cutting methods make it possible to localize some enzyme activities in cell components, which are often very well preserved. They can also be used for the examination of cell fractions under the light or electron microscope (REALE and LUCIANO, Histochemie *3*, 422—427, 1964). The cell homogenates are fractionated in a cellulose centrifuge tube, frozen quickly, cut on the cryostat (20—40 μ) and mounted on a collodion film. For examination under the light microscope the sections are stained — using one of the usual techniques — and subsequently treated by histochemical methods. For electron microscopic tests the sections are fixed in a buffered solution of osmium tetroxide and embedded in plastic. An alternative method is embedding after a double fixation (NOVIKOFF, BURNETT and GLICKMAN, J. Histochem. Cytochem. *4*, 416, 1956) in which a histochemical reaction is included. It seems that the various methods of pretreatment do not cause obvious damages to the organelles. By cutting frozen sections of homogenates we are able to investigate some enzyme activities (alkaline phosphatase, acid phosphatase, succinate dehydrogenase) in cell fractions.

Institut für Biophysik und Elektronenmikroskopie der Medizinischen Akademie, 4 Düsseldorf (Deutschland).

3.

M. VAN DER PLOEG and P. VAN DUIJN: A New Method for Demonstration of Peroxidases Studied on a Histochemical Model.

Studies were performed on a histochemical model in which proteins — with retention of their enzymic activity — are incorporated in polyacrylamide films. — Peroxidase-active proteins were found to cause rapid melanin formation in the films during their incubation in solutions containing 5,6-dihydroxy indole and hydrogen peroxide. Optimal conditions for this peroxidase reaction were determined. In leucocytes and erythrocytes, an incubation solution containing 5,6-dihydroxy indole and hydrogen peroxide, gives intensive, sharply localised melanin deposition. In kidney and liver sections from rats injected with lactoperoxidase, the incubation results in sharp staining of phagosomes. — The localized deposition of the melano-products during the classical DOPA reaction in tissue sections or smears is a direct result of an enzymic conversion of 5,6-dihydroxy indole by the peroxidases in leucocytes or methaemoglobin in erythrocytes.

Laboratory for Pathology, Biochemical Section, University of Leiden (The Netherlands).

4.

J.-P. PERSIJN and W. TH. DAEMS: Demonstration of Peroxidase Activity in Rat Spleen with the Electron Microscope.

A new histochemical reaction for peroxidase activity, based on the conversion of 5,6-dihydroxy-indole into melanin has been introduced for light-microscopy (VAN DER PLOEG and VAN DUIJN, this Congress). — Melanin, an organic polymer with a high mass, has a distinct electron density and is not attacked by OsO_4 or the embedding-media in electron microscopy. Experiments were therefore performed to test the applicability of this reaction for electronmicroscopical demonstration of peroxidase activity in spleen. — Frozen sections (50 μ) from rat spleen, fixed in glutardialdehyde, were incubated in a medium containing 5,6-dihydroxy-indole and H_2O_2. The composition of the incubation medium and further treatment, was according to DAEMS et al. (Histochemie *3*, 561, 1964). Sharply localized deposits were found in the cytoplasm of leucocytes throughout the section. The erythrocytes only give positive reaction in the periphery of the 50 μ sections. (Controls omitting substrate were negative.) Probably the H_2O_2 concentration diminishes towards the centre of the section until it falls below a critical value necessary for rapid melanin formation in erythrocytes. These observations are in agreement with the findings of VAN DER PLOEG that the critical value for erythrocytes is higher than for leucocytes.

Laboratory for Electronmicroscopy, University of Leiden (The Netherlands).

5.

CHARLES G. ROSA: Tetranitro-Blue Tetrazolium and Its Use in Light and Electron Histochemical Demonstrations of Succinic Dehydrogenase*.

The reduction of tetranitro-blue tetrazolium (TNBT) as a result of the action of the succinic dehydrogenase system (succinate tetrazolium reductase, SDH) has allowed for a more accurate localization of this enzyme in normal and pathologic tissues. In sections of normal mouse adrenal or adipose tissue crystalline artifacts of the reduced tetrazole characteristic of other procedures did not occur when TNBT was employed. The method which was used to study smear preparations of mouse sarcoma-37 ascites tumor cells also demonstrated better localizations of SDH in vaginal smear preparations of human squamous cell carcinoma of the uterine cervix. Previous investigations of SDH with the electron microscope in rat heart were extended to mouse liver. Mitochondria in hepatic cells did not appear as uniformly stained structures in any given plane of sectioning. In mitochondria where enzymatic activity was represented the extent of reduced TNBT observable always showed considerable variability ranging from large amorphous masses of formazan to very fine densities resident within the mitochondrial membranes. These intramembranous deposits (30 to 40 Å) were seen in blocks of tissue unfixed prior to incubation but later postfixed in osmium tetroxide, formalin or a combination of these fixing agents.

* Supported by U.S.P.H.S. Grant GM-06008.

Jefferson Medical College, Philadelphia, Pa. (USA).

6.

ERHARD FASSKE, INGEBORG STEINS und HERMANN THEMANN: Licht- und elektronenmikroskopische Darstellung der Glyzerinaldehyd-3-phosphat-Dehydrogenase-Aktivität in menschlichen und tierischen Zellen.

Nach einer Modifikation der Methode von HIMMELHOCH und KARNOVSKY (E. FASSKE und I. STEINS, Acta histochem. *16*, 302, 1963) wird die Aktivität der Glyzerinaldehyd-3-phosphat-Dehydrogenase dargestellt. Das glykolytische Enzym (GAP-DH) katalysiert die Oxydation von Glyzerinaldehyd-3-phosphat zu 1,3-Diphosphoglyzerinsäure. Das spezifische Substrat wird aus dem Glyzerinaldehyd-3-phosphat-diäthylazetal-Monobariumsalz durch hydrolytische Spaltung mittels *Dowex 50* und Abdestillieren des dabei frei werdenden Alkohols hergestellt. Als Wasserstoffakzeptor findet Nitro-BT Verwendung. Die lichtmikroskopischen Untersuchungen wurden an gesunden und kranken, lebendfrisch entnommenen menschlichen Gewebeproben ausgeführt (z.B. akute und chronische Entzündungen, gut- und bösartige Tumoren). — Für elektronenmikroskopische Untersuchungen wurden Kryostatschnitte in Hydroxyadipaldehyd, Bakers Formol, kaltem Azeton, Glutaraldehyd oder flüssigem Stickstoff vorfixiert. Nach der Inkubation in der spezifischen Lösung wurde in Osmiumtetroxyd nachfixiert und in Methacrylat, Vestopal oder Epon eingebettet. In einer simultanen Methode werden kleine Gewebewürfel in einem Gemisch aus Hydroxyadipaldehyd und der GAP-DH-Inkubationslösung bei 37° C inkubiert, nachfixiert, entwässert und für elektronenmikroskopische Untersuchungen in Methacrylat eingebettet.

Demonstration of Glycerolaldehyde-3-phosphate-dehydrogenase Activity in Human and Animal Tissue by Light- and Electron Microscopy. A modification of HIMMELHOCH and KARNOVSKY's method (E. FASSKE and I. STEINS, Acta histochem. *16*, 302, 1963) is used for the demonstration of the activity of glycerolaldehyde-3-phosphate-dehydrogenase. The glycolytic enzyme (GAP-DH) catalyses the oxidation of glycerolaldehyde-3-phosphate to 1,3-diphosphoglyceric acid. — The specific substrate is prepared from glycerolaldehyde-3-phosphate-diethylacetal monobarium salt by hydrolytic cleavage using *Dowex 50*, and distillation of the alcohol hereby released. Nitro-BT is used as hydrogen acceptor. — Normal and abnormal fresh human tissues (e.g. tissues taken from acute and chronic inflammations, malignant and non-malignant tumours) are used for the investigations carried out by light microscope. — For E.M. investigations cryostat sections are fixed in hydroxiadipaldehyde, Baker's formalin, cold acetone, glutaraldehyde, or liquid nitrogen. After being incubated in the specific solution, the sections are fixed for a second time in osmium tetroxide and embedded in Methacrylate, Vestopal, or Epon. — Using a simultaneous method, small pieces of tissue are incubated at 37° C in a mixture of hydroxiadipaldehyde and GAP-DH incubation solution, fixed, dehydratet, and embedded in Methacrylate for E.M. investigations.

Forschungsinstitut Borstel, Inst. f. experimentelle Biologie und Medizin, Pathol. Abtlg. 2061 Borstel (Bz. Hamburg) (Deutschland).
Institut für medizinische Physik der Universität Münster, 44 Münster/Westf., Hüfferstr. 68 (Deutschland).

7.

W. Schulze und A. Wollenberger: Elektronenmikroskopischer Nachweis einer ATPase-Reaktion in Mitochondrien, Zellmembranen und anderen Strukturelementen.

Biochemische Analysen der ATP-Spaltung isolierter Zellstrukturen unter den Inkubations- und Vorbehandlungsbedingungen, die für den elektronenmikroskopischen Nachweis angewendet werden, ergaben eine ausreichende ATPase-Aktivität auch in den Mitochondrien. Bisher konnte jedoch noch keine genaue Lokalisierung der ATP-Spaltung mit Hilfe des Elektronenmikroskops gezeigt werden. Durch Verwendung von 200 μ-Gefrierschnitten nach kurzer Vorfixierung in verschiedenen Aldehyden und in Osmiumtetroxyd (Caulfield) wird nach Inkubation in einem Medium, das 2,5 μM ATP-Natrium (Boehringer), 5 μM Magnesiumsulfat und 5 μM Bleiacetat enthält und mit Tris-HCl-Tartrat (pH 6,8) gepuffert ist, ein sehr feingranulärer Bleiphosphatniederschlag in den Membranen der Cristae mitochondriales und an den Außenmembranen der Mitochondrien aus adulten und embryonalen Herzmuskel- sowie aus Nierentubulizellen nachgewiesen. Ferner fanden sich feine Bleiniederschläge an dem Plasmolemm der Zellmembranen und an Pinozytosebläschen in Endothelzellen. Auch Teile des endoplasmatischen Retikulums zeigen eine positive ATPase-Reaktion. Versuche zur zytochemischen Differenzierung der verschiedenen ATPasen sind noch nicht abgeschlossen.

Electron Microscopic Demonstration of a ATPase Reaction in Mitochondria, Cellular Membranes and Other Structural Elements. A sufficient amount of ATPase activity was demonstrated in mitochondria by biochemical analysis of isolated cellular structures. The tests were carried out under the same conditions (in regard to pre-treatment and incubation) as used in electron microscopy. Fresh frozen sections (200 μ thick) were briefly fixed in various aldehydes and osmium tetroxide (Caulfield) and incubated in a medium which contained 2.5 μM ATP-Na (Boehringer), 5 μM Mg-sulphate and 5 μM lead acetate. The solution was buffered to pH 6.8 with Tris-HCl-tartrate. It was then possible to demonstrate a very delicate granulated precipitate of lead phosphate in the membranes of the cristae mitochondriales and on the exterior membranes of the mitochondria in adult and embryonic cells of heart muscle and kidney tubules. Furthermore there were delicate lead precipitates on the plasma membrane and on pinocytosis vesicles in endothelial cells. Parts of the endoplasmatic reticulum gave also a positive ATPase reaction. Tests for the differentiation of the various ATPases have not yet been completed.

Arbeitsstelle für Kreislaufforschung der Deutschen Akademie der Wissenschaften zu Berlin, Berlin-Buch (Deutschland).

8.

Joachim R. Sommer and Jacob J. Blum: Electronmicroscopic Localization of Acid Phosphatases in Euglena*.

An induced and at least one constituent acid phosphatase in *Euglena gracilis var. bacillaris* has recently been characterized by biochemical means. Parallel cytochemical studies with the light and electron microscopes employing the Gomori and/or the Burstone methods for the demonstration of acid phosphatases have shown the following results. In uninduced cells reaction product was found in the Golgi complex, in lysosome-like structures and in vesicles around the reservoir. In induced cells the reaction product was also found in well defined regions of the pellicle. In accordance with the biochemical studies no reaction product was seen in the pellicle following heat inhibition at 60° C for 20 minutes, whereas the reaction product in the Golgi complex and the perireservoir region remained essentially unchanged. In similar experiments with *Astasia longa*, a close relative of *Euglena gracilis*, no induced acid phosphatase was demonstrable biochemically. Cytochemically, reaction product was present in the Golgi complex and perireservoir region but not in the pellicle.

* This study was supported by grants GM 09730-02 and NSF GB 158 from the USPHS.

Depts. of Pathology and Physiology, Duke University Medical Center, Durham, N.C. (USA).

9.

Mary McCabe: The Latency of Leucine Amino-Peptidase Activity.

In controlled temperature frozen sections this enzyme has been demonstrated to occur in discrete particles which are not copper chelates. Such particles, with apparently only moderate activity, have been identified in the intestine, liver and kidney of the rat; the activity of this enzyme has been strikingly increased by suitable pre-treatment, suggesting that much of it is present in a latent form. Activation of this enzyme has been observed in the livers of mice fed with vitamin A; cyclic variation in its activity has been recorded in human endometrium. — The correspondence between the behaviour of this enzyme and that of those found in the lysosomes will be discussed.

Dept. of Pathology, Royal College of Surgeons of England, Lincolns Inn Fields, London W.C. 2 (Great Britain).

10.

AARON H. STOCK: Three New Extracellular Streptococcal Enzymes Demonstrated by Staining or Chemical Techniques*.

Three newly found extracellular streptococcal enzymes can be demonstrated following simple electrophoresis in agar gel by staining or chemical techniques. The samples consisted of 50-fold concentrates of streptococcal supernates. The extracellular enzymes are: streptococcal esterase (STOCK, URIEL and GRABAR), streptococcal beta glucuronidase and streptococcal leucine aminopeptidase. The β glucuronidase was identified by an agar overlay technique (STOCK and URIEL) by incorporating the substrate, phenolphthalein glucuronide, into the layer poured over the electrophorized enzyme. The β glucuronidase was produced by only certain strains of the groups and Lancefield group A types of streptococci but the production of esterase and aminopeptidase was more widely distributed. — The new staphylococcal enzymes to be described are extracellular carboxylic acid esterase and extracellular leucine aminopeptidase. Both are revealed by the histochemical staining methods used for demonstrating the streptococcal enzymes.

* Aided by grants from the U.S.P.H.S.

Dept. of Microbiology, School of Medicine, University of Pittsburgh, Pennsylvania (USA).

Fixierung · Fixation · Fixation
Verschiedenes · Divers · Divers

1.

G. SCHNEIDER und G. SCHNEIDER: Über Stoffverluste bei Formol-Fixierung von menschlichem Gehirngewebe.

Bei autoradiographischen Untersuchungen über die Synthese der Ribonukleinsäure in der Zelle treten nach verschiedenartiger Fixierung ein und desselben Gewebes starke Aktivitätsverluste (bis zu 86%) auf. Bei gleicher Fixierung verschiedener Organe findet man ebenfalls unterschiedliche Verluste. — Nach der Fixierung menschlichen Hirngewebes mit neutralem Formalin lassen sich in der Fixierungslösung zahlreiche Stoffe chemisch nachweisen. Erwartungsgemäß findet man in der Lösung größere Mengen niedermolekularer Substanzen, wie Aminosäuren und Kohlenhydrate. Indolderivate konnten unter den gegebenen Bedingungen nicht nachgewiesen werden. Überraschend ist die Tatsache, daß das Fixierungsmittel auch hochmolekulare Substanzen extrahiert. So lassen sich mit Hilfe der Hochspannungs-Papierelektrophorese mehrere Eiweißfraktionen nachweisen. Daraus ergibt sich, daß man wegen der möglichen Stoffverluste zumindest bei quantitativen autoradiographischen oder chemischen Untersuchungen von Geweben in der Wahl des Fixierungsmittels sehr kritisch sein muß.

Loss of Substances during Formalin-Fixation of the Human Brain. Autoradiographic investigations of the synthesis of Ribonucleic acid in the cell revealed serious loss of activity of the same tissue after fixation in different fixatives. The same method of fixation of different tissues also produces differential losses. — One can demonstrate chemically the presence of many substances in formalin used for fixation of human brain tissue. As is to be expected, one finds many substances of low molecular weight, such as amino acids and carbohydrates, in solution. The presence of indole-derivatives could not be established. Surprisingly, substances of high molecular weight are also extracted by the fixative. Thus one can demonstrate the presence, in solution, of several protein fractions by means of paper electrophoresis. Thus it is obvious that, because of possible losses of material, one has to be very critical in the choice of fixative in quantitative autoradiographic or chemical investigations of tissues.

Institut für Hirnforschung und allgemeine Biologie, 7828 Neustadt, Schwarzwald (Deutschland).

2.

LIISA RÄISÄNEN: Quantitative Determination of Esterase Activity in Readily and Poorly Soluble Tissue Fractions and the Effect of Fixation.

Previous work from the present laboratory has demonstrated that only poorly soluble desmo-enzymes can be histochemically demonstrated in fresh tissue sections, while readily soluble lyo-enzymes are lost during the histochemical procedure. In the present work, quantitative determinations of the enzyme activity were carried out to determine the esterase activity in fractions of different solubility, employing substrates commonly used for histochemical demonstration both as substrates and as competitive inhibitors towards other substrates. The effect of different types of fixation in immobilizing lyo-enzymes was also investigated by quantitative determinations. The observations as yet made suggest that a complete idea of the spectrum of esterases in a tissue can only be obtained by taking into consideration the solubility of various enzyme fractions.

Dept. of Anatomy, University of Helsinki, Siltavuorenpenger, Helsinki (Finland).

3.

DAVID T. JANIGAN and **WILLIAM K. GOURLEY: Aldehyde Fixation and Rat Liver Acid Phosphatases in Tissue Blocks and Fractions*.**

The effect of various aldehydes on the activities of B-glycerophosphatase (BGP'ase), phenyl phosphatase (PP'ase) and P-nitrophenylphosphatase (P-NPP'ase), of rat liver, all at pH 5.0, has been determined. On basis of recovery of activity after fixation the aldehydes could be divided into 3 broad groups: recovery of over 50% — hydroxyadipaldehyde, acetaldehyde; around 50% — formaldehyde, methacrolein; lower than 50% — crotonaldehyde, glutaraldehyde and acrolein. All 3 acid phosphatases were affected similarly except for p-NPP'ase activity which was depressed to a slightly greater degree. BGP'ase activity after hydroxyadipaldehyde fixation was largely in a soluble and non-sedimentable form while after glutaraldehyde it was completely sedimentable in water homogenates. — Formaldehyde fixation altered the pH activity curve of BGP'ase so that smaller declines in activity occurred at pH levels on the alkaline side of optimum as compared with unfixed liver. — Liver cell fractions, isolated by differential centrifugation, were subjected to formaldehyde fixation and it was found that microsomal acid BGP'ase activity was inhibited by only 25% while there was 75% inhibition of lysosomal and mitochondrial activity. Differences in the activity of the supernatant towards B-glycerophosphate and P-nitrophenyl phosphate were also observed, a feature already noted in guinea pig liver (NEIL and HORNER, Biochem. J. *84*, 32 p, 1962). — The discussion of these results will be directed towards an attempted correlation with the staining results for acid phosphatase as well as the choice of aldehydes for enzyme staining.

* Research supported by Public Health Service Research Grant No. AM-06626-01.
University of Kansas Medical Center, Kansas City, Kansas (USA).

4.

RICHARD M. TORACK: Differential Fixation of Adenosinetriphosphatase in Rat Brain. A Light and Electron Microscopic Study.

In rat brain, adenosinetriphosphatase activity at a neutral pH has been localized by means of electron microscopy to pinocytic vesicles of endothelial cells, vascular basement membranes, astroglial peri-vascular end feet, oligodendroglial plasma membranes, and glial-neuronal junctions at the surface of neuronal cell bodies, and at synapses. However some variation in the distribution of this enzyme activity was noted particularly at the periphery of the tissue section following different methods of prefixation. — In order to obtain greater reproduceability a more complete effect of fixation was produced either by perfusion fixation or by prolonged (18—24 hrs.) immersion of this coronal slices of rat brain. When glutaraldehyde, hydroxyadipaldehyde (HAA) and formaldehyde were used in this fashion the distribution of the reaction product differed distinctly. Following formaldehyde fixation all the previous sites of activity were again observed by means of light microscopy and electron microscopy. After HAA fixation the tissue evinced enzymatic activity only in vascular walls which was localized to the basement membrane by means of electron microscopy. Finally when glutaraldehyde was used as the fixative, the only activity was observed in the corpus callosum. The ultrastructural localization of this reaction product was to astroglial processes and to tubular structures in the axoplasm of myelinated fibers.

Dept. of Pathology, The New York Hospital—Cornell Medical Center, 1300 York Avn., New York 21, N.Y. (USA).

5.

PHILIP PIZZOLATO: Histochemical Demonstration and Identification of Calcium Oxalate Crystals by a Peroxide Silver Nitrate Procedure.

Calcium, when combined with carbonate or phosphate in animal tissue, gives a positive von Kossa silver nitrate reaction. Only some observers, however, have been able to show that calcium oxalate is blackened by silver nitrate. Because of these findings we undertook to investigate the cause of this variation. We were able to produce crystals of "pure" calcium oxalate in the rat kidney by the administration of ammonium oxalate or ethylene glycol. These crystals did not react with silver nitrate, hematoxylin or alizarin red S. However, when these kidney sections were treated with a mixture of concentrated hydrogen peroxide and silver nitrate, the crystals turned black. In order to prove that calcium carbonate or phosphate may not augment or catalyze the reaction, sections containing these salts, as well as, calcium oxalate were treated with dilute acetic acid until these was no staining with alizarin red S or hematoxylin. — We then found that the freed calcium oxalate did not react with the silver nitrate alone, but was blackened by the hydrogen peroxide-silver nitrate mixture.

Clinical Laboratory, Veterans Administration Hospital and Louisiana State University School of Medicine, New Orleans, Louisiana (USA).

6.

Kozo Okamoto, Osamu Midorikawa and Hidenori Kawanishi: Histochemical Demonstration of Zinc; Electron Microscopic Investigation on Zinc Granules and Comparative Studies on Dithizone and Silversulfide Methods.

Since Okamoto first developed in 1942 a histochemical method for detecting zinc, many papers on the zinc content and degree of granulation, especially in the islet cells of pancreas, have appeared, whereas no conclusive results have been obtained of the intracellular localization of zinc in relation to granule formation. The purpose of our present investigation, therefore, is to clarify the significance of the intracellular localization of zinc in the islet cells in the appearance of granule formation, storage and discharge under the various conditions of some experimental animals, using the electron microscopic technique. — Employing vital staining with dithizone and our improved method of dithizone-stained sections, the demonstration of intracellular sites of zinc in the islet cells of Langerhans has now been undertaken. The comparative studies have shown (1) that the two methods, dithizone and silver-sulfide method, do not always give concordant results on the content of zinc and its localization, (2) that the dithizone procedure is more specific than the silver-sulfide method when applied to the Langerhans islets and other tissues.

Kyoto University, Dept. of Pathology, Kyoto (Japan).

7.

Günther Geyer: Histochemische Anwendung der Carbodiimidreaktion.

Carboxylgruppen können mit Hilfe von Carbodiimiden in aktivierte Formen überführt werden. Diese aktivierten Carboxylgruppen reagieren mit NH_2-Gruppen unter Bildung von Peptidbindungen. Außerdem erfolgt eine Reaktion mit Sulfhydryl- und vielleicht auch mit Guanidogruppen. Aliphatische und phenolische OH-Gruppen, Indolreste und der Imidazolring sind dagegen reaktionsunfähig. — Die Carbodiimidreaktion mit Carboxylgruppen ist für die histochemische Erfassung von COOH-Gruppen und in einem zweiten Verfahren als Proteinnachweis verwendet worden. Dafür wurden folgende zwei Methoden ausgearbeitet. Beide Verfahren lassen sich ohne besonderen Aufwand als Routinemethoden ausführen. Die Aussagefähigkeit der Carbodiimidreaktion zum Nachweis von Carboxylgruppen entspricht weitgehend der Säureanhydridreaktion nach Barrnett und Seligman. Es reagieren vor allem COOH-Gruppen der Proteinseitenketten, außerdem weitere Carboxylgruppen, deren Zuordnung noch nicht erfolgen konnte. Die zweite Variante der Carbodiimidreaktion erfaßt die oben genannten reaktiven Gruppen, wobei Aminogruppen das Hauptkontingent darstellen. Sie wird deshalb am Paraffinschnitt als Proteinnachweis im allgemeinen Sinne verwendet. Die Befunde entsprechen weitgehend den Ergebnissen der Säureazidreaktion und anderen vergleichbaren Methoden.

The Histochemical Use of the Carbodiimide Reaction. With the help of carbodiimides carboxyl groups can be transformed into an activated form. These activated carboxyl groups will react with NH_2-groups; in the course of this reaction peptide-bonds are formed. In addition, they react with sulfhydryl and perhaps also with guanidino groups. Aliphatic and phenolic OH-groups, indole residues and the imidazole ring, however, do not react at all. The capability of carbodiimide to react with carboxyl groups is used for the histochemical determination of COOH-groups; in another process it is used for the determination of proteins. Two methods are described, which can easily be used as routine tests. The results obtained with the carbodiimide reaction correspond largely to the results of BARRNETT and SELIGMAN's acid anhydride reaction. The principal groups reacting are the COOH-groups of the protein side chains; some not yet classified carboxyl groups are also involved in the reaction. The second method includes the reactive groups mentioned above; the main contingent is formed by amino groups. It is therefore used as a protein reaction on paraffin sections. The findings correspond largely to the results of the acid azide reaction and other comparable methods.

Histochemische Abteilung des Anatomischen Instituts der Friedrich-Schiller-Universität Jena (Deutschland).

8.

H.-G. GOSLAR: Weitere Befunde zum histochemischen Verhalten freier Aminogruppen.

Zur kritischen Beleuchtung der Acrolein-Schiffreaktion auf Proteine im Vergleich zu der o-Diacetylbenzoreaktion wird über eine Reihe von Substanzen mit Doppelbindung sowie von Aldehyden mit und ohne Doppelbindung bezüglich ihres Verhaltens zu freien Aminogruppen berichtet und die Schwierigkeiten eines Spezifitätsbeweises der Acrolein-Schiffreaktion auch von dieser Seite aus dargestellt.

Further Results to the Histochemical Behaviour of Free Amino-Groups. For a critical consideration of the Acrolein-Schiff-reaction on proteins compared with the o-diacetyl-benzol-reaction it is reported on several substances with double-boundings as well as on aldehydes with and without double-boundings as to their behaviour to free amino groups. Besides the difficulties to prove the specifity of the Acrolein-Schiff-reaction are demonstrated.

Anatomisches Institut der Universität, 52 Bonn (Deutschland).

9.

JACOB Y. TERNER and ROBERT LEV: The Effect of Lactone Formation on the Basophilia of Acid Polysaccharides*.

Treatment of tissues with acetic anhydride:pyridine (15:25) will inhibit the basophilia and/or the metachromasia of a wide variety of epithelial mucins and ground substances. It is proposed that the mechanism of this inhibition is gamma lactone formation, involving the sialic acids of epithelial mucins and the glucuronic acid of the acid mucopolysaccharides. Alkaline hydrolysis will reverse this inhibition. If acetylated sections are treated with $LiAlH_4$, which reduces lactones to alcohols, or NH_3, which converts them to amides, the reversal of this inhibition by alkaline hydrolysis is prevented. The *in vitro* acetylation of a tetrasaccharide from hyaluronic acid results in the formation of an infra-red lactone peak with simultaneous loss of the normal carboxyl peak. The basophilia of gelatin models containing polygalacturonic or polymannuronic acids is essentially abolished by acetylation. The Alcian blue binding and mucicarminophilia of tissue sites containing hyaluronic acid (umbilical cord) and chondroitin sulfate A and C (cartilage) is completely abolished by desulfation and lactonization (acetylation). However, the staining of tissue sites containing significant concentrations of chondroitin sulfate B is not abolished by such treatment since the L-iduronopyranoside will not form a lactone.

* Aided by USPHS Research Grant AM-07228-02 and Training Grant GM-865-02.

Dept. of Pathology, College of Physicians and Surgeons, Columbia University, New York, N.Y. (USA).

10.

W. NIKLOWITZ, I. J. BAK and R. HASSLER: Electron Microscopical Observations of Pyramidal Cells in the Hippocampus after 3-Acetylpyridine Treatment.

The hippocampus is the main structure of the limbic system and known to be closely correlated with psychomotoric epilepsy. In recent studies it was shown that functional disturbances of the central nervous system could be induced by treatment with antimetabolites, resulting in epileptic seizures and morphological changes in the hippocampus. However the pyramids of the hippocampus can be divided into four different fields by cyto- and chemoarchitectonical methods. — In the present study rabbits were injected with various doses of 3-acetylpyridine (80—600 mg/kg) to examine the acute ultrastructural changes in the pyramids by electron microscopical and cytochemical methods. — It was found by lightmicroscopic observations that the pyramids, especially the field CA3, were the most affected area of the hippocampus. This effect was most clearly seen with the electron microscope. The cells in this field were found to undergo rapid and practically complete desintegration. Pathological changes occurred first in the mitochondria, and later in the ergastoplasm and Golgi apparatus. The mitochondria became swollen, the matrix disappeared, and the mitochondrial cristae were reduced in number. The ergastoplasm (Nissl-substance) showed a lysis of the membranes and the membrane-attached and free ribosomes, and only "spot-like" areas of ergastoplasm remained. From the atypical arrangement in the Golgi apparatus of the treated pyramids it was assumed that these cells have ceased to function. Further preliminary cytochemical observations with the electron microscope suggested that there were also abnormal metabolic processes resulting from the action of 3-acetylpyridine. This study agrees with previous findings but emphasizes the importance of considering the various fields of pyramidal cells.

Max Planck-Institut für Hirnforschung, Neuroanatomische Abteilung, Deutschordenstraße 46, 6 Frankfurt a. M.-Niederrad (Deutschland).

<h2 style="text-align:center">11.</h2>

A. MURESAN und G. SIMU: Histochemische und enzymochemische Veränderungen in den retikulo-histiozytären Zellen des zentralen Nervensystems gegenüber verschiedenen Antigenen.

Zum Studium des Verhaltens des retikulo-endothelialen Systems gegenüber Impftumoren wurde das Gehirn von 30 mit Guérin-Karzinom geimpften Ratten untersucht. Zum Vergleich wurden je 10 Ratten mit lebenden Mikroben und Freund-Antigen geimpft; ebenso wurden 10 unbehandelte Tiere unter Beobachtung gehalten. — Die stärkste Reaktion wurde bei den mit Mikroben geimpften Tieren beobachtet. Es traten im Gehirn wie in der Hirnhaut perivasculäre Infiltrate mit Histiozyten, Lymphozyten und Plasmazellen, manchmal auch Granulozyten auf. Weiter wurde eine Hyperplasie und Hypertrophie der Mikrogliazellen und der Endothelzellen der Gefäße beobachtet. Ebenso wurde eine Steigerung der Speicherfähigkeit für Kolloide der retikulo-histiozytären Zellen und eine Vermehrung der pyroninophilen Zellen festgestellt. In diesen Zellen ist auch die Aktivität der Adenosintriphosphatase und der alkalischen Phosphatase erhöht. — Bei den mit Guérin-Karzinom geimpften Tieren sind dieselben Veränderungen zu beobachten, aber in geringerem Ausmaß: es findet sich peri- oder endovasculär eine retikulo-endotheliale Hyperplasie, manchmal auch eine Hyperplasie der Mikroglia. Gegenüber den unbehandelten Kontrolltieren ist die Aktivität der ATP-ase und der alkalischen Phosphatase erhöht, aber in geringerem Maße als bei der vorhergehenden Gruppe. Die pyroninophilen Zellen und die Speicherfähigkeit für Kolloide im Plexus chorioideus sind erhöht. — Diese Ergebnisse zeigen, daß an der Reaktion des R.-E.-Systems gegenüber Impftumoren in einem gewissen Maße auch seine Zellen im Zentralnervensystem teilhaben. Diese Reaktion ist aber schwächer, verglichen mit der Reaktion auf andere Reize, wahrscheinlich infolge einer schwächeren Antigenwirkung des Tumorgewebes und dem Vorhandensein des Virchow-Robinschen Raumes.

Histochemical and Enzymochemical Changes in the Reticulo-histiocytic Cells of the Central Nervous System Caused by Various Antigens. In order to study the reaction of the reticuloendothelial system to transplanted tumours, Guérin carcinoma was inoculated into the brain of 30 rats. Ten rats, inoculated with living organisms, 10 with Freund's antigen, and 10 normal rats served as controls. The strongest reaction was found in the animals which had been inoculated with organisms: perivascular infiltrations containing histiocytes, lymphocytes, plasma cells and occasionally granulocytes were found in the brain and the cerebral meninges. A hyperplasia and hypertrophia of the microglial cells and the endothelial cells of the blood vessels was also observed. An increase in the storing capacity of colloids was observed in the reticulo-histiocytic cells. At the same time there was an increase in the number of pyroninophil cells. These cells showed a raised activity of ATP-ase and alkaline phosphatase. The same changes were observed in animals which had been inoculated with Guérin carcinoma. The changes were, however, not as marked as in the first group: a peri- and endovascular endothelial hyperplasia, and sometimes also a hyperplasia of the microglial cells, was found. If compared with the untreated control group the activity of ATP-ase and alkaline phosphatase was increased, though to a lesser extent than in the previous group. The number of pyroninophil cells and the storing capacity of colloids was increased in the Plexus chorioideus. — These findings show, that to a certain extent, the cells of the central nervous system take part in the reactions of the reticuloendothelial system to transplanted tumours. These reactions, however, are weak if compared with reactions to other stimulants. The reason may be found in the weaker antigenicity of the cancerous tissue and in the presence of the Virchow-Robin space.

Medizinische Fakultät, Institut für Pathologische Anatomie, Cluj (Rumänien).

Metachromasie · Metachromasia · Métachromasie
Mucopolysaccharide · Mucopolysaccharides · Mucopolysaccharides
Bindegewebe · Connective Tissue · Tissu conjonctif

1.

SAMUEL S. SPICER: Diamine Techniques for Characterizing Mucopolysaccharides Histochemically.

A mixture of the *meta* and *para* isomers of N,N-dimethyl-phenylenediamine (HCl) stains mucopolysaccharides selectively. Acid mucopolysaccharides whose mixed diamine staining succumbs to prior periodate oxidation and which are strongly PAS reactive are distinguished in this procedure from those whose staining resists prior oxidation and which have relatively weak or no PAS reactivity. Similarly, a sequence of periodate oxidation and exposure to the mixture of diamines, blocks somewhat variably affinity for Alcian blue, colloidal iron and azure A in some but not in other acidic mucins. — A selective method for acid mucosubstances employing the mixture of diamines with added $FeCl_3$ stains black most sulfated and a majority of sialic acid containing mucopolysaccharides. A modification with higher levels of diamines and $FeCl_3$ stains only sulfomucins. A sequence of the latter high iron diamine followed by Alcian blue distinguishes some epithelia containing only blue stained sialomucin, others with only black sulfomucin and others with both blue sialomucin and black sulfomucin either in different cells of the organ or often in the same cell. The basophilia of several autoradiographically proven sulfomucins is masked for Alcian blue, colloidal iron, and azure A but not for the high iron diamine or aldehyde fuchsin.

National Institutes of Health, Bethesda, Maryland, 20014 (USA).

2.

KARL-HEINRICH KNESE: Das metachromatische Färbungsmuster bei Verwendung verschiedener Farbstoffe am gleichen Objekt.

In allgemeinen Untersuchungen über die metachromatische Farbreaktion wurde eine Reihe verschiedener Farbstoffe angewandt. Wir haben nun nach Formolfixierung das Periost und Perichondrium von Rinderfeten von 17,5—850 mm SSL sowie Hühnerkeimscheiben von 24—120 Std Bebrütung mit den gebräuchlichsten metachromatischen Farbstoffen untersucht, unter anderem Toluidinblau, Azur A, B, Kresylechtviolett, Brilliantkresylblau, Janusgrün, Thionin. Diese Versuche wurden unternommen, um im Sinne einer „Färbung mit histochemischen Methoden" (SZIRMAI) eine „elektive" Darstellung verschiedener Elemente und Strukturen zu erreichen. Der pH der Farbflotte wurde variiert. Mit Erhöhung des pH ergibt sich die bekannte Steigerung der Farbstoffaufnahme. Das metachromatische Färbungsmuster ist bei den verschiedenen Farbstoffen recht unterschiedlich. Bei den osteoiden Säumen liegt ein Gefälle der Anfärbbarkeit vor; sie erscheinen von unterschiedlicher Breite. Diesem Färbungsmuster entspricht etwa die Ausreifung der Fibrillenstruktur des Knochengewebes nach Zahl der Fibrillen je Flächeneinheit des Schnittes und der Fibrillendicke, wie wir sie elektronenmikroskopisch beobachten konnten. Im Perichondrium reicht das Gefälle in einer Färbungsabnahme von der Knorpelbildungsfront in das Perichondrium hinein. Manche Befunde sprechen dafür, daß nicht nur ein einfaches Gefälle der Farbstoffaufnahme vorliegt. Über das beschränkte Ziel unserer Untersuchungen hinaus — die Chondrogenese und Osteogenese mit der Ausbildung spezifischer Bildungszellen — wird eine Reihe von Fragen aufgeworfen, die wir zur Diskussion stellen: Tritt bei den verschiedenen Farbstoffen eine metachromatische Reaktion bei einer unterschiedlich großen Zahl von elektronegativen Gruppen auf? Inwieweit hängt die Reaktion von der Größe und der Gestalt der Farbstoffmoleküle ab? Welche Rolle spielt die morphologische und chemische Struktur der untersuchten Gebiete bei der Farbreaktion?

The Metachromatic Staining Pattern of Various Dyes on the Same Object. A series of different dyes was used in a general investigation of metachromic colour reactions. After formalin fixation the periosteum and perichondrium of cattle foetus (17.5—850 mm C-RL) and chicken germinal discs (after incubation periods of 24—120 hrs.) were examined with some of the most usual metachromatic dyes (for example Toluidine blue, Azure A and B, Cresylecht violet, brilliant cresyl blue, Janus green, and Thionin). These experiments were carried out in order to achieve a "selective" staining of various elements and structures in the sense of SZIRMAI's "staining with histochemical methods". The pH of the dye-bath was varied. With rising pH the dye absorption increased, which was already a well-known fact. The metachromatic staining pattern varied markedly from dye to dye. Osteoid margins showed a gradient in colorability and appeared to be of varying width. This staining pattern was nearly in accordance with the maturation of the fibril structure in the bone tissue. The maturation was assessed on the number of fibrils per unit area of the section and on the diameter of the fibrils, as observed under the electron microscope. The perichondrium showed a diminution of stainability which reached from the marginal zone of cartilage formation to the perichondrium. Some findings supported the postulate that there is not only a simple decrease in the stainability. — Beyond the limited object of our investigations — chondrogenesis and osteogenesis and the formation of specific embryonic cells — a number of questions were raised which we would like to put forward for discussion: Do various dyes give a metachromatic reaction according to the number of negatively charged groups? How far does the reaction depend on the size and form of the dye molecule? — Is the morphological and chemical structure of the investigated tissue region of any importance for the outcome of the colour reaction?

Anatomisches Institut der Universität, Neue Universität, 23 Kiel (Deutschland).

3.

GABRIELE BATTAGLIA: Histophotometrical Demonstration of the Metachromasia of the Connective Tissue at Different pH on the Same Section.

It is shown that it is possible to study histophotometrically the metachromasia of the connective tissue stained with toluidin blue 1:10.000 buffered at different pH, on the same section instead of in different sections, by means of decolourisation with acidified alcohol and restaining with toluidin blue at the desired pH.

Istituto Anatomico, Università di Ferrara, Via Fossato di Mortara, 66 Ferrara (Italy).

4.

H. GALJAARD and J. A. SZIRMAI: The Staining Pattern of Hyaline Cartilage: A Histochemical and Interferometric Study.

The distribution of acidic glycosaminoglycans (mucopolysaccharides), neutral polysaccharides and proteins has been studied in the nasal septum cartilage of horses of various age groups. In addition, the concentration of the total dry mass in the different microscopic structures was determined by interference microscopy, using a double beam instrument (Leitz) and a photographic procedure. The results showed that the characteristic staining pattern of the intercellular substance is determined by variations in protein content rather than by that of the glycosaminoglycans; the latter showed a relatively uniform distribution. The variations in basophilia, observed when staining procedures are carried out at a low pH, could be explained by interference with the staining of glycosaminoglycans by the basic groups of proteins. This was also supported by the results of interferometric studies; furthermore, these revealed that the concentration of the dry mass varies markedly (from 25% to 50%, w/v) in the various regions around the cartilage cells, as well as in the different layers of the cartilage. Variations of the staining pattern and of the dry mass distribution in relation to age and development have also been observed.

Research Laboratories, Dept. of Rheumatology, University Hospital, Leiden (The Netherlands).

5.

JOHANNES LINDNER und MANFRED SCHMIDT: Die Histochemie der Grundsubstanz- und Fasersynthese im Bindegewebe.

Mit den bisher vorliegenden histochemischen Färbeverfahren sind die intrazelluläre Grundsubstanz- und Faserbildung nicht ausreichend nachzuweisen, die extrazellulären Synthese- und Differenzierungsschritte nur zum Teil. Auch die bisher verfügbaren ferment-histochemischen Verfahren erlauben keinen Einblick in die intrazelluläre Synthese der Interzellularsubstanz, sondern nur grobe Vergleiche mit den weitaus feiner differenzierten biochemischen Ergebnissen. Dagegen ist das histochemische Verfahren der Autoradiographie die einzige derzeitige Methode zur histotopologischen Lokalisation der Grundsubstanzbildung. Denn elektronenmikroskopisch ist bisher nur die intrazelluläre Faserbildung nachgewiesen. Die klassische Morphologenfrage, ob die Grundsubstanz- und Faserbildung in ein und derselben Zelle gleichzeitig stattfinden kann, ist schließlich ebenfalls nur autoradiographisch zu klären. Die zeitliche und örtliche Beziehung beider Synthesevorgänge zueinander wurde durch *in vitro*- und *in vivo*-Einbauzeiten mit getrennter und kombinierter Gabe von S^{35}-Sulfat und H^3-Prolin besonders an Knorpelgeweben geprüft (mit Variationen der Inkorporationszeit und -dosis, der Expositionszeit und -art, mit Anwendung von Filtern, Mehrfachbelegungen und weiteren technischen Hilfen des autoradiographischen Nachweises, sog. Doppelmarkierungen). Am sichersten ist dabei das Abwarten mehrerer Halbwertzeiten von S^{35}-Sulfat und der anschließende Vergleich der davor und der danach gewonnenen Folgeschnitte des gleichen (doppeltmarkierten) Materials (Blockes) miteinander. — Ergebnis: In einer Knorpelzelle können gleichzeitig Mucopolysaccharide (MPS) und Kollagen gebildet werden. Diese Syntheseschritte und ihre Abhängigkeit voneinander wurden vergleichend, sowie zur weiteren qualitativen Differenzierung und quantitativen Sicherung auch biochemisch und radiochemisch untersucht. Danach ergibt sich, daß im Gegensatz zu der synchronen und synonymen Beeinflußbarkeit des MPS- und des Proteinanteiles des MPS-Proteinkomplexes die Grundsubstanz- und die Faserbildung gleichsinnig und gegensinnig zu beeinflussen sind. Ihre Abhängigkeit voneinander ist also derzeit noch unübersichtlich.

The Histochemistry of the Synthesis of Ground Substance and Fibres in Connective Tissue. Up to now the formation of ground substance and fibres cannot be determined sufficiently with the usual stain methods, i.e. intra- and extracellular synthesis can be made visible only in part. Also the modern procedures of enzyme histochemistry do not yield sufficient findings on the processes of intracellular synthesis of ground substance, but only rough comparisons with the more subtle biochemical results. Autoradiography is, as a histochemical procedure, the only actual method for histotopological localization of synthesis of ground substance. In the electron microscope hitherto only the formation of fibres can be demonstrated. The classical problem of morphology, i.e. whether the formation of ground substance and fibres may occur in the same cell at the same time, can only be solved autoradiographically. The interaction of both synthesis processes concerning time and place has been investigated by separate and combined application of ^{35}S sulfate and of ^{3}H proline *in vivo* and *in vitro*, especially on cartilage tissue. The time of incorporation and exposure and the dosages have been modified. Further technical aids such as filters, double mounting procedures etc. have been used, and further autoradiographic detection methods for double labelling. The best way is the comparison of autoradiographs mounted immediately after tissue preparation and later on after several half lives of ^{35}S sulfate. — Result: In one chondroblast the synthesis of mucopolysaccharides (MPS) and collagen may occur simultaneously. These anabolic processes and their interdependence have been examined by biochemical and radiochemical methods in comparative studies to ascertain the qualitative and quantitative differentiation. The results show that the MPS and the protein portion of the MPS-Protein-complex is influenced in a similar way and at the same time. On the other hand the synthesis of the ground substance and fibres can be influenced in the same and in a contrary sense. Their interdependence cannot be sufficiently appreciated.

Pathologisches Institut der Universität, 2 Hamburg (Deutschland).

6.

GÖTZ FREYTAG: **Zur qualitativen und quantitativen Erfassung von Bindegewebsmucopolysacchariden.**

Der qualitative Nachweis und die quantitative Bestimmung saurer Mucopolysaccharide im Bindegewebe ist histologisch oder histochemisch trotz intensiver Bearbeitung der zahlreichen Probleme in den letzten Jahren nach wie vor schwierig, unvollständig und unbefriedigend. Vergleichende biochemische oder radiochemische Untersuchungen sind notwendig. Sie können aber auch die Probleme des Histochemikers vergrößern, vor allem dann, wenn zur Klärung biochemische und radiochemische Verfahren benutzt werden, welche selbst in einem Arbeitsschritt unklare Färbeverfahren verwenden und auf deren Ergebnissen aufbauen (wie z. B. bei der Chondroitinsulfatbestimmung nach DITTMANN und CREMER). Deswegen sind wir im Rahmen derartiger histochemischer, biochemischer und radiochemischer Vergleichsuntersuchungen bemüht, ihre kritischen Überschneidungs- und Berührungspunkte weiter aufzuklären. Da die Bestimmung der quantitativen Beziehungen zwischen den sauren Mucopolysacchariden und den zu ihrer Anfärbung besonders verwendeten Farbstoffen das gemeinsame Kardinalproblem dieser Vergleichsuntersuchungen darstellt, wurden diese Beziehungen *in vitro* und in der Größenordnung des histologischen Schnittpräparates geprüft. Zunächst wurden nach wie vor offene Fragen der Absorptionsspektren von metachromatischen kationischen Farbstoffen und von Kupferphthalocyaninen geprüft, weiterhin die Beeinflussung dieser Absorption durch bestimmte, im histologischen Schnittpräparat wesentliche und zu berücksichtigende Mucopolysaccharid-Eiweiß-Konzentrations- und -Bindungsverhältnisse sowie schließlich die Abhängigkeit dieser Absorptionsspektren von bisher bei derartigen Untersuchungen wenig berücksichtigten *in vitro*-Reaktionen. Die an charakteristischen Beispielen zusammengefaßten und zur Darstellung kommenden Ergebnisse zeigen, daß es bisher noch nicht möglich ist, in der Größenordnung des histologischen Schnittpräparates auch nur annähernd exakte Angaben über die im betreffenden Gewebsschnitt vorhandene Menge saurer Mucopolysaccharide zu machen oder ihre qualitative Zusammensetzung zu differenzieren.

On Qualitative and Quantitative Determination of Mucopolysaccharides of Connective Tissue. The qualitative identification and the quantitative determination of acid mucopolysaccharides of connective tissue is still a very difficult problem. In spite of the intensive histological and histochemical investigations in this field during the last years the results are incomplete and not satisfactory. Comparative biochemical and radiochemical work is necessary. On the other hand these subtle studies may intensify the problems of histologists working with histochemical methods, above all, if biochemical and radiochemical methods are applied which use in some detail uncertain staining technics. Therefore the conclusions of these results are not always valid (as for instance in the determination of chondroitin sulfate by DITTMANN and CREMER). Therefore we are occupied to investigate the critical factors of interaction and the common aspect of such comparative histo-, bio- and radiochemical studies. — The estimation of the quantitative correlation between the acid mucopolysaccharides and the dyes especially used to colour them is the main problem of such comparative investigations. Therefore these relations have been proved *in vitro* and on the level of histological sections. At first open questions concerning the absorption spectra of metachromatic cationic dyes and of cuprophthalocyanines have been examined, furthermore of this absorption is influenced, in histological sections, by special, important relations of concentration and binding of mucopolysaccharid-protein complexes. In addition to that, the dependence of the absorption spectra of *in vitro*-reactions until now have not been included sufficiently in those studies. — The results, summarized in characteristic examples, demonstrate the impossibility, until now, of determining the exact amount of acid mucopolysaccharides on the level of histological sections and to differentiate their qualitative composition.

Pathologisches Institut der Universität, 2 Hamburg (Deutschland).

7.

V. PRETO PARVIS and C. RAVETTO: Histochemical Study on Containing Sulfate and Sialic Acid Groups Mucins*.

The authors have studied cat submandibular gland, which is rich in mucous elements, and have noticed a partial action of neuraminidase, which attenuates but does not eliminate in cells the dye-binding capacity for toluidine blue. On the other side, by using the same dye at various pH levels, they remarked that the staining begins already at pH 0.4 and increases considerably at about pH 3. This finding has suggested the presence of acid sulfate groups masking the sialidase effect. Since it is known that the methylation can block the carboxyl groups by a reversible reaction and the sulfate groups by an irreversible reaction, authors have intended to exclude the last with such treatment. In fact, dye-binding capacity for toluidine or Alcian blue has disappeared after methylation, has been partially restored after demethylation and fully eliminated by further treatment with neuraminidase. — To confirm such result, by staining with Alcian blue at pH 0.5, a partial staining of mucous cells was obtained and further with Alcian yellow at pH 2.5 the cells have been stained green, partially blue or yellow. Positive answer to Bracco-Curti (with benzidine) and Geyer (with tetrazotised dianisidine) reactions has confirmed the presence of sulfate groups. — Same technical methodics have been successfully used on human submandibular gland and on other mucous glands.

* Research partially supported by Consiglio Nazionale delle Ricerche.

Istituto di Istologia ed Embriologia Generale dell'Università di Milano, Via Ponzio 7, Milano (Italia).

8.

F. BAUER: Die Rolle des Stoffwechsels in der Entstehung der Orthochromasie und Metachromasie im Hornhautgewebe.

Das Problem der Entstehung der Metachromasie und der Orthochromasie konnte bis heute noch nicht befriedigend geklärt werden. Die Ansichten von LISON und die von MICHAELIS stehen in Widerspruch zueinander. Die eigenen Untersuchungen hatten das Ziel, das Problem der Entstehung der Ortho- und Metachromasie im Spiegel des Metabolismus zu betrachten und die Entstehung der Farbtonveränderungen durch Komplexbildung des Farbstoffmoleküls mit verschiedenen Komponenten des Gewebes und der Stoffwechselprodukte zu erklären. Es wurde festgestellt, daß für die Entstehung der Metachromasie der Mucopolysaccharid-Farbstoffkomplex verantwortlich ist (die Anschauung LISONs), wobei sich ein Farbstoffmolekül an eine —SO$_2$OH-Gruppe bindet, d.h. bei hydrolytischer Abspaltung der chromotropen Gruppe (—SO$_2$OH) kann es zu keiner Komplexbildung des Farbstoffes mit dem Glukosaminmolekül kommen. Es entsteht vielmehr ein Schwefelsäure-Farbstoffkomplex mit der abgespaltenen schwefelsauren Estergruppe. Dieser Komplex entspricht einem kornblumenblauen Farbton. Dies bezieht sich auf bradytrophe Gewebetypen. Falls im Gewebe durch Vascularisation Eutrophie entsteht und durch Eintreten der oxydativen Phase der Glykolyse die mehrwertigen Säuren des Krebs-Cyclus Komplexe mit dem Farbstoff bilden, entsteht ebenfalls ein blauer Farbton. — Die Ergebnisse sind chemisch erklärlich und werden durch die Warburgsche Methode bestätigt.

Hornhaut	Q_{O_2}	Metachromasie
Epithel	6—8	blau (Orthochromasie)
Stroma	1	leuchtendrot (Metachromasie)
Endothel.	4—6	blau
Stroma (vascul.) .	4	blau (Orthochromasie)

The Influence of Metabolism on the Formation of Orthochromasia and Metachromasia in the Cornea. So far the problem of orthochromasia and metachromasia has not found a satisfactory solution. The two existing theories (LISON and MICHAELIS) contradict each other. Assuming that changes in the colour-tone are due to the fact that the dye molecule combines with various components of the tissue and metabolic products, my own investigations aim at solving the problem from a more metabolic point of view. It is found that a mucopolysaccharide-dye complex is responsible for the development of metachromasia (LISON's theory). In this case the dye molecule combines with a $—SO_2OH$ group, that is to say a hydrolytic cleavage of the chromotropic group ($—SO_2OH$) prevents the dye from combining with the glucosamine molecule. Instead, a sulphuric acid-dye complex is formed which includes the sulphuric ester group. The complex corresponds to a deep blue colour-tone. This applies to bradytrophic types of tissue. The same blue colour-tone is formed if by vascularisation eutrophia arises in the tissue and with the beginning of the oxidative phase of glycolysis the polyvalent acids of the Krebs-cycle form complexes with the dye. The results are explicable by chemical arguments and are confirmed by the Warburg method.

Cornea	Q_{O_2}	Metachromasia
Epithelium	6—8	blue (orthochromasia)
Stroma	1	brilliant red (metachromasia)
Endothelium . . .	4—6	blue
Stroma (vascul.) .	4	blue (orthochromasia)

Universitäts-Augenklinik, Debrecen (Ungarn).

9.

TAMÁS NEUMARK: The Relationship between the Collagen Fibres and the Amount of Acid Mucopolysaccharides Regarding the Development of Phenol Reaction.

In our experiments we have examined the relationship between the collagen fibres and the amount of acid mucopolysaccharides (MPS) in different connective tissues by examining the development of the phenol reaction by the polarisation microscope. According to the experiments of ROMHÁNYI, after sulphuration of collagen fibres, the phenol reaction does not occur. In consequence of this fact we supposed that the acid groups of the ground-substance may also have an inhibiting effect on the association of phenol molecules to the collagen fibres. We have demonstrated that in such connective tissue, where the amount of acid MPS is high, the optical inversion of the collagen fibres takes place only by an increased phenol concentration; but this does not happen where the amount of acid MPS is smaller. — By the treatment of connective tissues with testicular hyaluronidase or after methylation the birefrigence of collagen fibres is made negative by a lower concentration of phenol molecules. According to our experiments the acid MPS of ground-substance influences the development of the phenol reaction, and this observation indicates a close relationship between acid MPS and collagen fibres.

National Institute of Rheumatism and Medical Hydrology, Laboratory of Histopathology, Budapest (Hungary).

10.

CSABA HADHÁZY and BARNABÁS MÁNDI: Carbohydrate Histochemical Investigations on Cartilage Ground Substance in Formation and Resorption.

The cartilaginous material was obtained from regenerating articular surface; mature cartilage originated from intact articular surface and resorption was studied on cartilage grafts transplanted into muscle. The material was stained by PAS, Ritter-Oleson and Hale methods and by Alcian blue, toluidin blue and methylene blue buffer staining supplemented by various controls using digestion, blocking and extraction methods. — The hyaluronidase susceptible mucopolysaccharides produced during cartilage formation are found around the younger cells too (praechondroblasts) but their production is primarily due to the chrondroblasts. The ground substance surrounding the younger cells is less basophilic. The newly produced ground substance is generally Hale positive by Ritter-Oleson's method, the capsule remains Hale positive, whereas the rest of the ground substance becomes gradually PAS positive. The staining of the mature cartilage tissue is rhapsodic. The cartilage tissue originating from two different individuals may stain identically, and the neighbouring areas of the same articular surface may stain differently. In the vicinity of cartilage resorption the metachromasia of the ground substance decreased, the degree of basophilia was lower and strong PAS positivity ensued. — These characteristics refer to conditional changes of the mucopolysaccharides which may be explained by their protein binding, resp. depolymerisation.

Dept. of Anatomy, University Medical School, Debrecen (Hungary).

Tumoren · Tumors · Tumeurs

1.

C. J. LOUIS: Differential Staining of Neoplastic and Non-neoplastic Tissues.

WEILER (1956) reported that, coincident with aminoazo dye carcinogenesis, there was loss of organ specificity from rat liver as determined by complement fixation test. Using fluorescein *iso*cyanate-labelled organ specific antisera as histoserological stains, he demonstrated differential staining between rat liver and rat hepatoma. He regarded this lack of staining of tumour cells as confirmatory evidence of loss of their organ specific antigens. — All attempts to produce organ specific antisera by Weiler's or other methods were unsuccessful and fluorescein *iso*cyanate serum protein conjugates prepared from a wide variety of non-immunized animals gave identical differential staining. The mechanism of this staining will be discussed. — A large series of neoplastic lesions was investigated and it was found that malignant tumours uniformly showed greatly diminished fluorescence compared with the parent tissues. On the other hand innocent tumours and hyperplastic conditions uniformly fluoresced brightly and showed fluorescein-protein affinities similar to those of the tissue of origin. — More recently, other workers have reported results partly at variance with these; however they used fluorescein *isothio*cyanate instead of fluorescein *iso*cyanate. MILLER, WILLEY, LEESTMA and RIGGS (1963) investigated this problem and found that if fluorescein *iso*cyanate was utilized in preparing the stain they could produce an easily recognizable and consistent difference in cytoplasmic fluorescence between benign and malignant cells but if fluorescein *isothio*cyanate was used the results were equivocal and inconsistent.

Dept. of Pathology, University of Melbourne, Melbourne (Australia).

2.

P. J. MELNICK and S. H. ROSEN: Identification of Glycogen, Keratin and Mucin in Undifferentiated Lung Cancers*.

In more than 100 undifferentiated large cell carcinomas of the lung, 59 were identified as undifferentiated squamous cell carcinomas by: 1) their content of intracytoplasmic glycogen demonstrated with the PAS, Alcian blue-PAS, Gridley, and Gomori methenamine silver methods; and 2) keratin, identified by reactions for its sulfhydryl component with the alkaline tetrazolium, DDD, ferric ferricyanide and Bennett mercury orange methods, as well as the Kreyberg and Mallory phosphotungstic acid hematoxylin methods. 16 examples were identified as undifferentiated adenocarcinomas by their content of intracytoplasmic mucin droplets, identified by the PAS, Alcian blue-PAS, toluidine blue, Gridley, Gomori methenamine silver, and Kreyberg methods as well as the mucicarmine method. 26 cases were found to be undifferentiated mixed or adenosquamous carcinomas because of their content of all three, namely glycogen, keratin, and mucin. In 48 undifferentiated small cell and oat cell carcinomas no characteristic components could be demonstrated with the above methods.

* Aided by USPHS Research Grant No. CA-06987-01.

Dept. of Pathology, University of California Medical School, San Francisco, California (USA) and Pulmonary and E.N.T. Pathology Branch, Armed Forces Institute of Pathology, Washington, D.C. (USA).

3.

GILLES TREMBLAY: **Histochemical Study of Amylase Activity in Tumors of the Human Salivary Glands*.**

The tumors of the human salivary glands present a great variety of types and many conflicting views have been expressed regarding their histogenesis. The present work concerns the comparative histochemical study of amylase in normal salivary glands and in tumors of the salivary glands with the view of obtaining information on the histogenesis and the degree of differentiation of these tumors. Only surgical specimens are used. Amylase is determined by a starch film method (TREMBLAY, G. J., Histochem. Cytochem. *11*, 202—206, 1963). In this procedure a thin film of starch is exposed to a tissue section and, after incubation, is stained by the periodic acid-Schiff reaction. Clear areas are then observed in the film due to hydrolysis of starch by the amylase present in the tissue. Comparison of the unstained pattern in the starch film with the corresponding tissue section reveals the sites of amylase activity in the tissue. — This method was used to study the distribution of amylase in specimens of mixed tumours, cylindromas and papillary cystadenomas lymphomatosum.

* Supported by a grant from the National Cancer Institute of Canada.

Dept. of Pathology and Montreal Cancer Institute, Notre Dame Hospital and University of Montreal, Montreal (Canada).

4.

TADAO TAKEUCHI: **Histochemical Demonstration of Phosphorylase in Free Cancer Cells.**

Histochemical detection for phosphorylase in free cancer cells in the fluid of human ascites and various animal ascites tumors was easily made by improvement of Takeuchi and Kuriaki's technique. The materials were smeared on slides and dried quickly by a cool wind. Without fixation or after one minute fixation in methanol the smears were incubated in the substrate mixture containing 50 mg potassium glucose-1-phosphate, 20 mg muscle adenylic acid, 2 mg glycogen, 10 ml distilled water with 10 ml 0.2 M acetate buffer (pH 5.7), 2 mg sodium fluoride, 5 ml ethanol and insulin (about 10 units) at 37⁰ C for 1 hour. Occasionally 10 mg sodium adenosine-triphosphate, 2 mg magnesium chloride and 2 mg glucagon were added to the substrate mixture when weak reaction. — Cancer cells contained very often rich phosphorylase in almost all kinds of the tumors examined. It was, moreover, found by using the combination technique of histochemistry and radioautography that richer phosphorylase was demonstrable in cancer cells during synthesis of deoxyribonucleic acid and mitosis needing high energy metabolism, while glycogen deposit was decreased or lacking in these cells. Phosphorylase seemed to be localized in the cytoplasmic matrix to the exclusion of mitochondria, Golgi's regions and other organelles, by use of the Sasaki-Takeuchi's electronhistochemical technique using a glycogen-poor strain of Yoshida ascites hepatoma (AH 39).

Honjo-Machi 430, Dept. of Pathology, Kumamoto University School of Medicine, Kumamoto (Japan).

5.

F. PAPOUŠEK, J. ŠVEJDA and J. BRICHTA: **Quantitative and Qualitative Estimation of the Enzymes Activity during the Growth of the Rat Transplantable Tumor BŠ.**

The transplantable rat-tumor BŠ is a round cell carcinoma which in cell-free filtrates gives rise to transplantable leukaemia. The qualitative and quantitative estimation of the activity of some enzymes during the growth of the tumor appeared as a question of a great importance. The enzymes involved were alkaline and acid phosphatases, non specific AS-esterase, lipase, cytochromoxidase and beta-glucuronidase, as far as succinic-, lactic-, malic-dehydrogenase, and dehydrogenase of alpha-glycerophosphate. The activity of these dehydrogenases was estimated also quantitatively with the aid of spectrophotometry in respect to the quantity of formazan calculated on 1 mg of fresh tissue. The time intervals were 7th, 11th, 15th, 21st and 28th day after the transplantation. Parallel measurements of tissue respiration of the tumor in Warburg apparatus were made. The greatest activity of the enzymes and the tissue respiration was proved on 15th day after transplantation. These findings are of great value for the research of the biological activity of the tumor, namely as far as the further growth, metastases and tissue cultivation of the tumor is concerned.

1. Dept. of Path. Anat. Med. Fac. University J. E. Purkyně, Brno, Pekařská 53 (Czechoslovakia).

6.

N. T. RAIKHLIN: Comparative Histochemical Investigations of Changes in Redox Enzymes in Virus, Chemical, and Plastics Cancerogenesis and in Spontaneous Malignancy of Fibroblasts in Monolayer Cultures.

The activity of cytochromoxidase, diaphorase, and DPN- and TPN-linked dehydrogenases were investigated at different stages of cancerogenesis. The changes at different stages of experimental cancerogenesis were similar and independent of the nature of the acting cancerogenic factor. It was found in the earliest stages of cancerogenesis that the permeability of the mitochondrial membranes was affected and the activity of all the enzymes studied was enhanced. The enzymatic activity in different cells, in the presence of pre-tumour proliferation is rather polymorphous, but it is still maintained at a sufficiently high level. A redistribution of localisation among a number of enzymes takes place in the tumour cells between individual structural units of cells. — The activity is decreased in enzymes that take part in the electronic transport (cytochromoxidase, TPN-diaphorase), as well as of succindehydrogenase, dehydrogenase of α-glycerophosphate, and at times of DPN-diaphorase. The activity of dehydrogenases of lactic acid, glutamate, glucose-6-phosphate and 6-phosphogluconate in the tumour is markedly increased. Special experiments have shown that changes of enzyme activity are due to a great extent to the permeability of the cellular and mitochondrial membranes.

Institute of Experimental and Clinical Oncology, Academy of Medical Sciences of USSR, Moscow (USSR).

7.

G. R. N. JONES: Quantitative Histochemistry: Respiratory Enzymes in Experimentally Induced Cancer*.

Histological control of experimental material is a fundamental problem in the biochemical study of pathological material, since most biochemical procedures involve the disruption of tissue into subcellular components as an initial step, and the histological architecture is thereby destroyed. Histochemical investigations can yield relevant and helpful information at a qualitative level since both biochemical activity and histological structure are simultaneously preserved. Quantitative histochemical methods are potentially yet more useful, since biochemical function can then be studied under standardised conditions at the level of the unhomogenised cell. — Cytochrome oxidase and succinate-neotetrazolium reductase have been studied by quantitative histochemical techniques in secondary liver tumours found in the lung of a rat which had received 4-dimethylaminoazobenzene in the diet. In mitotically active material the levels of these enzymes were of the same order as those of normal rat liver. In tissue in which there was little or no mitotic activity, and in which the cells showed varying degrees of damage, succinate-neotetrazolium reductase activity was somewhat reduced, while cytochrome oxidase was only slightly less active than in normal rat liver. These results will be discussed in relation to the findings of other workers.

* This work is supported by the British Empire Cancer Campaign.

Dept. of Pathology, Royal College of Surgeons of England, Lincoln's Inn Fields, London W.C.2 (Great Britain).

8.

W. MÜLLER, Köln/Deutschland: Cytophotometrische Untersuchungen an Neurinomen.

9.

ÉVA GÁTI: Histochemische Untersuchungen an bilateral beimpften degranolempfindlichen bzw. degranolresistenten Mäuselymphomen.

Wir impften Mäusen jeweils subkutan auf eine Seite Geschwulstzellen, die aus degranolempfindlichen (DE), auf die andere Seite solche, die aus degranolresistenten (DR/NK) Lymphomascitestumoren stammten. 10—12 Tage nach der Überimpfung gaben wir den Mäusen einmalig eine semiletale Degranoldosis (100 mg/kg). Vor der Behandlung bzw. 24, 48, 72, 96 Std danach entnahmen wir die Tumoren und fertigten Kryostatschnitte an. — Die Succinodehydrogenase-Aktivität blieb im DR-Tumor konstant, sank im DE-Tumor 24 Std nach Degranolgabe auf ein Minimum ab, um sich 72 Std später wieder langsam zu erhöhen. Der Inkubationslösung zugegebenes 0,001 M Cystein behebt die Hemmung der Succinodehydrogenase-Aktivität. — Die Aktivität der DPN-Diaphorase und der Milchsäuredehydrogenase ist im unbehandelten DR-Tumor stärker als im unbehandelten DE-Tumor. — Unter der

Degranolwirkung vermindert sich die Milchsäuredehydrogenase-Aktivität im DE-Tumor, um im DR-Tumor konstant zu bleiben. Die Aktivität der Apfelsäuredehydrogenase ist im DE- und im DR-Tumor gleichermaßen gehemmt, während die der DPN-Diaphorase und der Alkoholdehydrogenase in beiden Tumoren keine Änderung erfährt. — Wir untersuchten die Aktivität der Phosphatasen nach GOMORI, wobei uns Nukleotide, Kohlenhydrate und Glyzerinphosphate — bei alkalischem, saurem und neutralem pH — als Substrat dienten. — Unter der Degranolwirkung nimmt die Spaltung von Glyzerinphosphaten und DPN bei alkalischem Milieu schon nach 24 bzw. 48 Std im DE-Tumor stark ab, im DR-Tumor dagegen zu.

A Histochemical Investigation of Bilaterally Inoculated Degranol-Sensitive and Degranol-Resistant Lymphoma in Mice. Mice were injected with Degranol-sensitive (DS) or else Degranol-resistant (DR/NK) lymphoma ascites tumours in two subcutaneous sites. 10—12 days after the inoculation a semi-lethal dose of Degranol (100 mgs/kg) was applied. Prior to the application of the drug and 24, 48, 72 and 96 hours afterwards specimens were taken from the tumours and sections cut on the cryostat. The activity of succinate dehydrogenase remained constant in DR-tumours; 24 hours after the application of Degranol a marked decrease was found in the activity of DS-tumours, which, after 72 hours, was followed by a slight increase. If 0.001 M cysteine was added to the incubation solution the inhibition of succinate dehydrogenase activity was removed. The activity of DPN-diaphorase and lactic acid dehydrogenase was more distinct in untreated DR-tumours than in untreated DS-tumours. After Degranol treatment lactate dehydrogenase activity was reduced in DS-tumours while it remained constant in DR-tumours. The activity of malate acid dehydrogenase was inhibited in both DS- und DR-tumours, whereas DPN-diaphorase and alcohol dehydrogenase were unaffected. We tested the phosphatase activity (Gomori method) using nucleotide, carbohydrate and glycerol phosphates as substrates. The tests were carried out at alkaline, neutral and acid pH. 24—48 hours after Degranol treatment the cleavage of glycerol phosphates and DPN (alkaline pH) was markedly reduced in DS-tumours, whereas it was increased in DR-tumours.

Aus dem Onkopathologischen Forschungsinstitut, Budapest XII (Ungarn).

10.

FRITZ FEY: Phosphatasen bei virusinduzierten Leukämien.

Mittels der Azokupplungsmethode wurde die Aktivität von alkalischer und saurer Phosphatase an virusinduzierten Leukämien der Maus und an normalen Mäusen verschiedener Inzuchtstämme untersucht. Im einzelnen wurden Kryostatschnitte von Thymus, Milz und Lymphknoten, sowie Tupf- und Ausstrichpräparate der Leber, des Knochenmarks und des Blutes von normalen und leukämischen Tieren getestet. Während bei den Normaltieren nur eine geringe zellgebundene Phosphatasen-Aktivität nachgewiesen werden konnte, ergaben sich bei den Leukämien folgende Befunde: Die lymphatischen und Paramyeloblasten-Leukämien besitzen durchweg eine starke alkalische und schwache saure Phosphatase-Aktivität. Bei den übrigen myeloischen Leukämieformen ist das Auftreten der Phosphatasen heterogen, vermutlich besteht eine Korrelation zwischen Reifegrad der Leukämiezellen und Phosphatase-Aktivität. Die Retikulumzell-Leukämien zeigen hinsichtlich der alkalischen Phosphatase-Aktivität den geringsten Effekt, während die saure Phosphatase in den meisten Fällen erhöhte Werte aufwies. — Die zytochemisch ermittelten Resultate bezüglich der alkalischen Phosphatasen konnten biochemisch verifiziert werden.

Phosphatases in Virus-Induced Leukaemias. By means of the coupling azo dye method the activity of alkaline and acid phosphatases was investigated in various inbred strains of normal and leukaemic mice. Tests were carried out on cryostat sections of thymus, spleen and lymph nodes as well as on imprints and smears of bone marrow, liver, and blood of the normal and the leukaemic animals. Little cell-bound phosphatase activity was found in the normal animals. The findings in the leukaemic animals were as follows: lymphatic and paramyeloblastic leukaemias showed a high activity of alkaline and a low activity of acid phosphatase. In all other forms of myeloid leukaemia the activity of phosphatases was heterogeneous; presumably there is a correlation between the grade of maturity of the leukaemia cells and phosphatase activity. Reticulum cell leukaemias showed the lowest effect as far as the activity of alkaline phosphatase was concerned; whereas the acid phosphatase was raised in most cases. As far as alkaline phosphatase was concerned the cytochemical results were in agreement with biochemical findings.

Institut für experimentelle Krebsforschung der Deutschen Akademie der Wissenschaften, Berlin-Buch (Deutschland).

Donnerstag, Thursday, Jeudi, 20. August 1964

Haut und Epithel · Skin and Epithelium · Peau et tissu épithélial
Verschiedenes · Divers · Divers

1.

MAFFO VIALLI: **The Granulous Glands of the Skin of Bombinator pachypus.**

In the skin of *Bombinator pachypus* fixed with formol can be observed two types of granulous glands containing either large or small granules which differ both histochemically and histophysically. The acetone extracts of the skin contain a high concentration of 5-hydroxytryptamine and also another substance which couples with diazonium compounds. In material prepared by freeze-drying, the large granules show a complex structure, being composed of numerous minute granules immersed in a homogeneous mass of different reactivity. In addition to the properties already mentioned, these granules react positively in the p-dimethylaminobenzaldehyde reaction, the Morel and Sisley reaction and the blockable PAS reaction. — The fluorescence excited by Wood's light reveals differences between granular types, between granules themselves and also variations dependent on the fixing process. The spectral curves have been obtained using a Zanotti universal histophotometer. The large granules show strong white-blue primary fluorescence which is increased by post-fixation but is more intense than that of the formol-fixed material; the spectral curves vary in the same way. The small granules show a slight yellow-brown primary fluorescence; post-fixed material and that fixed directly in formol show a fairly intense yellow-gold fluorescence and modified spectra. The fixing process appears to influence the intensity and peak distribution in the UV absorption spectra of the two granular types which accords with the data obtained from fluorescence. The results indicate the complexity of the secretion.

Institute of Comparative Anatomy and Center of Histochemistry of C.N.R. University of Pavia (Italy).

2.

R. W. GOLTZ and A. M. HULT: **Microspectrophotometric Quantitative Estimation of Elastic Fibers in Human Skin*.**

A method has been developed, permitting the estimation of elastic fibers of human skin, in normal and pathological states. This method, based on the property of orcein to bind stoichiometrically with elastin at a maximal absorption peak of 589 μm, permits measurements to be made, in the visible light range, with a Zeiss microspectrophotometer, under conditions which obey Beer's law. A ratio is obtained, which reflects a relative amount of elastic fibers. This procedure was applied to skin from various anatomical sites of normal individuals of different age, sex and race; and to skin from patients with such systemic connective tissue diseases as pseudoxanthoma elasticum, Ehler-Danlos Syndrome, lupus erythematosus, scleroderma, and cutis laxa. Results of these measurements will be briefly summarized.

* Supported by N.I.H. Grant No. GM 10148.

Division of Dermatology, University of Minnesota Medical School, Minneapolis, Minnesota (USA).

3.

ANNA MARIA BOLOGNANI FANTIN and LORENZO BOLOGNANI: **Observations on the Histochemistry and Biochemistry of the Cutaneous Mucus of Anguilla vulgaris.**

Histochemical research of the epidermis of *Anguilla* has revealed the presence of three types of cell: — principal cells, club cells and mucous cells. The principal cells are intensely PAS positive only in certain layers. The young and more mature club cells do not show the reactions of polysaccharides but do show those of proteins, the more mature cells giving more positive results. The mucous cells which may also be differentiated into young and more mature cells, can produce either a secretion containing only neutral polysaccharides (only PAS positive), or one containing neutral polysaccharides and non-sulphated acid polysaccharides, one component being sialic acid (the secretion is partly digested by sialidase and hydrolysed chemically, according to QUINTARELLI et al.). The characteristics are not always correlated with the degree of maturation. — In order to clarify the composition of the cutaneous mucus of *Anguilla*, the mucus was scraped manually from the skin and subjected to digestion by papain. It was then possible to isolate a polysaccharide fraction without sulphated groups. The principal constituents are hexosamine (galactosamine), sialic acid, and neutral sugars; only traces of uronic acids were found. These findings confirm and illustrate the results of histochemical investigations of the cellular elements concerned with the production of mucus and of the principal cells which also contain mucopolysaccharides.

Institute of Comparative Anatomy and Center of Histochemistry of C.N.R., and Institute of Biochemistry, University of Pavia (Italy).

4.

ALDO UGO: Histochemical Observations on Some Cutaneous Elastopathies.

Further to previous histochemical researches on normal elastic membranes and fibers (PRETO PARVIS and UGO 1962—1964) the characteristic fibers of some cutaneous elastopathies (Favre-Racouchot disease, Pseudoxanthoma elasticum, Cutis hyperelastica) have also been studied. In particular, attention has been paid to the lipid component of these fibers by applying standard histochemical methods (Sudan black, Nile blue sulfate, Liebermann reaction, Baker's test for lipine, PFAS reaction and its control with bromination, 3-4 benzpyrene, phosphine 3 R, primuline, rhodamine). Comparison of the results from normal elastic and collagen, and those from more or less dystrophic fibers which are characteristic of the elastoses, favours the derivation of the latter from elastic rather than from collagen. Collateral use of methods for aminopolysaccharides, especially for acid mucopolysaccharides (staining with methylene blue at various pH levels and observation of metachromasia, Alcian blue, Hale's reaction, aldehyde-fuchsin, PAS reaction, etc.), demonstrates in general a concomitant increase of these substances in the elastotic derma. This finding is discussed, taking in account the altered circulation in cutaneous zones which present the observed elastotic picture.

Istituto di Istologia ed Embriologia Generale dell'Università di Milano, Via Ponzio 7, Milano (Italia).

5.

COLIN JOHN SMITH: Epithelial Cell Lysosomes in Malignancy.

Lysosomes have been disclosed in hamster cheek pouch epithelium by staining fresh-frozen, unfixed, cold-microtome sections for acid phosphatase and non-specific esterase. Apart from staining by these techniques the particles demonstrate other properties necessary for designation as lysosomes (BITENSKY, L., in Ciba Foundation Symposium "Lysosomes", p. 362, J. & A. Churchill, London, 1963). Striking changes in size of the lysosomes have been found in the course of carcinogenesis, produced by regular applications of a 0.5% solution of 9:10-dimethyl-1,2-benzanthracene in liquid paraffin. Prior to the onset of malignancy the pouch epithelium passed through a series of well-defined stages: (1) an initial inflammatory reaction in which an occasional cell of the basal layer displayed enlarged lysosomes; (2) a recovery phase in which the normal fine, discrete pattern of lysosomes was evident; and (3) a period of epithelial hyperplasia and papilloma formation with enlarged lysosomes again appearing in the basal cells. In the established malignant tumours large lysosomes were apparent in cells throughout the lesion, though not all cells were affected. This appearance in experimentally induced malignancy was similar to that seen in sections of human oral carcinoma stained by the same methods. Other pathological conditions of the oral mucosa have been subjected to these techniques for comparison of their lysosomes pattern.

Dept. of Dental Science, Royal College of Surgeons of England, Lincoln's Inn Field, London W.C.2 (Great Britain).

6.

HOWARD H. CHAUNCEY, JOSEPH H. KRONMAN and GEORGE YERGANIAN: Enzyme Histochemistry of Oral Tissues in the Normal and Diabetic Chinese Hamster*.

Histochemical characteristics of the cheek pouch were determined in normal Chinese hamsters *(Cricetus griseus)* and in animals with spontaneous Diabetes Mellitus. Classification of the severity of the disease process was dependent on the time of onset of symptomology and daily dosage of insulin required to maintain sugar-free urine. — Tissue samples were obtained immediately after sacrifice. Representative specimens were fixed in neutral buffered formalin, processed, and stained with hematoxylin and eosin. Other specimens were placed in chloral hydrate formalin solution for 18 hours (4⁰ C), washed, and gelatin embedded. Cryostat cut sections, 10 μ thick, were placed free-floating in buffered substrate solutions for the localization of acid phosphatase and nonspecific esterases. Succinic dehydrogenase activity was localized in unfixed tissue sections. — Morphologic changes noted in the diabetic animals were reduction in number and size of mucosal papillae, accompanied by a loss of normal epithelial dimensions. Histochemical alterations were: reduction in acid phosphatase activity of basal layers of epithelium; decreased nonspecific esterase activity in all zones of interpapillary epithelium; and increased esterase activity in the basal region of the papillary epithelium. Succinic dehydrogenase activity did not appear to be quantitatively altered. — These findings imply that in the diabetic animal epithelial metabolism is altered and rate of keratinization decreased. The thinning of keratin layer and change in acid phosphatase activity tend to substantiate this hypothesis.

* Aided by grants from the National Institutes of Health and the National Science Foundation.

Tufts University School of Dental Medicine and Children's Cancer Research Foundation, Boston, Massachusetts (USA).

7.

GOTTFRIED GEILER: Histochemische Untersuchungen am Fibrinoid subkutaner Rheumaknoten bei Rheumatoidarthritis.

Da die fibrinoide Veränderung des Bindegewebes bei ätiologisch sehr unterschiedlichen Erkrankungen und bei der Alteration verschiedener Gewebe vorkommt, ist es nicht berechtigt, sie a priori als eine einheitlich definierte Alteration aufzufassen. Gezielte Untersuchungen sind darum von Fall zu Fall notwendig. — Für die vorliegende histochemische Analyse des Fibrinoid subkutaner Rheumaknoten wurden Reaktionen zum Nachweis von Kohlenhydraten, Eiweißsubstanzen und Nukleinsäuren durchgeführt. Außerdem kamen histoenzymatische Untersuchungen sowie allgemeine Bindegewebs- und Faserfärbungen zur Anwendung. — Nach dem Reaktionsausfall der Kohlenhydratnachweismethoden enthält das Fibrinoid reichlich α-Glykole. Freie saure MPS liegen nicht vor, sondern nur geringe Mengen, die fest an Proteine gebunden sind. — Die Ninhydrin-Schiff-Reaktion weist auf einen hohen Gehalt an α-Aminosäuren hin. Im einzelnen lassen sich Tyrosin, Arginin, Cystin und Cystein erfassen. — Das Verhalten der Nukleinsäurefärbungen spricht für das Vorhandensein von RNS. — Durch Trypsin und Pepsin wird das Fibrinoid vollständig verdaut, gegen Kollagenase ist es resistent. — Bei den Fibrinfärbungen nach WEIGERT und MALLORY sowie bei den Allgemeinfärbungen verhält sich das Fibrinoid wie Fibrin. — Die Faserfärbungen zeigen abgestufte Schweregrade der Bindegewebsfaserschädigung bis zur Nekrose der kollagenen und elastischen Fasern. — Zusammenfassend läßt sich sagen, daß Exsudatfibrin am Aufbau des Fibrinoid der Rheumaknoten einen wesentlichen Anteil besitzt, daß aber auch andere Substanzen (Mukoproteide, Glykoproteide, geringe Mengen fest an Proteine gebundene saure MPS, Kernabbauprodukte, nekrotische Bindegewebsbestandteile) an seiner Zusammensetzung beteiligt sind.

A Histochemical Examination of the Fibrinoid of Subcutaneous Rheumatoid Nodules in Cases of Rheumatoid Arthritis. Fibrinoid transformation of the connective tissue occurs in diseases of varying aetiology, accompanying the alteration of various tissues. It is therefore unjustified to consider it, *a priori*, as a homogeneous alteration. Systematic investigations are therefore necessary for each particular case. The present histochemical analysis of the fibrinoid of subcutaneous rheumatoid nodules comprises tests for the determination of carbohydrates, proteins and nucleic acids. Histoenzymatic tests and general staining methods for connective tissue and fibres have been used additionally. According to the results of the carbohydrate determination large quantities of α-glycols are present in the fibrinoid. Free acid MPS is not demonstrable but small quantities of protein-bound MPS are present. The ninhydrin-Schiff-reaction indicates a high proportion of α-amino acids. Tyrosine, arginine, cystine and cysteine can be demonstrated separately. The results of the staining reactions for nucleic acids seem to indicate the presence of RNA. Fibrinoid is broken down completely by trypsin and pepsin whereas collagenase has no effect at all. Fibrinoid reacts like fibrin when stained according to the methods of WEIGERT and MALLORY or in general staining procedures. Special staining methods for fibres reveal various stages of lesions in the connective tissue which gradually lead up to the necrosis of collagen and elastic fibres. According to the results it can be concluded that exudate fibrin contributes in a high proportion to the formation of the fibrinoid of rheumatoid nodules. Other substances involved are mucoproteins, glycoproteins, small quantities of protein-bound acid MPS, nuclear debris and necrotic components of the connective tissue.

Pathologisches Institut der Karl-Marx-Universität, Liebigstr. 26, Leipzig C 1 (Deutschland).

8.

M. A. DINA and R. POZZUOLI: Histochemical Findings in Mummified Malherbe's Epithelioma.

Some histochemical findings of amorphous material wich characterizes Malherbe's epithelioma have been reported. The main consituent of the amorphous masses seems to be keratin relatively poor of SH groups. The distribution pattern of histochemical properties shows a preminent reactivity in the peripheric growing layers being the central, amorphous, scarcely reactive. No modification is noted with the aging of the neoplastic process. Quite parallel findings have been found in subcutaneous epidermoid cysts.

Istituto di Anatomia e Istologia Patologica, Viale Mancini 5, Sassari (Italy).

9.

HANS H. PFEIFFER: Rheooptical Behaviour of a Protein from Arthropod Ligaments.

Resilin isolated by WEIS-FOGH is a protein secreted to Arthropod elastic ligaments just as elastin in walls of vessels of Vertebrates. In dry state resilin is glass-like, but in polar solvents, like water, formamide, ethylene glycol or glycerol, it assumes the solidity of cherry-gum. Its swelling depends on the dissociation of ionizable groups of agents. It is insoluble in all solvents not breaking peptide bonds, but is digested by proteases, acids and bases. Stretched pieces fail to show, if subjected to X-ray diffraction, any regular leptonic orientation. But just like other isotropic substances, e.g. glass, gelatine, cherry-gum etc., the protein acquires, if squeezed within the sphere of elasticity, negative anisotropy to the direction of pressure. The intensity of birefringence increases with the rate of pressure (photo-elastic effect). After some seconds, however, the negativity shifts suddenly into positivity because the leptones place themselves transversely to the direction of pressure. The behaviour against Barrnett's reaction with 2,2'-dihydroxy-6,6'-dinaphthyldisulfide, which was synthesized especially for this purpose, establishes that SH and S-S groups are absent. Resilin is still a protein relating to elastin, and the protein character results from Gomori's staining (1950) with aldehyde-fuchsin (5 minutes, purple colour) or from the deep violet colouring after staining 15 minutes with orcinol-new fuchsin according to FULLMER and LILLIE (1956).

Aus dem Labor f. Polar.-Mikrosk., Wilhelmstr. 7, 28 Bremen I (Deutschland).

10.

HANS LUPPA: Histochemische Studien an den Müllerschen Riesenzellen von Lampetra planeri.

Die unmittelbar unter dem Ependym gelegenen Müllerschen Riesenzellen des Bachneunaugengehirnes besitzen im Cytoplasma granuläre Einschlüsse unterschiedlicher Größe und Lokalisation. Im Rahmen von Untersuchungen über die Funktion der Müllerschen Riesenzellen durchgeführte Färbungen und topochemische Nachweise zur Klärung der chemischen Zusammensetzung der sekretartigen Granula bzw. kolloidartigen Cytoplasmaeinschlüsse brachten folgende Ergebnisse: Die Granula färbten sich bei Larven und adulten Tieren kräftig mit Paraldehydfuchsin, blieben jedoch mit Chromalaunhämatoxylin-Phloxin ungefärbt. Der Nachweis von Proteinen bzw. Aminosäuren ergab die Anwesenheit von freien Aminogruppen, Tyrosin, Sulfhydrylgruppen, Disulfidgruppen, Histidin und Arginin, wobei der Anteil der Disulfidgruppen nach den Ergebnissen der Alcianblaufärbung (nach Voroxydation mit saurer Permanganatlösung) etwa 4% beträgt. Übereinstimmende Reaktionsergebnisse wurden in den Nucleolen erzielt. Während der larvalen Entwicklung traten in den Granula Schwankungen der Protein- bzw. Aminosäurekonzentration auf. Eine bemerkenswerte Ausnahme bildet der Argininnachweis, der nach Anwendung alkoholfreier Fixierungsgemische in den Granula sämtlicher Zellen von Larven und adulten Tieren gleichmäßig stark ausfällt. In topischer Übereinstimmung mit den Sekretgrana wurden regelmäßig Lipide (Phospholipide?), neutrale Mucopolysaccharide und säurefestes Lipofuscin nachgewiesen. Saure Phosphatase und (Naphthol-AS-D-Acetat-) Esterase waren positiv, Bernsteinsäuredehydrogenase offenbar negativ. Die Ergebnisse der Färbungen und histochemischen Nachweise scheinen auf eine neurosekretorische Tätigkeit der Müllerschen Riesenzellen von *Lampetra planeri* hinzudeuten.

A Histochemical Investigation of the Müller Giant Cells in Lampetra planeri. In the brain of *Lampetra planeri* the Müller giant cells are to be found directly under the ependyma. Their cytoplasm shows granulated inclusions of various sizes and variable localization. In the course of research into the function of the Müller giant cells we tried to obtain details about the chemical structure of these secretory-like granules or the colloid-like inclusions of cytoplasm. The results of staining methods and topochemical determinations are as follows: In larvae and adult animals the granules stain intensely with paraldehyde fuchsin, but remain unstained with chromalum haematoxylin phloxine. Protein or amino acid determinations yield positive results for free amino groups, tyrosine, sulphydryl groups, disulphide groups, histidine and arginine. According to the Alcian blue staining (after preliminary oxidation in acid permanganate solution) the number of disulphide groups amounts to 4%. Corresponding results are obtained in the nucleoli. During the chrysalis state fluctuations in the concentration of proteins or amino acids are observed in the granules. A noteworthy exception is arginine; after fixation in alcohol-free fixatives the granules in all the cells of adult animals and larvae give equally strong results. In topical agreement with the secretory granule lipids (phospholipids?), neutral mucopolysaccharides and acid-fast lipofuscin are regularly found. Acid phosphatase and (naphthol-*AS-D*-acetate-) esterase are positive; succinate dehydrogenases gives negative results. The results of staining and histochemical determinations seem to point at a neurosecretory function of the Müller giant cells in *Lampetra planeri*.

Zoologisches Institut der Karl-Marx-Universität, Leipzig (Deutschland).

Physikalische und quantitative Methoden · Physical and quantitative Methods
Méthodes physiques et quantitatives
Autoradiographie · Autoradiography · Autoradiographie

1.

MARTIN RITZÉN: **Microspectrophotometry of Monoamines.**

A recently developed method for the detection of monoamines in tissues has been studied by means of a modified Caspersson-Zeiss Universal Microspectrophotometer. This histochemical reaction involves a condensation of the monoamines with formaldehyde vapour to an isoquinoline derivative, that shows an intense fluorescence when under ultraviolet or blue violet light. In this study a model system is used, consisting of microdroplets of various catecholamines and indolalkylamines in a protein solution. Fluorescence and excitation spectra, and fading of the emitted light during irradiation, has been used for identification of the different amines. The fluorescence maximum at a neutral pH for catecholamines is around 480 mμ, for 5-hydroxytryptamine (5-HT) around 525 mμ. Together with the rapid fading of 5-HT and the fact that 5-HT — but not catecholamines — has an additional excitation maximum at around 300 mμ, this can be used for the certain identification of and differentiation between very small amounts of these two monoamines. These findings are valid also in tissues which contain a known catecholamine (e.g. sympathetic nerve endings and ganglia, adrenal medulla) and 5-HT (enterochromaffin cells, mast cells from rat and mouse) and also in tissues which contain both, but in different cells (brain stem of rat). The quantitative aspects of the reaction are investigated by determining the dry mass of a microdroplet by interference microspectrophotometry, and measuring the total amount of fluorescence at the fluorescence peak of the very same droplet.

Dept. for Cell Research, Karolinska Institutet, Stockholm (Sweden).

2.

GÖSTA GAHRTON: **Quantitation of the Periodic Acid-Schiff (PAS) Reaction by Means of Combined Microinterferometry and Microspectrophotometry.**

The periodic acid-Schiff reaction has been studied quantitatively in glycogen microdroplets and human neutrophil leukocytes by means of combined microinterferometry and microspectrophotometry. Various modifications of the McManus PAS reaction have been performed. The color development as a function of periodic acid oxidation time and Schiff reaction time was faster in neutrophil leukocytes than in glycogen microdroplets. The PAS positive substance in the leukocytes was removed by digestion with α-amylase and was therefore considered to be glycogen. A mean value of PAS reactive glycogen per neutrophil leukocyte was estimated to be 6×10^{-12} g, based on comparison of results of complete PAS reaction in both microdroplets and leukocytes. The number of nuclear lobes in the leukocytes had no influence on the amount of PAS reactive glycogen.

Institute for Cell Research, Karolinska Institutet, Stockholm (Sweden).

3.

LEON CARLSON: **A Modified Principle for an Interference Microscope.**

A shearing interference microscope has been constructed, using conventional microscope objectives as condenser and objective. The beam is split and re-combined by Wollaston prisms which are imaged in the back focal plane of the condenser and objective respectively by the aid of simple lens systems.

Dept. for Cell Research, Karolinska Institutet, Stockholm (Sweden).

4.

D. J. Goldstein: Interference Microscopy of Histological Sections*.

Accurate and reproducible interference microscopy of unstained sections requires (a) the use of a shearing system, preferably used with a half-shade eye-piece; (b) complete removal of embedding wax from the specimen (the usual treatment with xylene is inadequate); (c) complete replacement of one mounting medium by any successive medium; (d) accurate measurement of the refractive index (RI) of the mounting media at the temperature and wave-length used for microscopy. — Optical path differences between histological objects and a reference area are consistently lower when the section is mounted in water, than would be expected from measurements in non-polar fluids. This anomalous result is attributed to swelling of tissue components in water. — It has been found that some mucins, cellulose and especially glycogen have a lower RI than most other tissue components. Cytoplasmic RNA and various secretory granules have a relatively high RI. The highest RI observed was in calcified cartilage matrix. The "effective thickness" of many tissue components, particularly certain epithelial mucins, is considerably less than the thickness of the section; this is attributed to the presence in such components of substantial spaces containing mounting medium. — The results will be presented in detail, together with a discussion of their correlation with the permeability of tissue components to histological dyes.

* This work was performed in the Department of Zoology, Oxford University, during the tenure of a Nuffield Dominion Travelling Fellowship.

Dept. of Anatomy, University of the Witwatersrand, Johannesburg (South Africa).

5.

Ulf Friberg and John F. Burke: Automatic Contact Microradiography with Ultrasoft X-Rays*.

Through the use of contact microradiography with ultrasoft x-rays it is possible to visualize the distribution of dry mass in soft tissues or in demineralized hard tissues. This technique can also be utilized for the study of, e.g., early calcification at the tissue level. Previous instrumental limitations on the feasibility of extensive biological studies with ultrasoft microradiography have largely been overcome through the design of an automatic apparatus for the exposure of the microradiograms.

* Aided by grants from the National Institutes of Health (DE 01777; E 2392), Public Health Service, U.S. Department of Health, Education and Welfare, and the John A. Hartford foundation, Inc.

Karolinska Institutet, Stockholm (Sweden) and Harvard Medical School, Boston, Mass. (USA).

6.

Gy. Rappay* and P. van Duijn: Aminoethylcellulose Membranes as Model for a Quantitative Cytochemical Study of the Ninhydrin-Schiff Reaction.

Aminoethyl groups can be bound covalently to cellulose by reaction with 2-aminoethylsulfuric acid. According to results reported in the literature and own experiments, it was established that the cationic groups attached to the cellulose are primary amino groups. The high stability of the aminoethylcellulose membranes (AE-cellulose) renders it possible to carry out quantitative investigations of the presently known histochemical colour reactions of the NH_2 group by application of special quantitative colorimetry (van Duijn et al., J. Histochem. Cytochem. *10*, 473, 1962). — In the present work, the reaction conditions of the ninhydrin-Schiff reaction (NSR) were investigated in detail. In order to investigate the stoichiometry of the NSR, the number of amino groups in AE-cellulose per mg cellulose was determined by titration as well as by the dinitrofluorobenzene reaction. — It appears that AE-cellulose is a suitable model for investigations of other colour methods for amino groups in protein histochemistry.

* Recipient of a grant from the Netherlands Organisation for the Advancement of Pure Research (Z.W.O.).

Institute for Experimental Medical Research, Dept. of Morphology, Budapest (Hungary) and Pathological Laboratory, University of Leiden, Leiden (The Netherlands).

7.

P. G. Bosque und G. Werner: Eine Methode zur radioautographischen Darstellung niedermolekularer und wasserlöslicher Stoffe in Organen.

Es wird über eine Methode berichtet, die es gestattet, niedermolekulare und wasserlösliche Radioisotopen-markierte (z.B. ^3H, ^{14}C) Substanzen radioautographisch in Organen und Geweben darzustellen. Das Verfahren besteht im wesentlichen aus folgenden Schritten:

Einfrieren der Organe in Isopentan bei —180° C, Gefriertrocknen und Einbetten in Paraffin im Hochvakuum, Anfertigung der Schnitte und deren Entparaffinierung in Xylol, radioautographische Exposition, Trennen des Schnittes von der trockenen Fotoemulsion (Ilford G 5) und Färbung bzw. photographische Entwicklung. — Als Beispiel für die Leistungsfähigkeit der Methodik werden Radioautographien der Schnitte von Dünndarm, Niere, Leber, Magen, Pankreas und Speicheldrüse der Maus gezeigt, der ³H-Atropin injiziert worden war. Die photographischen Schwärzungen der Radioautographien können an demselben Schnitt histologisch dargestellten Organstrukturen exakt zugeordnet werden. — Die Methode eröffnet wegen ihrer allgemeinen Anwendbarkeit auf jedwede Radioisotopen-markierte Substanzen vielseitige Möglichkeiten für viele Forschungsprobleme.

A Technique for Radioautographic Demonstration in Organs of Low-Molecular and Water-Soluble Substances. A technique is described for radioautographic demonstration in tissues and organs of low-molecular and water-soluble substances marked with radioactive isotopes (e.g. ³H, ¹⁴C). The technique consists in the following steps: freezing of the organs in isopentane at —180° C, freeze-drying and embedding in paraffin, both in vacuum, cutting of the organ and removing of the paraffin with xylol, radioautographic exposure, detachment of section and dry photo-emulsion (Ilford G 5), the first one is stained and the second one developed. — Radioautographs of small intestine, kidney, liver, stomach, salivary glands (submaxillary) of the mouse are shown; Atropine marked with tritium was previously injected to the mice. The dark spots of the radioautographs correspond exactly with the organic structures of the stained sections. — The technique can be applied for the demonstration of any substance labeled with radio-isotopes and therefore creates new possibilities to solve many research problems.

Max Planck-Institut für Hirnforschung, Arbeitsgruppe Neurochemie, 6 Frankfurt a. M. (Deutschland) und Anatomisches Institut der Universität in Valladolid (Spanien).

8.

N. G. Khrushchov: Autoradiographic Study of Sulfate Metabolism in Different Structures of the Loose Connective Tissue.

At a different time (from 2 min to 24 hrs) after the intraperitoneal injection of radiosulfate its incorporation into the cells and intercellular substance of subcutaneous connective tissues of albino mice was studied. — The incorporation of the radiosulfate into the molecule of sulfomucopolysaccharide was shown to proceed usually within the cells. The production of sulfomucopolysaccharides of the intercellular substance (chondroitin-sulfate acids) is realized mainly by the fibroblasts. — Mast cells possess a particular ability to accumulate sulfate. Its binding by mastocytes proceeds through sulfation of earlier formed mucopolysaccharides. — A conclusion is drawn on the possibility of the transformation of a polyblast into a mast cell. This process is related to an increased accumulation of sulfate groups by the macrophage. It is suggested that an excess of sulfate is the direct cause of the appearance of mast cells in the loose connective tissue. — An idea of the mast cells as a system regulating the amount of free sulfate in the loose connective tissue is grounded.

Institute of Animal Morphology, Academy of Sciences of USSR, Vavilov street 12/2, Moscow B-133 (USSR).

9.

Jochen Kunz: Autoradiographische Untersuchungen der Mastzellen in Eisendextran-induzierten Sarkomen.

Zur Frage der Beziehungen zwischen Mastzellproliferation und Tumorwachstum wurden bei Ratten durch langfristige subcutane Applikation von Eisendextran (Ursoferran) Spindelzellsarkome erzeugt. Das Auftreten von Mastzellen im Tumorgewebe wurde autoradiographisch nach Na₂³⁵SO₄-Gabe mit der Strippingfilm- und Emulsionstechnik sowie mittels Toluidinblaufärbungen vergleichend untersucht. Die Auszählung definierter Tumorgewebsbezirke ergab einen signifikant erhöhten Gehalt an Mastzellen in den Autoradiogrammen im Vergleich mit den Präparaten nach Toluidinblaufärbung. Bei Anwendung der Toluidinblau-pH-Reihe wiesen die Mastzellen der Sarkome mit fallender Wasserstoffionenkonzentration eine deutliche Zunahme auf, während die vergleichsweise untersuchten Mastzellen der Rückenhaut eine derartige pH-abhängige Anfärbbarkeit ihrer Granula vermissen ließen. Diese Befunde werden als Ausdruck einer Neubildung von Mastzellen im Tumorgewebe angesehen. Das Fehlen einer Mastzellanreicherung in den Tumorrandzonen und in den Vorstadien der Sarkome sowie das Auftreten einer diffusen Mastozytose im Zentrum der voll ausgebildeten Sarkome spricht gegen die vielfach vertretene Ansicht einer defensiven Rolle der Mastzellen gegen das Tumorwachstum.

Autoradiographic Examination of Mast Cells in Ferrodextran Induced Sarcomata. In an attempt to solve the problem of the relationship between the proliferation of mast cells and tumour growth, spindle cell sarcomata were produced in rats by injecting them subcutaneously with ferro-dextran (Ursoferran) over a long period. The presence of mast cells in the cancerous tissue was demonstrated by autoradiography after $Na_2{}^{35}SO_4$ had been applied (stripping film and emulsion technique). Comparative studies were done on preparations stained with Toluidine blue. A count of mast cells in defined tumour areas showed a significant rise in the autoradiographs compared with the number of cells in the Toluidine blue preparations. Employing Toluidine blue in a descending pH series, the mast cells originating from sarcomata showed a distinct increase in staining with falling hydrogen-ion concentration. The granules of the mast cells from the skin of the back, which were tested for comparative reasons, did not show the same pH-dependence, however. These findings seemed to indicate that mast cells were formed in the cancerous tissue. On the one hand, mast cells did not accumulate in the marginal zones of the tumour, or in the preliminary stages of sarcoma formation. On the other hand there was a diffuse mastocytosis in the centre of the fully developed sarcomata. These facts do not support the view that mast cells take a defensive part in tumour growth.

Pathologisches Institut der Humboldt-Universität zu Berlin, Schumannstr. 20/21, Berlin N 4 (Deutschland).

Enzyme · Enzymes · Enzymes

1.

ZDENĚK VESELÝ, JAROSLAVA MIKULÁŠKOVÁ and ZDENĚK LOJDA: Substrates for the Histochemical Demonstration of β-Glucuronidase.

8-hydroxyquinoline glucuronide and glucuronides of naphthols AS-BI, AS-BO (the phosphates of which proved to be very useful substrates in studies of the intracellular localisation of acid phosphatase (LOJDA et al., Histochemie *3*, 428, 1964) were prepared by a modification of chemical (KATO et al., Chem. Pharm. Bull. *8*, 239, 1960) and biological (WILLIAMS, Biochem. J. *37*, 329, 1943, and FISHMAN and BAKER, J. Histochem. Cytochem. *4*, 570, 1956) procedures. The purity of prepared glucuronides as well as of the commercial 8-hydroxyquinoline glucuronide (Light) were controlled by elementary analysis, UV and infrared spectroscopy and chromatography. These substrates were then used for the histochemical detection of β-glucuronidase in histological sections using the Fishman and Baker's method and our own modification of the azocoupling method in differently treated sections of rat kidney and liver. The authors came to the conclusion that the results obtained by histochemical reactions are very much dependent on the purity of substrates.

Research Institute of Natural Drugs, Prague, and Angiological Laboratory and 1st Institute of Pathology, Faculty of General Medicine, Charles University, Prague (ČSSR).

2.

WILLIAM H. FISHMAN, S. GREEN, Y. NAKAJIMA and S. GOLDMAN: Positive Correlations between the Fishman-Baker Method and the Naphthol AS-BI β-Glucosiduronic Acid-Post-Coupling Technique for β-Glucuronidase.

The Fishman-Baker method produces staining reactions which have now been unequivocally proven to be caused by the enzyme, β-glucuronidase and not by unspecific iron uptake as argued by JANIGAN and PEARSE. Experimental evidence is based on quantitative measurement of glucuronic acid liberated into the incubation medium and on the use of control conditions employing competing substrate, alternate substrate or change of pH. The desirability of using a non-chelate procedure for comparison with the Fishman-Baker chelate one has led to the synthesis of naphthol AS-BI β-D-glucosiduronic acid and the demonstration of its suitability as a substrate for β-glucuronidase activity. Factors of fixation, substrate concentration and pH are important variables as well as the choice of coupling agent (Garnet G.B.C., Fast Dark Blue RR, hexazonium pararosanilin). Post-incubation coupling gave the most consistent results. A comparison of a number of representative tissues by both chelate and non-chelate methods shows identical cytoplasmic and tissue sites of β-glucuronidase activity. Particularly pertinent is the demonstration by the non-chelate procedure of β-glucuronidase activity in rat muscle; a β-glucuronidase-poor tissue of high affinity for iron. The general application of biochemical criteria for evaluating enzyme staining reactions is reviewed.

Tufts University School of Medicine, Boston, Mass. (USA).

3.

Masando Hayashi: Histochemical Localization of β-Glucuronidase*.

The distribution of β-glucuronidase activity has been studied in many tissues of the adult Wistar rat with the method employing naphthol AS-BI glucuronide as substrate and hexazonium pararosanilin as diazo reagent (Hayashi, Nakajima and Fishman: 14th Histochemical Society Meeting, Washington, D.C., 1963). Tissues known to be the enzyme-rich by homogenate assay showed a rapid and intense reaction in a great number of cells (e.g., preputial gland, spleen, liver, epididymis, ovary, lymph node, thymus, kidney and thyroid). In enzyme-poor tissues, the staining was limited to a small number of specific cells (e.g., nervous tissues, eye, testis and muscle). Cells responsible for the intense reaction in these organs were various types of phagocytic cells; liver and kidney parenchymal cells; and several kinds of mucous membrane cells. In each cell the staining appeared usually as discrete and mostly spherical granules localized in the cell cytoplasm and none in the nucleus. Size, number and location of the granules were variable, depending on the type of cells and large granules were especially abundant in macrophages. Similarities in localization of β-glucuronidase with those of other lysosome enzymes were noted in several tissues studied with, however, variations in activities between the enzymes.

* Aided in part by a grant from the National Institutes of Health, NB-03015.

Howe Laboratory of Ophthalmology, Harvard University Medical School, 243 Charles Street, Boston, Massachusetts (USA).

4.

Zdeněk Lojda, Zdeněk Lodin, Arnošt Bass and Jiří Pilný: Cytospectrophotometric Determination of Acid Phosphatase.

The possibility of the cytospectrophotometric determination of acid phosphatase demonstrated in histological sections with Gomori and azodye methods using phosphates of α-naphthol, naphthol AS-TR, AS-BI, and AS-BO and hexazotized p-rosanilin was investigated. Cryostat sections (unfixed or briefly fixed in cold formol) and frozen sections from cold formol fixed rat liver and kidney were used. The results were compared with the microchemical determination of acid phosphatase performed on parallel sections using the same material and solutions as for the histochemical detection and also solutions where lead nitrate or hexazotized p-rosanilin were omitted. Results obtained with azodye methods were comparable with the microchemical determination and were influenced primarily by the treatment of material before the histochemical test was performed.

Angiological Laboratory and Iˢᵗ Institute of Pathology, Faculty of General Medicine, Charles University, Prague, Institute of Physiology of Czechoslovak Academy of Sciences, Prague (ČSSR).

5.

L.-D. Leder: Experimental Contributions to the Exceptional Position of the Naphthol AS-D Chloroacetate Esterase in Contrast to Other Histochemically Demonstrable Hydrolases.

Naphthol AS-D chloroacetate, when used as chromogenic substrate for the histochemical demonstration of esterases, provides results which strikingly differ from those obtained with the simple naphthol esters. Furthermore it is possible to demonstrate the naphthol AS-D chloroacetate esterase in routine paraffine embedded tissue without any detectable loss of activity. This complete resistance of the activity to the influences of routine paraffine embedding arises the question, whether the obtained reaction really is due to an enzyme or not. In order to clear this problem the reaction has been performed both on fresh frozen tissue and on paraffine embedded material and the effects of some inhibitors have been studied. The results are compared and the exceptional position of the naphthol AS-D chloroacetate esterase is pointed on.

Pathologisches Institut der Universität, 23 Kiel (Deutschland).

6.

Arno Hecht: Zur Technik des histochemischen ATPase-Nachweises.

Bei fermenthistochemischen Untersuchungen zum experimentellen Herzinfarkt der Ratte beobachteten wir unter anderem als eine der ersten Veränderungen bereits nach 1 Std Infarktdauer eine erhöhte Aktivität der mitochondrialen ATPase. Diesen Befund deuteten wir vorerst als Folge einer *Entkopplung der oxydativen Phosphorylierung.* Zur Sicherung dieser Annahme wurden fixierte und unfixierte Schnitte zur Entkopplung der oxydativen Phosphorylierung mit 2,4-Dinitrophenol vorbehandelt, außerdem wurde DNP direkt in die Inkubationslösung gegeben. Mit dem gleichen Ziel wurden Herzen einer Autolyse von 1 und

3 Std bei 37⁰ C unterworfen. Neben der mitochondrialen ATPase (pH 7,5 nach PEARSE u. REIS) interessierte uns das Verhalten der myofibrillären ATPase des Herzmuskels (pH 9,4 nach PADYKULA u. HERMANN) unter gleichen Versuchsbedingungen sowie der Einfluß von Fixierungsdauer und Inkubationszeit. Die Ergebnisse bestätigen, daß es sich bei den im Infarkt beobachteten Veränderungen tatsächlich um die Folgen einer Entkopplung der oxydativen Phosphorylierung handeln muß. Durch DNP wie Autolyse läßt sich gleichfalls eine Steigerung der mitochondrialen ATPase-Aktivität erzielen. Die myofibrilläre ATPase läßt dagegen keine Wirkung erkennen. Entscheidend ist eine vorherige Formalinfixierung der Schnitte, ohne die sich die erhöhte Aktivität nicht feststellen läßt. Unsere Untersuchungen zeigen, daß der histochemische ATPase-Nachweis unter Berücksichtigung entsprechender methodischer Voraussetzungen durchaus geeignet erscheint, Störungen der oxydativen Phosphorylierung zu erfassen. Um Fehldeutungen zu vermeiden, sind exakte Versuchsbedingungen, vor allem hinsichtlich der Fixierung und der Pufferbehandlung, unerläßlich.

The Technique of the Histochemical Demonstration of ATPase. During enzyme-histochemical investigations of experimental myocardial infarction in rats, we observed as one of the first changes a raised ATPase activity. We interpreted this finding as a result of an uncoupling process in the oxidative phosphorylation. To confirm this hypothesis, fixed and unfixed sections were treated with 2,4-Dinitrophenol in order to uncouple the oxidative phosphorylation. For the same reason some rat hearts were exposed to 1—3 hours autolysis at 37⁰. Apart from the mitochondrial ATPase (pH 7.5, PEARSE and REIS) we were interested in the behaviour of the myofibrillar ATPase of the heart muscle (pH 9.4, PADYKULA and HERMANN), under the same experimental conditions. We also wanted to find out whether the period of fixation and incubation has any influence on the results. The results obtained confirmed that the changes observed are in fact due to the uncoupling of the oxidative phosphorylation. DNP treatment and autolysis resulted both in an increase of mitochondrial ATPase activity. No effect was observed on the myofibrillar ATPase. Formalin fixation of the sections was crucial for the outcome of the test as no increase of activity could be observed in unfixed material. The results of our investigations revealed that the histochemical demonstration of ATPase is suited to show malfunction of oxidative phosphorylation. To avoid misinterpretations exact experimental conditions — especially regarding fixation and buffer treatment — are essential.

Pathologisches Institut der Humboldt-Universität, Abt. für Histochemie, Schumannstr. 20/21, Berlin N 4 (Deutschland).

7.

L. KALEVI KORHONEN: Histochemical Demonstration of Inorganic Pyrophosphatases.

In hypertonic saccharose solutions homogenized rat kidneys showed inorganic pyrophosphatase (PPase) activities at pH 5.5, 8.5 and 7.0 (PPase I). PPase I is a soluble enzyme. Fixation in absolute acetone caused no or very little inactivation, but the nuclei were intensively stained. Fixation in acetone containing 0.5% trichloro acetic acid caused a considerable inactivation, but the nuclei were unstained. Higher aldehydes (acetaldehyde, glutaraldehyde, hydroxyadipaldehyde) inactivated only under 10 per cent and nuclear staining was absent or very weak. Enzymatic activity of PPase I is a function of pH and concentration of the substrate ($Na_4P_2O_7$) and Mg^{2+}. According BLOCH-FRANKENTHAL the actual substrate is the complex anion $(MgP_2O_7)^{2-}$. The Mg^{2+} can be replaced by Co^{2+} which also forms a labile soluble complex anion $(CoP_2O_7)^{2-}$ with the pyrophosphate. The stability of the solution depends on the ionic strength, temperature and pH. A solution of 0.01 M $Na_4P_2O_7$ and 0.005 M $Co(NO_3)_2$ at pH 7.2—7.6, in which saccharose 0.25 M and KCN 0.005 M has been added, was ideal for the histochemical purposes. Enzymatically liberated ortho P is precipitated as cobaltous phosphate, and the cobalt is visualized as its sulphide. This histochemical method gives results superior those described by KURATA and MAEDA (1956).

Dept. of Anatomy, University of Turku, Turku 3 (Finland).

8.

SIBAJYOTI GUHA and RAYMOND WEGMANN: On the Histochemistry of Phosphorylase: A Critique and Evaluation.

Phosphorylase, the enzyme of glycogen mobilization, aroused much interest since the publication of a histochemical technique by TAKEUCHI and KURIAKI (1955). This enzymatic activity was found present in various organs. Further progress was made by the demonstration of phosphorylase activity by activation of phosphorylase-kinase. GUHA and WEGMANN (1960) also differentiated the activity of preexisting active phosphorylase and of total (active and inactive form) phosphorylase. The demonstration of active phosphorylase permitted the evaluation of the dynamic equilibrium existing between the active and the inactive phosphorylase of different tissues. Moreover, the determination of the active phosphorylase activity permits a valid comparison with the results obtained by biochemical methods. — Recent biochemical findings indicate that phosphorylase may be responsible only for the degradation of glycogen while the synthesis is done by other enzymatic systems. Though the histochemists seem to have accepted this interpretation, yet it is not substantiated by most of the histochemical findings. This discrepancy between the findings of histochemists and biochemists can be explained by the difference of techniques used by the workers of these two disciplines. The protective action of the activators of phosphorylase as protamine, polyvinyl pyrrolidon (PVP) and ethylenediamine tetraacetic acid (EDTA) was studied. The possibility of the demonstration of the auxiliary enzymes of phosphorylase as branching enzyme and phosphorylase-kinase is discussed.

Institut d'Histochimie Médicale, Faculté de Médecine, Paris-6e (France).

9.

ROBERT HESS: Observations on the Specificity of α-L-Glutamyl Arylamidase Activity.

The description by GLENNER et al. (Nature *194*, 867, 1962) of a Ca^{++}-activated "aminopeptidase A" hydrolysing α-L-glutamyl naphthylamide (GNA) prompted us to study the possible relationship of this activity to angiotensinase of serum and kidney. Exopeptidase activity with a relative specificity for N-terminal L-dicarboxylic amino acids can be assumed to initiate the degradation of angiotensin II and analogues of an α-L-Asp^1 configuration (REGOLI et al., Biochem. Pharmacol. *12*, 637, 1963). — Studies with rat serum have shown that GNA-hydrolysis is competitively inhibited by various isomers of the octapeptide Val^5-angiotensin II, more strongly by those of the α-L-configuration than by α-D, β-L and β-D analogues. A similar inhibitory pattern, however, was observed for the hydrolysis of L-leucyl-β-naphthylamide with the exception of β-L-angiotensin II which interferred solely with GNA hydrolysis. Studies with degradation products showed that, in contrast to the dipeptide H-Asp-Arg-OH (α-L and β-L), the hexapeptide H-Val-Tyr-Val-His-Pro-Phe-OH was strongly inhibitory with both substrates. — Kinetic studies with rat kidney microsomes were carried out using α-L-glutamyl p-nitroanilide (GpNA) and L-leucyl p-nitroanilide (LpNA) as substrates. Reaction velocities were determined by direct observation of absorbance at 405 mμ. With the exception of α-L-Asp^1-angiotensin II and of α-L-Asp^1-(NH$_2$)-angiotensin II no striking differences in K_i values were observed when hydrolysis rates of the two substrates were compared in the presence of various angiotensin derivatives. Apparent K_m values were very similar (GpNA: 1.4×10^{-4} M; LpNA: 1.7×10^{-4} M). — These and other observations (including studies with purified leucine aminopeptidase) suggest a considerable overlap in the specificity to aryl substrates of peptidases implicated in inactivating angiotensin. Criteria for the usefulness of the glutamyl arylamidase reaction as applied to altered physiological conditions will be discussed.

Research Laboratories, Pharmaceutical Dept., CIBA Limited, Basle (Switzerland).

10.

VÄINÖ K. HOPSU: Further Studies on the Enzymes Hydrolyzing Acetyl Naphthylamides.

A variety of substrates and tissue homogenates and fractions has been studied in an attempt to characterize further the enzymes in tissues responsible for hydrolysis of acetyl naphthylamides first described by GOMORI. By means of starch gel electrophoresis of Guinea pig kidney homogenates, it was found that in mutually exclusive bands, in which acetyl, mono, and dihalogen (AMD) acetyl naphthylamides were hydrolyzed and the others in which trihalogen acetyl naphthylamides were hydrolyzed, hydrolysis of 1-naphthol and Naphthol AS acetates occurred in all. The activities in both groups were inhibited by fluoride, DFP and E 600. Further investigation revealed that the trihalogen derivative hydrolyzing activity was predominantly in the particulate fraction, showed pH optimum 8.3 and was relatively heat stable. The AMD acetyl naphthylamidase activity was predominantly in the supernatant fraction, had pH optimum 6.6 and hydrolyzed acetanilide, the O-acetyl group of O,N-diacetyl tyrosine and several other esters. It appears that there exist in Guinea pig kidney enzymes with esterase activity hydrolyzing two different classes of acetyl naphthylamides. Enzyme activity hydrolyzing the same substrates was also found in commercial hog kidney acylase preparations and was separated from acylase I by electrophoresis and DEAE-cellulose chromatography, but no separation into specific acetyl naphthylamide groups was obtained.

Dept. of Anatomy, University of Turku (Finland).

11.

E. K. Pletchkova: Cholinesterase Activity of Parasympathetic and Sympathetic Neuro-Muscular Junctions in Smooth Muscles of Visceral Organs, Vessels and in the Myocardium.

In our laboratory by the thiocholine method of Koelle the activity of the total and non-specific cholinesterase in different organs innervated by cranial and sacral parasympathetic nerves in dogs, cats and rabbits has been studied. To verify the adequacy and accuracy of the method there were used not only biochemical means of inactivation of the enzyme but as well the morphological ones — cholinesterase activity of the nerve fibres in organs innervated only by sympathetic nerves was examined. — In muscles of the organs with sympathetic innervation, namely in the nictitating membrane and the uterus, no nerve fibres possessing cholinesterase activity have been revealed. However, in these organs one could observe cholinesterase activity in the nerve fibres innervating the blood vessels. — The degree of cholinesterase activity of peripheral plexuses of parasympathetic nerve fibre in different organs is not the same. The highest cholinesterase activity is found in nerve elements of the digestive tube and the heart; a relatively low activity is shown by peripheral nerve plexuses of the organs innervated by the sacral part of the parasympathetic. — The character of distribution of the active cholinesterase suggests that terminal plexuses of nerve fibres in smooth muscles of various organs and vessels and the myocardium represent extensive mediatory fields.

Laboratory of Neurohistology by the Name of B. I. Lawrentjew, Institute of Normal and Pathological Physiology, Academy of Medical Sciences, Moscow (USSR).

Nervensystem · Nervous System · Système nerveux

1.

Arnold Schabadasch: The Histochemical Peculiarities and Equivalents of Functional Properties of the Nervous Tissue.

New methods of histochemical detection of different types of nucleoproteins, glyco-lipoproteins and phosphoproteins, developed by us, allow distinction of many peculiarities of the specialized structures of neurocytes. The visual pattern of proposed reactions competes with results of classical neurohistological techniques in precision. The series of investigations leads us to essential altered conceptions about the nature and reactivity of mitochondria, nucleolus, axons and synaptical terminals in normal or experimental conditions. Ribonucleoproteins with different isoelectric points are the main tissular anions in mitochondria and tigroid granules; a part of mitochondria contains also glycoproteins. The concentration of glyco-lipoproteins, similar to gangliosides, rises in axons; it increases considerably in unmyelinated fibers and is maximal in terminal ramifications. Each special portion of the neuron is thus characterized by a typical electrocolloidal charge; a gradient of different values of charges corresponds to inequality of spatial distribution of the mentioned polyelectrolytes, along the neuron axis. The degree of I.E.P. negativity rises in the following succession: synapses, axon, cytoplasm, tigroid, mitochondria; depending upon the functional state, these differences may diminish or increase. — Histochemically detectable changes of "fixed anions" are followed by uptake, binding or release of various cationes, the neuromediators included; thus they can serve as concrete signs of local physiological processes, especially in terminal ramifications.

Institute of Biophysics, Academy of Sciences of USSR; Moscow (USSR).

2.

Ambrosius Ábrahám: Histochemical Investigations on the Neurosecretory System of the Swimming Beetle (Dytiscus marginalis).

In the protocerebrum of the swimming beetle two cell groups of greater size, were found the members of which show a remarkable neurosecretory activity. In the cellbodies and in the thick and straight processes granules of smaller and greater size could be seen which using the Gomori-Bargmann chromhaematoxylin-phloxin technique appear dark violet, pale violet and in some cases red. To get a real picture concerning the function of the cells and the factors influencing them, the animals were enligthed with electric light, treated with ultrasonic, with alternating current and some of them became blind. On the basis of the examinations made with different staining and impregnation methods was established, that the cells produce the secretum in great quantity, the greater part of which passes on the cellprocess and other places of the cellsurface into the neuropil of the brain, without any special pathways. The smaller part of the secretgranules enter into the nervus cardiacus via the two pathways originated from the secretory cell groups. The productivity of the cells is increased by the ultrasonic as well as by the electric excitement but the highest hypersecretion could be seen when the animals were exposed to a constant and strong electric light. The neurosecretum appearing in the same cell — dark violet, pale violet or red — is a glycoprotein which shows different reactions in the different phases of the transformation.

Jószef Attila University, Zool. and Biol. Institute, Szeged (Hungary).

3.

H. HAGER and G. KREUTZBERG: Light- and Electron Microscopical Identification of Lysosomes in Nerve Cells of the Cerebral Cortex by Means of the Acid Phosphatase Reaction.

The concentration, modification and storage of foreign and metabolically inert materials can take place in organelles of nerve cells that are morphologically identical to the so-called microbodies, or cytosomes, as was shown by experimental tellurium poisoning (HAGER 1960, 1962). The designation of these organelles as lysosomes (DE DUVE and coworker 1950, DE DUVE 1963) was the object of the present study. Demonstration of the acid phosphatase activity both for the light, and also for the electron microscope, employed a slightly modified form of the method described by MILLER (1962). A histochemical method was also used to try to demonstrate non-specific esterases (CREVIER and BÉLANGER 1955). — Acid phosphatase activity appeared in the light microscope as granules of fairly uniform size which were found especially numerous in nerve cells. Electronmicroscopically the reaction product was identified in the perikaryon of nerve cells as a fine precipitate on the membrane and within the matrix of the microbodies (cytosomes), even after relatively short incubation times. The activity of the enzyme was confined to these bodies and occasionally to other sharply defined, smaller intracytoplasmatic bodies. Particular attention was given to their relation to the Golgi zone. If it is correct that the acid phosphatase reaction is a "Lysosome-marker" (MILLER 1962, MILLER and PALADE 1963), then a large number of microbodies in nerve cells may be spoken of as lysosomes. The meaning of this finding for the understanding of liquifactive necrosis and the striking types of nerve cell necrosis in the central nervous system is apparent.

Deutsche Forschungsanstalt für Psychiatrie (Max Planck-Institut), 8 München 23, Kraepelinstraße 2 (Deutschland).

4.

GERARD M. LEHRER and MURRAY B. BORNSTEIN: The Quantitative Enzyme Histochemistry of the Rat Cerebellum during Development*.

Five enzymes have been studied by quantitative micro methods in the granularis of developing rat cerebellum. Cerebelli were removed from litter mates at various intervals postnatally. They were quick frozen. Cryostat sections were dried without thawing and stored under vacuum at —60° C. Hexuplicate samples of granularis, dry weight 0.01 to 0.04 mcg., were dissected, weighed on quartz fiber balances, and analyzed for hexokinase, glucose-6-phosphate, lactic and malic dehydrogenases, and Beta-D-glucuronidase. — Hexokinase increases steadily from 2 moles per kg. dry weight per hour (MKH) to an adult level of about 5 MKH by the 19th day. Lactic dehydrogenase shows a gradual increase from about 20 MKH at birth to a nearly adult level of about 60 MKH by the 19th day. Malic dehydrogenase increased from about 80 MKH at birth to about a nearly adult level of 300 MKH by the 19th day. Glucose-6-phosphate dehydrogenase, 1.5 MKH at birth, reaches a peak of about 4 MKH on the 8th day, receding to 2.5 MKH, a nearly adult level, by the 14th day. Glucuronidase data were as described by ROBINS *et al.* The significance of these data as well as correlation with development will be discussed.

* Supported in part by grant No. 293-2 from The National Multiple Sclerosis Society and by grant No. CA-05766-03 from The National Cancer Institute, U.S.P.H.S.

Laboratory of Neurochemistry, The Mount Sinai Hospital, New York, N.Y. (USA).

5.

P. KÁSA and B. CSILLIK: Histochemical Studies on the Effect of Nerve Degeneration in the Cerebellar Cortex.

Mossy fiber apparatuses in the archicerebellar cortex exhibit a strong acetylcholinesterase activity under normal conditions, whereas the perikarya of Purkinje cells are devoid of it. In acutely deafferentated animals the enzyme activity of the mossy fiber apparatuses decreases, while the corresponding Purkinje cells gain activity (5th—8th postoperative day). From the 15th day, the original distribution of the enzyme starts to be restituted. These observations suggest that: 1. In a latent form, acetylcholinesterase is present in Purkinje pericarya under normal conditions. Experimental alterations in the amount of cellulipetal impulses result in a reactivation of this latent enzyme (that used to present in an active form during ontogenetical development). 2. As a result of the degeneration of the mossy fibers that contained acetylcholinesterase under normal conditions, deafferentation leads to the abolition of this enzyme. Restitution of the latter after a short silent period may be the result of a latent enzyme similar to that in the Purkinje cells. 3. Archicerebellar mossy fiber apparatuses appear to contain a sophisticated multisynaptic mechanism, including both excitatory and inhibitory synapses.

Dept. of Anatomy, University Medical School, Szeged (Hungary).

6.

M. BRANCA: Succinodehydrogenase im Gehirn von Meerschweinchen mit Brandverletzungen.

Bei anästhesierten Meerschweinchen (Äther) mit einem durchschnittlichen Gewicht von 250 g wurden durch Eintauchen der Tiere für 15 sec in 70⁰ C warmes Wasser Verbrennungen auf $^1/_3$ der Körperoberfläche gesetzt. Die Tiere wurden 2, 6, 16, 22 und 36 Std nach der Verbrennung getötet. Untersucht wurde das Vorkommen der Succinodehydrogenase in den motorischen Regionen unter Verwendung von Nitro-BT und MTT als Elektronenakzeptor. Zur genaueren Ortsbestimmung der Enzymreaktion wurden die Gewebsschnitte nach der Photographie nach UNNA-PAPPENHEIM gefärbt. Die Succinodehydrogenase kommt im Gehirn normaler Kontrolltiere vor allem im Protoplasma der Nervenzellen und hauptsächlich am Neurohilus vor. Werden die Ergebnisse der nach verschiedenen Zeiten der Verbrennung getöteten Tiere miteinander verglichen, so zeigt sich eine kritische Abnahme der Enzymaktivität um die 2.—6. Std (bis zum vollständigen Verschwinden). Um die 16. Std tritt wieder eine schwache positive Reaktion auf. Zwischen der 22. und 26. Std ist die Intensität der Reaktion wieder normal. Die Ergebnisse stehen in enger Beziehung zur Stresslage der Tiere.

The Activity of Succinate Dehydrogenase in the Brain of Guinea Pigs with Burns. Ether anaesthetized Guinea pigs of an average weight of 250 gms were immersed for 15 seconds into 70⁰ C water. Thus burns were produced on $^1/_3$ of the body surface. The animals were killed 2, 6, 16, 22 and 36 hours afterwards. The presence of succinate dehydrogenase in the motor regions was investigated; Nitro-BT and MTT were used as electron acceptors. Photographs were taken and the sections then stained (UNNA-PAPPENHEIM) in order to make possible an accurate localization of the enzyme reaction. In normal control animals succinate dehydrogenase was found above all in the protoplasm of nerve cells and mainly in the neurohilus. When the various results from the test animals were compared a significant decrease (leading up to complete absence) of enzyme activity was found between the 2nd and 6th hour. After 16 hours the reaction became weakly positive. Between the 22nd and 36th hour the intensity of the reaction became normal. The results were closely connected with the condition (stress) of the animals.

Istituto di Anatomia ed Istologia Patologica, Via Aristide Gabelli 35, Padova (Italia).

7.

PAUL J. ANDERSON and NICHOLAS CHRISTOFF: Chromatographic and Histochemical Evaluation of Acid Phosphatase in Central Nervous System Autolysis and Necrosis*.

Total, bound, and unbound acid phosphatase activities were determined in homogenates and centrifugates of rat brains that were autolyzed for periods up to 7 days. Chromatographic separation of isozymes and proteins was performed on DEAE cellulose using a parabolic gradient of NaCl and continuous flow analysis of the eluate was recorded with an Autoanalyzer. Histochemical localization was obtained with the naphthol AS-TR phosphate-hexazonium pararosanilin technic. These results were compared with similar observations on embolized and ligated brains. — Autolysis was associated with early increases in unbound activity and corresponding decreases in bound activity. Bound activity showed a progressive decline over the course of several days in autolyzing brains, but structure-bound activity was histochemically detectable in neurons and pericytes even after 48 hours. Bound activity in necrotizing brains, however, was predominantly localized to macrophages. The chromatographic pattern of the isozymes was not altered but significant differences in peak amplitudes and specific activities were recorded. Enzymatic release, chromatographic separation, and histochemical localization were unaffected by pretreatment with agents that are postulated to exert a stabilizing or labilizing influence on lysosomes.

* Aided by a grant from the National Institute of Neurological Diseases and Blindness (NB 03236).
Mount Sinai Hospital, New York, N.Y. (USA).

8.

J. Escolá und E. Thomas: Elektronenmikroskopische Beobachtungen über die Lokalisation der sauren Phosphatase im Reaktionsbereich experimentell erzeugter Hirngewebsnekrosen.

Auf Grund des histochemisch nachweisbaren starken Anstiegs der sauren Phosphatase im Randgebiet von Hirngewebsnekrosen (durch Koagulation an Rattengehirnen erzeugt) wurden elektronenmikroskopisch die reaktiven Zellvorgänge in diesem Bereich unter Anwendung der Gomori-Bleimethode zum Nachweis der sauren Phosphatase untersucht. In den Makrophagen konnten wir eine starke Enzymreaktion feststellen; manchmal fand sich das Reaktionsprodukt in umschriebenen Körperchen (Lysosomen) oder in der Wandung von Phagozytosevakuolen, außerdem sahen wir Niederschläge gleichmäßig und herdförmig im ganzen Cytoplasma und dessen Ausläufern. Andere, den Makrophagen unmittelbar benachbart liegende Zelltypen, wie Plasmazellen und Fibroblasten, zeigten eine negative Reaktion. Auch in Axonauftreibungen liegen rundliche dichte Körperchen mit einem Durchmesser von etwa 0,2—0,5 μ, in denen man eine positive Reaktion in Form feinkörniger Niederschläge beobachten kann.

Electron Microscopical Studies on the Localization of Acid Phosphatase in the Reactive Area of Experimental Brain Necroses. In the marginal area of brain necroses (caused by coagulation in rat brains) a considerable rise of acid phosphatases can be demonstrated by histochemical methods. The reactive process in the cells of this area was investigated by means of electron microscopy and Gomori's lead nitrate method for the demonstration of acid phosphatases. We were able to demonstrate a strong enzyme reaction in macrophages; sometimes the reactivity product could be seen in circumscribed particles (lysosomes) or in the walls of phagocytosis vacuoles. In addition we saw homogeneous and circumscribed precipitates in the cell protoplasm and its processes. Other cells, which adjoined the macrophages directly — as for example plasma cells and fibroblasts — gave a negative reaction. Dense, roundish particles — of a diametre of 0.2—0.5 μ — could be demonstrated in the axons. They also gave a positive reaction by showing a fine granular precipitate.

Max Planck-Institut für Hirnforschung, Neuropathologische Abteilung, Deutschordenstr. 46, 6 Frankfurt a. M.-Niederrad (Deutschland).

9.

Aulikki Kokko: Histochemistry of Esterases in the Spinal Ganglion of the Rat.

Esterases have been histochemically demonstrated in sections of the spinal ganglion of the rat, using several substrate-inhibitor combinations to discriminate between activities of acetylcholinesterases, non-specific cholinesterase, E 600 sensitive non-specific esterase and E 600 resistant non-specific esterase. While non-specific cholinesterase activity was mainly observed outside the ganglion cells, each of the other three types of esterase activity was predominant in the cytoplasm of the neurons. The activity level greatly varied from one individual cell to another, ranging from intense to weak. Acetylcholinesterase and E 600 resistant non-specific esterase activities were particularly variable from one cell to another. Efforts have been made to correlate the activity levels of the different types of esterases in individual cells (1) by examining neighbouring sections in which different esterases were demonstrated and (2) by first demonstrating in a section the distribution of cholinesterase activity and, after this was registered by photomicrography, treating the same section to obtain a superimposed reaction for non-specific esterases.

University of Helsinki, Dept. of Anatomy, Siltavuorenpenger, Helsinki (Finland).

10.

J. M. Sosa and H. M. Savio de Sosa: Cytochemical Characteristics of Aging in Spinal Ganglion Nerve Cells.

In previous communications, with another co-worker, the senior author has found that multiangular spinal ganglion cells do represent an aging expression of neurons in normal spinal ganglions during the various rabbit life periods. Through the use of different histochemical methods, the present authors find free aminoacids inside cytoplasm of multiangular ganglion cells, while typical round shaped spinal ganglion cells show no positive aminoacid reactions. This fact is here interpreted as an indication of some cytoplasmic protein desintegration during the normal aging process. This histochemical aging feature runs parallel with other cytological aging variations in spinal ganglion nerve cells.

Instituto de Histología, Facultad de Medicina, Universidad de Los Andes, Apartado 149, Mérida (Venezuela).

Nucleinsäuren · Nucleic Acids · Acides nucléiques
Verschiedenes · Divers · Divers

1.

G. Yasuzumi: Cytochemical and Electron Microscopic Analysis of Transformation of Nucleic Acids into Periodic Acid-Schiff-Positive Substances.

In the development of the atypical spermatozoa of *Cipangopaludina malleata* Reeve, the nucleus fails to differentiate and simply shrinks in volume until only a remnant, devoid of deoxyribonucleic acid (DNA), is left. The cytoplasm shows numerous vesicles containing small DNA-positive granules. The cytoplasm of the atypical spermatids of *Melania libertina* Gould is characterized by the presence of ribonucleic acid (RNA)-positive granules in the vesicles. Such two kinds of granules increase in number and size as spermiogenesis proceeds. Many of them are used in the formation of the sheath of middle piece of mature atypical spermatozoa in the former, but in the latter in the formation of the head sheath of the mature. At that time the DNA- and RNA-positive granules are transformed into periodic acid-Schiff (PAS)-positive substances. Such process is pursued by several fixation techniques, cytochemical analysis employing Feulgen-, Unna-Pappenheim-, and nucleosidediphosphatase-reactions, and utilizing autoradiography. It is concluded that the Golgi complex functions as an important cell organelle in the process of transformation of nucleic acids into PAS-positive substances.

Electron Microscope Research Laboratory, Dept. of Anatomy, Nara Medical University, Kashihara city, Nara (Japan).

2.

Vinod C. Shah and Walter L. Hughes: Autoradiographic Evidence of the Incorporation of I[125]-Deoxyuridine in the DNA of Chloroplasts of Euglena gracilis[*].

Isotopically labeled Iododeoxyuridine (IUdR) has been recently found as a suitable precursor for the study of DNA metabolism. I^{125}-UdR was selected for this study because it can be prepared with higher specific activities and would be better for the detection of small amounts of DNA than tritiated thymidine. The *Euglena* cells growing in logarithmic phase were incubated with 0.5 to 1 $\mu c/ml$ I^{125}-UdR in the medium. Autoradiograms were prepared using Kodak-NTB$_2$ emulsion after Carnoy fixation and several washings of 5% TCA. — After two hours of incubation 16% of the cells were found labeled. The percentage of labeled cells increased with increasing incubation periods up to eight hours. All the cells which were in the process of division were found labeled and the label was both in cytoplasm as well as in nucleus. Of the nondividing labeled cells 76% had only cytoplasmic label while the rest had both cytoplasmic as well as nuclear label. Treatment with DNase (2 mg/ml) for two hours at pH 6.5 and 37° C removed 69% of cytoplasmic and 78% of nuclear grains. Chloroplasts isolated in a sucrose gradient from cells incubated for five to eight hours, were also found labeled. The percentage of labeled chloroplasts varied from twelve to twenty one in different preparations. Treatment with DNase resulted in loss of grains in the chloroplasts also. Significant difference in the incorporations was found in the cells incubated in dark and light as well as the normal and U.V. mutants. — The above results support the existence of DNA in the chloroplasts of *Euglena*.

* Supported by U.S. Atomic Energy Commission.

Medical Research Center, Brookhaven National Laboratory, Upton, N.Y. and Physiology Dept., Tufts University School of Medicine, Boston, Mass. (USA).

3.

C. Crăciun, Florica Motzoc, Smaranda Constantinescu and C. Tasca: Quantitative Variations of DNA in Conditions of Hyperoxygenated Guérin Tumor Rats.

Using histophotometric methods we attempted to compare optical density/DNA units (OD/DNA) with karyometric, interkaryometric and pathomorphic findings in Guérin tumors from rats kept in a hyperoxygenated 30% atmosphere (O_2 Rats). This last condition we previously found to induce changes in enzymes related to the Krebs cycle. In spite of the strictly identical conditions, important individual differences were present. — A relative increase to OD/DNA = 17 was found, for instance, in O_2R346 for 6.8 μ large nuclei (69%) and 1.8 μ internuclear interval (60%), — while 8.6 μ nuclei (24%) at 8.5 μ interval (5%) had OD/DNA = 19.5 — associated with vigorous cancer growth. Control Rat 344 showed OD/DNA = 8.5—13.5 for 4—7—8 μ nuclei (7—52—35%) at 0—1.6—2.4 μ interval (14—44—35%). The liver of O_2Rat346 with 65% of 6.8 μ nuclei at practically equal 7.0 μ intervals and OD/DNA = 11 has to be compared with OD/DNA = 5.5 for the same nuclear size and intervals. — DNA content, pathomorphic picture, nuclear size, and intervals, do not appear clearly related. Dystrophic lesions (karyo and cytopycnosis and lysis etc.) with unequal growth might produce a "feverish" graph of quantitative values and appear, therefore, to express gradients of malignancy. — Investigation by indirect methods (DNA content etc.) of the main metabolic problem (protein synthesis), should be connected with routine pathomorphic slides in order to follow quantitatively the actual receptivity of each cancer patient to powerful chemo- and actinotherapy. Herewith we attempt to make out experimentally the theoretical basis of such a dynamic morphofunctional practical test.

The Victor Babeş Institute and the University Dept. for Pathological Anatomy, Bucarest (R. P. Roumania).

<h3 align="center">4.</h3>

FREDERICK H. KASTEN and PAUL GERBER: Cytochemical Demonstration of Viral DNA in Infected Cells*.

Simian Vacuolating Virus 40 produces cytopathogenic effects in susceptible monkey kidney cells in tissue culture. By use of the sensitive fluorescent-Feulgen reagent, Auramine $O—SO_2$, progressive DNA alterations are followed. Nuclear inclusions develop which are demonstrated to contain viral DNA. At 48 hours prominent inclusion bodies are seen which appear glassy and homogeneous whereas host DNA is granular. Inclusions fluoresce green while host DNA fluoresces yellow to yellow-green. Viral DNA in the inclusion resists DNase digestion apparently due to its protein coat and is selectively visualized in the central portion of the nucleus, often surrounding a prominent nucleolus. Detailed observations will be presented.

* This investigation was supported in part by Public Health Service Research Grant CA 07017-01 from the National Cancer Institute.

Pasadena Foundation for Medical Research, 99 N. El Molino Ave., Pasadena, California, and Division of Biologics Standards, National Institute of Health, Bethesda 14, Maryland (USA).

<h3 align="center">5.</h3>

E. TAKÁCS-TOMITY, S. ÖRDÖGH and Ö. TAKÁCS: Histochemistry of Nucleic Acids in A-Avitaminosis.

It has been shown in our previous studies that vitamin A has an important part in nucleic acid metabolism. One group of Wistar rats has been kept on a vitamin A deficient diet, another group has been kept on the same diet plus vitamin A. The content of DNA and RNA has been estimated in both groups in the kidneys, liver, spleen, adrenals. There occurred a significant increase in the DNA and RNA content in the vitamin A-treated group. At the same time, also the incorporation of P^{32} was enhanced. To obtain a more detailed analysis of this phenomenon, the organs of the same rats have been submitted to a histochemical assay. By means of the Feulgen reaction and the methylgreen-pyronin staining, virtually the same results could be obtained histochemically as those described previously on the basis of biochemical estimations. The most conspicuous difference between the two groups has been observed in the reactive of the Ferrein columns in the kidney cortex.

Dept. of Anatomy and Histology, Dept. of Physiology, Medical University, Szeged (Hungary)

<h3 align="center">6.</h3>

M. HOLLMANN und H. ZIMMERMANN: Histochemie und Morphologie der HeLa-Zellen während der Alterung*.

Unterschiedliche Angaben im Schrifttum über Histochemie und Morphologie stationärer HeLa-Zellen lassen vermuten, daß diese Unterschiede durch Alterungs- und/oder Degenerationsvorgänge bedingt sein können. Diese Vermutung wird durch die zumeist im Schrifttum fehlenden Angaben über Alter und Zellkonzentration im Nährmedium (Zelldichte) bestärkt. Es wurden HeLa-Zellen in verschieden hohen Konzentrationen von 50000—220000 Zellen pro Milliliter Nährlösung explantiert und histochemisch untersucht. Die Aktivität oxydativer Enzyme — Succinodehydrogenase, Isocitricodehydrogenase, endogene Reduktasen — sowie der sauren Phosphatase nahm bei niedriger Zelldichte leicht, bei höherer Zellkonzentration stark ab. Die Reaktion der alkalischen Phosphatase nahm mit steigender Auspflanzungszeit zu. Histochemische Fettreaktionen wurden mit zunehmender Explantationszeit intensiver, zunächst Sudanschwarz und nach dessen Abfall Scharlachrot. Aminopeptidase- und PAS-Reaktion fielen nur wenig ab, Glykogen war zu keinem Zeitpunkt nachweisbar. Morphologisch bleiben die HeLa-Zellen bei hoher Zellkonzentration klein, bei geringer Zelldichte wurden sie zunächst größer und schrumpften schließlich. Die Beobachtungsdauer betrug 10 Tage. Die Veränderungen waren in allen Versuchsreihen gleichsinnig, bei hoher Zelldichte wesentlich schneller als bei geringer. Sie werden als Alterungs- und Degenerationserscheinungen infolge Nährstoffverbrauches gedeutet. Die Zunahme der alkalischen Phosphatasereaktion bleibt problematisch. Zum Vergleich wurden alle Reaktionen an Suspensionszellen ausgeführt, die unter optimalen Ernährungsbedingungen lebten. Wir fanden keine wesentlichen Unterschiede zu frisch explantierten stationären Zellen.

Histochemistry and Morphology of HeLa-Cells during the Aging Process. Varying information given in the literature about the histochemistry and morphology of stationary HeLa-cells suggest that these discrepancies are caused by aging and/or degeneration processes. This conjecture finds support in the fact that there is hardly ever any information given about the age of the cells and the cell concentration in the culture-medium (cell density). HeLa-cells were explanted in different concentrations (50.000—220.000 cells per ml culture-medium) and histochemically analysed. The activity of oxidative enzymes — succinate dehydrogenase, isocitrate dehydrogenase, endogenous reductases — decreased slightly in low and markedly in high cell concentrations. The activity of alkaline phosphatase as well as histochemical fat reactions were enhanced with the extension of the transplantation period. First Sudan black became more intensive and after its decrease, Scharlach R. There was a slight decrease only in aminopeptidase and PAS reaction; glycogen could not be determined at any time. In high cell concentrations the HeLa-cells remained morphologically small, whereas in low cell concentrations the cells first grew big and finally shrunk. The observation period comprised 10 days. The variations were identical in all series of tests, essentially faster in cases of high cell concentration than in low cell concentrations. They were interpreted as an aging and degeneration process owing to a consumption of the culture-medium. The increase of the alkaline phosphatase reaction cannot be explained. For comparative reasons all reactions were carried out on cell suspensions, which were grown under optimum conditions. There was no essential difference with freshly explanted stationary cells.

* Mit Unterstützung der Deutschen Forschungsgemeinschaft.

Pathologisches Institut der Universität, 6 Frankfurt a. M. (Deutschland).

7.

H. P. LANGE und H. G. SCHIEMER: Interferenzmikroskopische Untersuchungen an HeLa-Zellen*.

An etwa 1700 HeLa-Zellen (Deckglas-Kulturen) wurden interferenzmikroskopische Untersuchungen (Interferenzmikroskop der Fa. Zeiss) durchgeführt und am 1., 3., 5., 7. und 10. Tag nach dem Umsetzen Volumen, Trockengewicht und %-Trockengewicht bestimmt. Mit Hilfe der elektronischen Rechenmaschine (IBM 7090, Deutsches Rechenzentrum Darmstadt) konnten die Werte getrennt für Cytoplasma, Golgi-Zone, Kern und Nucleolus ermittelt werden. Dabei zeigten sich charakteristische Schwankungen der Ergebnisse: Wir fanden einen Anstieg für Volumen und Trockengewicht am 3. Tag, eine rückläufige Phase am 5. und ein erneutes Zunehmen dieser Werte vom 7. Tag an. Das %-Trockengewicht verhält sich am 3. und 5. Tag gegenläufig, ab 7. Tag bleiben die Werte konstant oder steigen leicht an. Konformes Verhalten ergab die DNS-Untersuchung (s. Vortrag SCHIEMER und HÜBNER). Diese Befunde an Gewebekulturzellen ähneln dem Verhalten der Zellen bei postischämischer Regeneration (ROTTER, LAPP, ZIMMERMANN, HÜBNER) sowie den Befunden an Plattenepithelien der Mundschleimhaut bei Angina (SCHIEMER). Das Umsetzen der Gewebekulturen (Trypsinisierung etc.) wirkt also synchronisierend auf die Regeneration, die wie bei anderen gereizten Zellsystemen (z. B. durch Ischämie oder Toxine) einen mehrphasischen Ablauf zeigt.

An Interference Microscopic Investigation of HeLa-Cells. 1700 HeLa-cells (cultured on coverslips) have been investigated by interference microscopy (Zeiss interference microscope). The volume, dry weight and %-dry weight was estimated on the first, third, fifth, seventh and tenth day after transplantation. With the help of an electronic computer (IBM 7090, Deutsches Rechenzentrum Darmstadt) the values for cytoplasm, Golgi-zone, nucleus and nucleolus were calculated separately. The results obtained showed the following characteristic pattern: An increase in volume and dry weight was observed on the third day, a recurrent phase on the fifth day, which was followed by another increase from the seventh day. The %-dry weight showed an inverse ratio on the third and fifth day; from the seventh day, however, the values remained either constant or increased slightly. Corresponding results were obtained from DNA estimations (see lecture by SCHIEMER and HÜBNER). These findings on cultured cells resemble the behaviour of cells in postischaemic regeneration (ROTTER, LAPP, ZIMMERMANN, HÜBNER) as well as findings on the pavement epithelium of the mucous membrane of the mouth in angina (SCHIEMER). — It can therefore be said that the transplantation (trypsinization etc.) has a synchronizing effect on the regeneration which, as it is known from other stimulated cell systems (ischaemia/toxins, for example) shows a polyphase course of development.

* Mit Unterstützung der Deutschen Forschungsgemeinschaft.

Pathologisches Institut der Universität, 6 Frankfurt a. M. (Deutschland).

8.

H. G. SCHIEMER und K. HÜBNER: Cytophotometrische Untersuchungen an HeLa-Zellen*.

An über 250 HeLa-Zellen wurden Nucleinsäurebestimmungen getrennt für Kern, Kernkörperchen und Cytoplasma durchgeführt. Die cytophotometrischen Untersuchungen erfolgten mit dem Universalmikrospektrophotometer der Firma Zeiss. — Die Substanz- und Volumenwerte sowie die prozentuale Substanzmenge am jeweiligen Volumen wurden mit Hilfe eines datenverarbeitenden Programmes (Rechenanlage des Deutschen Rechenzentrums, IBM 7090) berechnet. Auf diese Weise konnten die Kernsäurewerte für Kern, Kernkörperchen und Cytoplasma getrennt bestimmt, sowie auch Zellen mit mehreren Kernen und Kernkörperchen einer Auswertung zugänglich gemacht werden. Die Untersuchungen wurden an HeLa-Zellen, und zwar am 1., 3., 5., 7. und 10. Tage einer Gewebekulturpassage durchgeführt. Es zeigt sich, daß der Kernsäuregehalt von Kern, Kernkörperchen und Cytoplasma während der 10tägigen Gewebekulturpassage gleichartige Veränderungen erfährt, wobei ein doppelgipfliger Verlauf zu beobachten ist. Einer initialen Kernsäurezunahme (1.—3. Tag) folgt ein Substanzabfall (3.—5. Tag), an den sich ein erneuter Kernsäureanstieg mit einem Maximum am 7. Tage anschließt. Schließlich wird am 10. Kulturtage der Ausgangswert annähernd wieder erreicht. Die Ergebnisse finden in interferenzmikroskopischen Parallelversuchen ihre Bestätigung. — Die Untersuchungsergebnisse zeigen, daß für quantitative histochemische Untersuchungen an Gewebekulturzellen die Angabe über die Dauer bzw. das Alter der jeweiligen Passage von entscheidender Bedeutung ist.

Cytophotometric Investigations of HeLa-Cells. Nuclei, nucleoli and cytoplasm of more than 250 HeLa-cells were examined separately for their content of nucleic acids. The „Universalmikrospektrophotometer" (Zeiss) was used for the cytophotometric investigations. The data obtained for volume and substance as well as the percentage quantity of substance for the respective volume were calculated with an electronic computer (Rechenanlage des Deutschen Rechenzentrums, IBM 7090). In this way it was possible to calculate the amount of nucleic acid for nuclei, nucleoli and cytoplasm separately and to include cells in the investigation, which contained several nuclei and nucleoli. The tests were carried out on HeLa-cell cultures on the 1st, 3rd, 5th, 7th and 10th day of the cell-passage. It appeared that during the 10 days' passage the content of nucleic acid in nucleus, nucleolus and cytoplasm was subject to diurnal variations. Showing two peaks, the curve obtained was much alike for the three cell components in question. An initial increase of nucleic acid (1st—3rd day) was followed by a decrease (3rd—5th day) which again was followed by another rise, which had its maximum on the 7th day. On the 10th day the initial position was regained. The findings were in agreement with interferometric investigations carried out in parallel. — The results obtained show that an exact statement of the age, or the length of passage, is absolutely crucial for the outcome of quantitative histochemical tests on tissue cultures.

* Mit Unterstützung der Deutschen Forschungsgemeinschaft.

Pathologisches Institut der Universität, 6 Frankfurt a. M. (Deutschland).

9.

MASAOKI YAMADA: Isolation of the Egg Cell Envelope of Teleost Fish.

The cortical envelope was isolated from egg cells of the teleost fish (*Cyprinus auratus* L.) by the Tris-Mg method of the author. A 1/500 M-MgCl₂ solution buffered with 1/50 M-Tris at pH 7.4 was used for the isolation. — Morphological manifestations were observed under an interference microscope. The envelope has a meshlike structure with square and pentagonal openings less than 1 micron in diameter. The widths of the openings are due to the effects of different cations. — By chemical analyses, P is lacking in the envelope, DNA is indetectable, and in the hydrolysate the peculiar aminoacid, hydroxyproline, is present. — By spectrophotometry a yellow pigment is revealed in the internal side of the envelope. The Emax is about 450 mμ and the absorption in the UV zone is not prominent.

Dept. for Cell Research, Karolinska Institutet, Stockholm (Sweden).

10.

D. S. Robinson: Distribution of Carbonic Anhydrase in Egg Shell Membrane.

The outer shell membrane, obtained by decalcifying the shells of whole avian eggs, consists of four distinct fibrous layers. The outermost layer of the membrane contains aggregates of mucoprotein which are believed to represent the organic matrix of the inorganic mammillae first described by Nathusius. Treatment of this layer according to Häusler's modification of Kurata's cobalt bicarbonate method for the detection of carbonic anhydrase has shown that such enzymatic activity was concentrated at the organic mammillae. Further support for this observation was afforded by the observed inhibition of the staining reaction with 2-acetylamino-1,3,4-thiadiazole-5-sulphonamide sodium and the failure of the membrane to retain cobalt either in the absence of the bicarbonate substrate or after boiling. Further studies of the outer shell membrane obtained from premature eggs show that inorganic crystal formation commences at the organic mammillae during shell formation. It therefore seems that carbonic anhydrase might initiate shell deposition at these sites by localising high concentrations of carbonate ion. — Pretreatment of the membrane with solutions of either 8 Molar urea or inorganic salts has little effect on the staining characteristics of the mammillae, whereas pretreatment with 8 Molar urea solutions containing mercaptoethanol caused dissolution of the mammillae and subsequent loss of enzymatic activity. Disulphide bonds may thus be necessary for the integrity of the organic mammillae.

Low Temperature Research Station, Cambridge (Great Britain).

Magen—Darm · Gastro-intestinal Tract · Tube digestif

1.

S. Capurro, D. Zaccheo and C. E. Grossi: Differentiation of the Salivary Glandular Epithelium and Mucopolysaccharides during the Development of Man.

When applying histochemical methods for the mucopolysaccharides, the *glandulae labiales, palatinae* and *linguales* were studied in human embryos and foetuses at different stages of development, till birth. — Our research has brought evidence that the mucopolysaccharides appear in the buds of the glands considered, during the process of canalisation of the solid epithelial columns, in which such substances are to be found in superficial sites of the still undifferentiated cells and in the spaces under formation between them. It is a matter of acid mucopolysaccharides which presumably are mixed with other neutral ones. — The product of the first typical mucous cells contains this same substance while mucocytes with an exclusively neutral secretion appear, as a rule, only in a further stage and are in small numbers, mixed with the others. In the earlier stages of development, the perceptible acid products are represented, at least predominantly by sialic acids; in the 5th month and in the following ones, the presence of sulphomucopolysaccharides is also demonstrated. — In the *glandulae linguales* the cells of the demilunes and the mucoserous ones of the tubules contain, from the 5th month onwards, neutral mucopolysaccharides and in a smaller extent, sialomucins. — From the present research it follows that already during intrauterine life the mucopolysaccharide production of the epithelia considered reaches the same histochemical characteristics which are met in the adult.

Istituto di Anatomia Umana Normale, d'Istologia e di Embriologia Generale dell'Università di Genova (Italia).

2.

P. Mangiante and M. Schippers: Enzymatic Reactions for the Differentiation of Salivary Glandular Types in Birds.

In the epithelium of the buccal glands of some birds of various ages, different enzymatic activities have been looked for by means of histochemical methods. The buccal glands are represented in the same species, by elements of different type. The reactions have demonstrated that in these different cells, different enzymatic activities display themselves, as in *Gallus* for example. The differences are related not so much to single elements of the secretory units but more to complete glandular masses which have or not a determined enzymatic activity and show a very slight activity as compared to other glandular groups which on the contrary have a very high activity. Previous research of other authors on enzymatic activities had not pointed out this peculiar characterization of the secretory apparatus of the buccal mucosa of birds. — It has been possible to establish that this different enzymatic inheritance is correlated to the type of secretory elements which are comparable to the salivary elements, mucous or "amphoteric", of certain mammals. For example, the elements of the "amphoteric" type have been seen to react strongly with substrates for the acid phosphatase and the non-specific esterase; mucous cells did not stain or exhibited a weak reaction. This fact which has arisen for the first time in the field of birds, is well known for more evoluted animals and represents therefore interesting evidence which will have to be considered also in relation to the type of feeding and of environment.

Istituto di Anatomia Umana Normale, d'Istologia e di Embriologia Generale dell'Università di Genova (Italia).

3.

MARIA GABRIELLA MANFREDI ROMANINI and ANNUNZIA FRASCHINI: Histochemical Studies on the Functional Differentiation of Gastric Glandular Cells during the Development of Salamandra atra and Salamandra maculosa.

Salamandra atra and *Salamandra maculosa*, although both defined as viviparous species, differ in that *Salamandra atra* remains in the maternal oviduct even after metamorphosis, whereas *Salamandra maculosa* is born at an advanced larval stage able to live freely only in the water. We have studied in these species the stages in the functional differentiation of the gastric glandular "multipotent" cells using histochemical methods designed to demonstrate the pepsinogenesis (by means of specific reactions of amino acids containing derivatives such as tyrosine and tryptophan), the accumulation of ribonucleic acid and of lipides, either free or masked. These substances appear in both species at a very early larval stage immediately after completion of the morphological differentiation of the gastric epithelium and before complete differentiation of the intestinal epithelium. At this stage, the larvae still have yolk and well developed gills. The histological situations are very similar in *Salamandra atra* precocious larvae, which still live free for a long time in the maternal oviduct, and for those of *Salamandra maculosa* at the birth. In both species, during the successive phases of development there is a rapid increase in pepsinogen-like material, the content of which per cell reaches a maximum just before metamorphosis and then diminishes, possibly as a result of the increased length of the glandular tubes. The lipides increase after metamorphosis and the two species differ especially as regards the ratio between the various lipidic fractions.

Institute of Comparative Anatomy and Center of Histochemistry of C.N.R., University of Pavia (Italy).

4.

C. E. GROSSI, S. CAPURRO and D. ZACCHEO: Characteristics of the Gastric Epithelial Mucins in the Pre-Natal Development of Man.

The elaboration of the mucous secretion of the stomach has been followed, by histochemical methods for the polysaccharides, in a series of human embryos and foetuses. Both the surface epithelium and the glandular one have been considered, bearing in mind also their histo- and cytogenetic modifications during the development. — In the two months of the embryonic life the surface epithelium, formed by many layers of cells, contains exclusively glycogen which appears uniformly distributed. But already at the end of the 2nd month, corresponding to the formation of the gastric pits. appears a product which collects in the apical region of the cells and which is comparable to the mucopolysaccharides. In the following months, with the differentiation of the mucous elements, the cellular product increases and assumes definite histochemical characters. Two types of polysaccharides are present in the surface epithelium: neutral mucoproteins and sialomucins. The former are very abundant and distributed in all the surface cells with a certain uniformity. On the contrary, sialomucins do not have constant standards both as regards their appearance and their localisation. The first traces of carboxylic acid mucopolysaccharides have, however, appeared at the 5th month. — The gastric glands have not shown conspicuous regional differences in the course of pre-natal life. Very poor in glycogen and devoid of mucopolysaccharides in the initial stages of their buds (3rd month), the glandular cells later enrich themselves of neutral mucopolysaccharides (4th—5th months) while carboxylic acid mucopolysaccharides also appear in the neck cells. Proofs of the presence of sulpho-mucopolysaccharides have not been noticed in any of the epithelial sites.

Istituto di Anatomia Umana Normale, d'Istologia e di Embriologia Generale dell'Università di Genova (Italia).

5.

D. ZACCHEO, C. E. GROSSI and S. CAPURRO: Characterization of the Mucopolysaccharide Material during the Development of the Human Intestinal Epithelium.

The present data concern the appearance and localisation of the epithelial mucopolysaccharides in the human intestine during the ontogenesis. — In the first months of the prenatal life, the epithelium of the intestine shows conspicuous modifications in relation to morphogenetic processes which are known, even if still discussed: the process of villus formation, the formation of duodenal glands, the buds of the intestinal crypts in the small intestine and furthermore, in the large intestine, the disappearance of the villi and the final arrangement of the mucosa. — During the two months of the embryonic life, the polysaccharides evidenced in the intestinal epithelium are formed exclusively of glycogen. Already at the end of the 2nd month however, a mucopolysaccharide product appears in some epithelial cells. All the same, it was not possible to establish a relationship between the appearance of such material and the numerous phenomena connected with the histogenesis of the epithelium. In spite of this, the qualification data of the mucopolysaccharide material have permitted the observation of certain differential aspects in the various segments. It is particularly referred to the goblet cells, the secretion of which has presented characters of progressive, increasing, acidity proceeding in a cranio-caudal direction. The mathematic-statistic elaboration of the data obtained through the goblet cells count, done in the different parts of the intestine, has permitted the schematic representation of the existing differences. While, during all the prenatal life, in the small intestine, the secretion of the muciparous cells contains neutral and acid carboxylic mucopolysaccharides, without any significant differences, in the colon and particularly in the rectum cells containing also sulpho-mucopolysaccharides appear.

Istituto di Anatomia Umana Normale, d'Istologia e di Embriologia Generale dell'Università di Genova (Italia).

6.

ZDENĚK LOJDA and PŘEMYSL FRIČ: **Histoenzymatic Study of Jejunal Biopsies in Malabsorption Syndromes.**

Hydrolytic and oxidative enzymes were followed histochemically in 62 jejunal biopsies obtained with a Crosby capsule: 25 normal, 12 non-tropical sprue (primary malabsorption syndrome — PMS) and 25 patients with malabsorption of other etiology (secondary malabsorption syndrome — SMS). Enzymes were detected with various methods in differently treated specimens. In SMS (with the exception of Whipple's disease) the enzymatic equipment of enterocytes, goblet cells, Paneth cells and Kultschitzky cells was not altered. In enterocytes of patients with PMS enzymes confined to microvilli (various phosphatases, aminopeptidase), "lysosomal" enzymes (acid phosphatase and E 600 resistant esterase), and mitochondrial enzymes (dehydrogenases) in the supranuclear zone were affected to various degree depending on the stage of the disease. The enzymatic equipment of goblet and Paneth cells was not altered. In Kultschitzky cells the activities of acid phosphatase and E 600 resistant esterase were increased in untreated PMS. Implications of these results are briefly discussed.

Laboratory of Angiology, 1st Institute of Pathology and 2nd Research Unit of Gastroenterology, Faculty of General Medicine, Charles University, Prague (ČSSR).

7.

A. LINDNER, T. KUTKAM, F. LINDNER and W. O. SMITH: **Cellular Changes in Gastric Mucosa of Potassium Depleted Rats: A Correlation with Function*.**

The aim of this study was to correlate gastric functional changes with changes in mitochondria, nucleic acids, protein and enzymes in cells of stomachs of potassium deficient rats. The functional changes found in gastric secretion consisted of a decrease in concentration of HCl and decreased total four hour secretion of HCl, Cl, K, Mg and pepsin (SMITH and BAXTER, Proc. Soc. Exp. Biol. and Med., *112*, 859, 1963). — Groups of rats were fed a potassium deficient diet while pair fed controls received the same diet with potassium added. Rats were sacrificed at the end of two weeks. Small blocks of tissue were processed by freeze-substitution or were rapidly frozen and cut in a cryostat for enzyme studies. Changes in mitochondria, nucleic acids, protein and enzymes were studied using histochemical methods. Results: a decrease of about 40% in concentration of RNA and protein and 20% of DNA was found in parietal and chief cells, as determined by photographic photometry; some swelling of mitochondria and a decrease in activity of succinic dehydrogenase, cytochrome oxidase and carbonic anhydrase was seen in parietal cells. The changes were noted in the potassium deficient animals as compared with pair fed controls. — The decrease in HCl might well be explained by the cellular changes found in the stomach.

* Supported in part by a grant from the National Institutes of Health.

Dept. of Pathology, University of Oklahoma School of Medicine and Veterans Administration Hospital, Oklahoma City, Oklahoma (USA)

8.

UMBERTO A. BIANCHI: The Histochemistry of the Intestinal Mucosa in Sigmoid and Rectal Artificial Bladders by an Estimation of Some Normal Histophysiological Aspects.

In patients who had undergone pelvic exenteration for advanced cancer of the cervix, biopsies were made of the mucosa of sigmoid and rectal artificial bladders. The experiment was controlled by the study and comparison of the corresponding normal intestinal tracts, the bladder and, in some cases, the corresponding foetal organs. — The data also depend on the time lapse after operation. Metaplastic phenomena are not observed; the mucosa retains its normal structure. The findings can be summarized as follows: appearance of glycogen, especially in the surface epithelium; some modification in the nature of the mucus; permanence of enterochromaffin cells; disappearance of pseudomelanin pigmented cells; oscillation of the mitotic index; variation in DNA distribution depending on the rate of cellular turnover; reduction in the RNA underlying the striated border; slight reduction of proteins. — The following conclusions were drawn: the epithelium of the intestinal segments under consideration has a well defined intrinsic adaptability; it can therefore support, without structural alteration, the radical change of environment caused by the urine; there may also be a revival of foetal characteristics. In more detail: the enterochromaffin cells have no direct connection with digestive phenomena; the pseudomelanin pigment is formed only under the influence of the neutral environment; the cellular turnover, a fairly constant value, is an intrinsic characteristic correlated with the variation in DNA; also in the terminal tract of the intestine there is some functional protein synthesis; the major part of the protein is synthesised during the proliferative process and is of remarkable stability.

Institute of Comparative Anatomy and Center of Histochemistry of C.N.R., University of Pavia; Autonomous Obstetric School of Brescia, University of Milan (Italy).

9.

G. RONCORONI, F. CISOTTI, and A. ALLARA: Comparison between S^{35} Localizations and Findings from Other Histochemical Methods for Gastric Sulfomucins.

In a previous research the authors have proved, with an autoradiographic method, the presence of a sulfated component in aminopolysaccharide secretions of various tracts in the cock's gastric apparatus (glandular stomach, isthmus, muscular stomach). In the present study the authors have checked on these tissues the answer to histochemical reactions which are specific for sulfate groups: Bracco-Curti and Geyer reactions, staining with Alcian blue at pH 0,5, methylation-demethylation test. — In the glandular stomach (surface epithelium), there is agreement of positivity between S^{35} localization and Bracco-Curti, Geyer reactions, and staining with Alcian blue at pH 0,5. In glandular lobules, where S^{35} appeared in low amount as delicate diffuse dotting, Geyer and Bracco-Curti reactions were negative, while Alcian blue at pH 0,5 has shown a slight azure staining. In the isthmus and in surface sections of the muscular stomach mucosa (on the boundary of covering cuticle) parallel to strong autoradiographic demonstration of sulfur appears a positivity to Geyer, Bracco-Curti and Alcian blue at pH 0,5, more intense than in the glandular stomach. — In all considered regions, on the contrary, after demethylation, the dye-binding capacity for toluidine blue is restored. This fact should testify the concomitant presence of carboxylic acid along with sulfate groups. The green answer of various tonalities to staining with Alcian blue at pH 0,5, followed by staining with Alcian yellow at pH 2,5, would agree in this sense.

Istituto di Istologia ed Embriologia Generale dell'Università di Milano, Via Ponzio 7, Milano (Italy).

10.

G. C. BUDD: Intestinal Mucus Synthesis Studied by High Resolution Autoradiography.

The mucus present in mammalian intestine is secreted mainly by goblet cells on the villi and in the intestinal crypts. The acid nature of this secretion product is associated with the mucoitin sulphate component of the complex. — Using sulphur-35 sodium sulphate to label newly synthesised mucoitin sulphate in the goblet cells of the mouse, the intracellular site of sulphate binding has been studied by light and electron microscope autoradiography. In addition the sequence of stages between synthesis of the labelled material and its eventual discharge into the intestinal lumen have been followed and timed. — After intraperitoneal administration of radioactive sulphate, the label is incorporated rapidly into 700—800 Å diameter granules of mucigen which are situated between parallel agranular lamellae. From their position and appearance these membranes appear to represent part of an extensive Golgi apparatus. — Within a few minutes after labelling, granules are aggregated into larger droplets, each of which is bounded by an agranular limiting membrane derived from the

Golgi lamellae. The droplets are shed into a swollen storage zone together with small amounts of other cytoplasmic components. — Near the apical part of the cytoplasmic storage zone, the granular content of mucigen droplets changes in appearance prior to discharge of mucus from the cell. The significance of these observations to an understanding of the secretory process will be discussed.

Royal Free Hospital School of Medicine, 8, Hunter Street, London, W.C.1 (Great Britain).

11.

M. SCHIPPERS: Further Contribution to the Knowledge of the Enzymatic Activities during the Embryonic Development of Man.

The study of two human embryos of 12 mm and 22 mm CR respectively, by various histochemical methods, has filled some gaps in the knowledge about enzymatic activities which occur during the ontogenesis. These evidences confirm the variability of the enzymatic value in relation to the differentiation of the buds; the research has permitted to better specify the time of appearance or of disappearance of some enzymatic activities in determined sites; it was even possible to observe in some differentiating buds, the succession of different enzymatic activities. — From the comparison of the various enzymatic reactions it appears also, that in this period the carboesterase activities are very weak; the reaction comes to light especially in the condensed mesenchyma, when in this site the alkaline phosphatase activity has disappeared. — In some mesenchymal condensations and also in certain epithelia where the alkaline glycerophosphatase activity decreases in the period under consideration or where it is no longer to be noticed, the 5-nucleotidase activity manifests itself *ex novo* or it appears with a greater intensity. The appearance of the specific phosphatase activity instead of the non-specific one is however a phenomenon localized to certain sites. — In certain buds, finally, acid phosphatase activity appears or increases gradually as the alkaline one exhausts itself in relation to a higher stage of differentiation of the buds in question. The acid phosphatase is the most common enzyme in the considered stages and manifests itself in some sites also simultaneously to the alkaline phosphatase activity.

Istituto di Anatomia Umana Normale, d'Istologia e di Embriologia Generale dell'Università di Genova (Italia).

Freitag, Friday, Vendredi, 21. August 1964

Urogenitalsystem · Urogenital System · Système urogénital

1.

OTTO V. DEIMLING, GÜNTER BAUMANN und HARALD NOLTENIUS: Hormonabhängige Enzymverteilung in Geweben: Geschlechtsunterschiede der alkalischen Phosphatase, unspezifischen Esterase, Glukose-6-phosphatase und Succinodehydrogenase in der Niere der Maus.

In der Niere der Maus lassen sich Geschlechtsunterschiede der Aktivitäten von alkalischer Phosphatase, unspezifischer Esterase, Glukose-6-phosphatase und Succinodehydrogenase histochemisch an Kryostatschnitten gut darstellen. Da die Aktivitätsunterschiede auf bestimmte Abschnitte des Nephrons beschränkt sind, so ergibt sich das Bild einer unterschiedlichen Fermentverteilung. Dies ist besonders gut am Beispiel der unspezifischen Esterase zu erkennen: In der männlichen Niere findet sich in allen Anteilen der Hauptstücke eine starke Esteraseaktivität. In der weiblichen Niere findet sich eine Esteraseaktivität gleicher Intensität nur in den gestreckten Hauptstücken. Die gewundenen Hauptstücke sind esterasearm. Umgekehrt zeigen beide Geschlechter eine vergleichbare Succinodehydrogenaseaktivität in den gewundenen Hauptstücken, während die Aktivität der gestreckten Hauptstücke beim weiblichen Tier deutlich geringer ist als beim männlichen Tier. Die alkalische Phosphataseaktivität ist schließlich in der weiblichen Niere wesentlich mehr verbreitet als in der männlichen Niere. — Durch Kastration von männlichen Tieren entsteht in den Nieren der weibliche Fermentverteilungstyp bei allen vier Fermenten. Die Kastration weiblicher Tiere verändert deren Fermentverteilung nicht. Nach Testosterongaben tritt in weiblichen Nieren die typisch männliche Fermentverteilung auf. Östradiolgaben verändern die Fermentverteilung der männlichen Niere nicht. Demgegenüber sind bei analogen Versuchen an der Ratte Orchektomie und Testosteron wirkungslos, Ovarektomie und Östradiol aber von sehr starker Wirkung.

Hormone Dependent Enzyme Distribution in Animal Tissues: Sex Difference of Alkaline Phosphatase, Unspecific Esterase, Glucose-6-Phosphatase, and Succinic Dehydrogenase in the Kidney of Mice. Sex differences of the activities of alkaline phosphatase, unspecific esterase, glucose-6-phosphatase, and succinic dehydrogenase are easy to demonstrate histochemically in cryostat sections of mouse kidney. As the differences of activity are limited to certain segments of the nephron, the result is a picture of a varying enzyme distribution. This is particularly easy to notice in the example of the unspecific esterase: In the kidney of the male mouse there is a strong esterase activity in all segments of the proximal tubules. In the kidneys of the female mice an esterase activity of equal intensity is only in the descending limb. The convoluted proximal tubules are poor in esterase. Inversely both sexes have a comparable succinic dehydrogenase activity in the convoluted proximal tubules, whilst the activity of the descending limb of Henle is considerably less in the females than the males. Finally, the alkaline phosphatase activity is very more widespread in the female kidney than in the male. — Castration of male mice results in the female type of enzyme distribution with all four enzymes. Castration of female mice does not alter their enzyme distribution. Following testosterone, the typical male enzyme distribution occurs in the female kidney. Oestradiol does not alter the enzyme distribution of the male kidney. On the other hand orchidectomy and testosterone are without effect on the kidney of rat in analogous experiments. Ovarectomy, oestradiol, however, have a very strong effect.

Pathologisches Institut der Universität, 78 Freiburg (Deutschland).

2.

JERZY KAZIMIERCZAK: Glucose-6-phosphate Dehydrogenase (G-6-PDH) and α-glycerophosphate (Menadione Reductase) Dehydrogenase (α-GPDH) in Juxtaglomerular Apparatus (JGA) of Various Species*.

G-6-PDH and α-GPDH, as revealed by histochemical methods, have been found to be particularly active in JGA (G-6-PDH in macula densa and α-GPDH in JG-cells) of the kidney of animals made hypertensive by help of Goldblatt clamps. Since the JGA is the place of renin formation, it was suggested that these enzymes may participate in some synthetic processes leading to renin production and secretion. — Most of these experiments have been performed, however, only on rats, and it was of interest to see if the pattern and intensity of staining reaction for these enzymes in JGA of the kidney of various species reflects the renin concentration in their kidney. — The present examination of rat, rabbit, mouse, Guinea pig, and cat kidneys have shown, that while the overall gradiation of the staining reaction for G-6-PDH in macula densa of various species were found to be somewhat parallel to the reported concentration of renin, the reaction for α-GPDH had a different pattern in each examined species. The enzyme was found in different parts of the JGA in different species, and the intensity was without any connection with the concentration of renin. In hypertensive animals both features, indicating increased metabolism in JGA may, however, be found. That was observed also in the surgically treated hypertensive patient, where the stenosed kidney revealed prominent increase of staining reaction for both G-6-PDH in macula densa and α-GPDH in JG-cells in contrast to the opposite kidney where the reaction for these enzymes in the JGA was very weak.

* These investigations were carried out in the University Institute of Experimental Medicine, Copenhagen.

Rheumatic Research Laboratory, University Institute of Pathological Anatomy, Copenhagen Ø (Denmark).

3.

ALFRED FARAH: Effects of Antidiuretic Hormone on Protein-Bound Sulfhydryl and Disulfide in Renal Cells*.

Protein-bound sulfhydryl (PBSH) and disulfide (PBSS) were determined by cytophotometric methods based on the Barrnett and Seligman procedures for PBSH and PBSS (CAFRUNY, DI STEFANO and FARAH, J. Histochem. and Cytochem. *3*, 354, 1955; TEIGER, FARAH and DI STEFANO, Ibid. *5*, 403, 1957). Injections of antidiuretic hormone, 24—48 hours dehydration, or injection of hypertonic NaCl into a carotid artery of rats produced a maximal reduction of 20 to 25% in PBSH groups in all types of renal cells studied. Persistence of this change depended upon the dose of antidiuretic hormone and correlated with the antidiuretic action of the hormone. Changes in PBSS are not observed during the period of antidiuresis but a marked reduction in PBSS is seen at a time when PBSH is recovering. These effects on PBSH and SS groups due to dehydration or hypertonic salt administration are not seen in hypophysectomized or diabetes insipidus rats. Synthetic lysine or arginine vasopressin produced quantitatively similar effects while oxytocin in relatively high doses did not produce these PBSH changes in renal cells. It can be shown that the effects on PBSH are not due to a stoichometric binding of this antidiuretic hormone with renal cellular PBSH.

* Supported by a grant from the United States Public Health Service.

State University of New York, Upstate Medical Center, 766 Irving Ave., Syracuse, New York (USA).

4.

MARGHERITA RIZZOTTI: Glycogen and Mucopolysaccharides in Renal Pelvis Glands in Horse.

In the horse's renal pelvis there are tubulo-alveolar glands with mucous secretion, which are not well known histochemically. — The transitional epithelium, very rich in glycogen in this species, causes introflections in places in which the epithelium at first keeps its own characters, then becomes a cuboidal compound and finally a true secretory epithelium. Both in cells of tissues with compound epithelium and in cells of glandular tract, glycogen and mucopolysaccharides can be observed; the former prevailing in the highest tracts of the glandular recesses, the latter in the glandular "fundus". However elements still full of glycogen can be traced somewhere also in this "fundus". — Chromotrope, Alcianophil mucopolysaccharides are constituted essentially of sialopolysaccharides on the basis of acid hydrolysis after QUINTARELLI, TSUIKI, HASHIMOTO and PIGMAN. — The peculiar histochemical characteristics of these glands and their localization in a typically glycogenic epithelium give rise to the question of a metabolic glycogen-mucopolysaccharides relationship in the genesis of the latter.

Istituto di Anatomia degli Animali domestici — Istologia ed Embriologia dell'Università di Milano, Via Celoria, 10 (Italia).

5.

A. J. MATTY: Histochemistry and Fine Structure of the Toad Bladder.

The bladder of toads (*Bufo marinus* and *Bufo bufo*) of recent years have proved convenient membranes for investigational work by physiologists and pharmacologists, particularly for those investigating the effects of hormones such as neurohypophysial hormone and thyroid hormone on ion and water permeability. — This study is an attempt to correlate physiological and histochemical findings. The distribution of enzymes which have been implicated in active transport processes is described. Enzymic activity is concentrated in the mucosal layer, little enzyme activity occurring in the serosal layer. Thyroxin ánd vaso-pressin do not cause swelling of the mucosal cells of *Bufo bufo*, but thyroxine analogues appear to affect the E.M. picture of the mucosal cells by increasing the intercellular spaces at a time when there is known to be rapid water movement across the membrane. The serosal cells have a characteristic vacuolated appearance and much pinnacytosis appears to be taking place in these cells.

Wellcome Laboratories of Pharmacology, The University, St. Andrews (Scotland).

6.

UBALDO FILOTTO: Cytotopochemistry and Histochemical Properties of Prostate Gland Secretion in Equus caballus and Bos taurus.

Prostate glands of several mature animals of various ages have been put in proper fixing solutions to study their morphology and to carry on the appropriate test for polysaccharides (PAS, PAS-diastase, Iodine stain, Alcian Blue 8 GN, metachromasia with toluidine blue, methylation-demethylation reaction), for lipids (Sudan Black, Nile blue sulfate, plasmal reaction, BAKER's test), for proteins (Bromophenol blue) and for RNA (BRACHET's method). The horse secretory cells contain small, uniform granules distributed in the whole cytoplasm. Considering the response of the material contained in the cells and in the ducts to the histo-chemical methods, the author thinks that the secretion consists of glycogen, glycoproteins, ribonucleoproteins and lipids (neutral fats and phospholipids). In the bull the secretory epithelium of the gland is characterized by a prominent cell polymorphism. Apparently serous, serous-mucous, mucous elements are readily observed. These cells probably represent func-tionally differentiated phases of a single cell type which elaborates, by different modalities, various substances such as glycogen, neutral and acid polysaccharides (with carboxyl and sulfate groups), ribonucleoproteins and lipids (neutral fats and phospholipids).

Istituto di Anatomia degli Animali Domestici—Istologia ed Embriologia dell'Università di Milano, Via Celoria n⁰ 10, Milano (Italy).

7.

MIKKO NIEMI: Esterase Activity of Rat Testis during Various Developmental and Functional Stages.

Non-specific esterase activity is one of the qualitative histochemical characteristics of the steroid hormone producing cells in the ovary, testis and adrenal gland. Previous studies have suggested that esterase activity of these organs might parallel the steroid production quantitatively. Histochemical studies were carried out on rat testis on three developmental phases (prenatal, postnatal prepubertal and adult animals); hypophysectomized adult rats were studied as well. Besides three histochemical techniques (α-naphthyl acetate, Naphthol-AS-D acetate and 5-bromoindoxyl acetate) quantitative chemical estimation of esterase activity was also done. Isozymes of esterase were characterized by starch gel and disc electro-phoresis. Esterase activity was visible in the interstitial cells of the testis from the 15th intrauterine day onwards. The number of positively reacting cell groups increased rapidly toward the 21st fetal day, but after birth their number decreased to very low figures within a couple of days. At the beginning of puberty (around 21st postnatal day) a very rapid appearance of esterase positive cells took place in all the interstitial tissue of the testis. Hypophysectomy caused a marked decrease of histochemically detectable esterase activity which could be, however, partially restored with treatment with chorionic gonadotrophin. Chemical estimates paralleled the volume of interstitial tissue and its histochemically demonstrable esterase activity. Due to the great changes in the volume of the tubular tissue, quantitative activity measurements did not reveal the true status of the interstitial cell, which could adequately be studied by histochemical methods.

University of Helsinki, Dept. of Anatomy, Siltavuorenpenger, Helsinki (Finland).

8.

ZOLTÁN PÓSALAKY: Acid Phosphatase Activity in Spermatogenesis.

In the course of spermatogenesis in the rat, acid phosphatase activity presents itself in the spermatogonia and primary spermatocytes, i.e. in the resting stage, the leptotene and the zygotene, as delicate granules scattered over the cytoplasm. In the next-following develop-mental stage, in the early pachytene, enzymic activity is absent, but in the late meiotic pro-phase, the pachytene, it reappears in the form of a single clearly delimited granule. In the spermatids it shows as one larger and numerous minute granules. In this stage of maturation, the reaction increases in intensity both in the detaching plasma portions and around the heads of spermatids within the Sertoli cells. The acid phosphatase-positive granules display also sulphatase activity. — The lysosomal nature and the role of the acid phosphatase activity are discussed and related to findings under the electron microscope.

Institute of Experimental Medical Research of the Hungarian Academy of Sciences, Morphological Dept., Budapest VIII., Szigony-u. 43 (Hungary).

9.

MARTTI IKONEN: Enzymic Activity of the Interstitial Tissue of the Early Development of Human Fetal Testes.

Human fetal testis was studied using several histochemical techniques for the demonstration of enzymatic activity. Quantitative chemical determinations were performed simultaneously. Special emphasis was paid on the activity of steroid-3β-OL-dehydrogenase. To confirm the validity of this reaction testis tissue was also incubated with tritiated progesterone and 17-hydroxypregnenolone. The generated steroids were separated using thin-layer chromatography and the radio-activity measured using gas-counting of tritium. The youngest fetus in the series was 7 weeks old (CR-length 2,4 cm). The testis contained 25% of interstitial tissue which was composed of large polyhedral cells showing a weak tetrazolium reductase activity towards NADH, NADPH and lactate. These oxidative enzymic activities increased during the subsequent 15 weeks, when also the volume of the interstitial tissue steadily grew. Steroid-3β-OL-dehydrogenase, β-hydroxybutyric acid dehydrogenase and non-specific esterase activities became visible at the age of 10—12 weeks, when also some sudanophilic lipid droplets were first time observed.

University of Helsinki, Dept. of Anatomy, Siltavuorenpenger, Helsinki (Finland).

10.

CARLO BIGNARDI and GIUSEPPE AURELI: Morphological Evolution and Steroidogenic Activity of the Interstitial Cells in the Mammal's Testis during Ontogenesis.

Remarkable differences among the various species are present in the evolution of the interstitial cells during ontogenesis. 1) Interstitial cells may remain entirely differentiated from the beginning of their fetal life up to adult age (e.g. *Sus scrofa*). The steroid-3 β-ol-dehydrogenase activity is always demonstrable with a maximum in the adult life. 2) An interstitial cell involution may occur already during fetal life until cells become apparently dedifferentiated in the prenatal phase and then reappear in the prepuberal phase (e.g. *Bos taurus*). The steroid-3 β-ol-dehydrogenase activity is demonstrable in all phases in which the cells are morphologically differentiated. 3) In the *Equides*, the development of interstitial cells is prominent in the middle of fetal life with evident morphological characters of secretory activity and with the appearance of numerous capillaries. Cells successively undergo lipochromic degeneration ("xanthochrome" cells according to ANCEL and BOUIN) and in the prepuberal age they are completely absent. The steroidogenic activity, which is very low or absent in the interstitial cells during the most active phase of fetal development, is moderate shortly before the birth and disappears in the testis interstitial tissue during prepuberal age.

Istituto di Anatomia degli Animali Domestici—Istologia ed Embriologia dell'Università di Milano, Via Celoria, 10 (Italia).

11.

SARDUL S. GURAYA*: A Histochemical Study of Lipids in the Mammalian Testis.

Lipids have been studied histochemically in testicular material of the adult rabbit, Guinea-pig, sheep, goat, buffalo, marmoset and chimpanzee. Frozen gelatine sections of material fixed in 10% formalin and formaldehyde-calcium, with and without postchroming, were used. Two types of sudanophilic lipids are found. The first type of lipids are granules and spheres of various sizes; the minute granules are lost during fixation and postchroming. Generally they are found in the Sertoli cells, late spermatids and Leydig cells and show a great variation in their amount in different mammals. Their amount is least in rabbit and Guinea pig. They react moderately for phospholipids. Some of them in the seminiferous tubules of sheep, goat, buffalo, marmoset and chimpanzee, and Leydig cells of Guinea pig, marmoset and chimpanzee develop neutral fats (triglycerides). Since the primates remained in the captivity for unknown time, the spermatogenesis in some individuals was stopped as determined from degenerative changes in the various stages of spermatogenesis as well as in the epididymis. The lipid droplets in the Leydig cells of such individuals also contained cholesterol and its esters. The second type of lipids are in the form of heterogeneous or fenestrated lipid bodies of various size. They are irregularly distributed among the spermatogenic cells of seminiferous tubules. They show a great variation in their amount in different mammals. Comparatively their number is more in the rabbit and Guinea pig. They were not observed in those primates whose testis showed advanced degenerative changes in spermatogenic cells and epididymis. They react more intensely for phospholipids which seem to differ from those of the first type in that they give a strong reaction with acid-haematein and are preserved only when the material is fixed in formaldehyde-calcium and postchromed in dichromate-calcium. The significance of these lipid bodies will be discussed elsewhere.

* Population Council, Post Doctoral Fellow.

Dept. of Obstetrics and Gynecology, University of Kansas Medical Center, Kansas City, Kansas (USA).

Immunologie · Immunology · Immunologie
Amyloid · Amyloid · Amyloïde
Verschiedenes · Divers · Divers

1.

B. B. Fuchs, I. V. Konstantinova, S. G. Kolaeva, A. P. Tsiegankov, V. A. Schulga, P. M. Krass and L. F. Maximovsky: **Anti-BSA Formation Initiated in vivo and in vitro by Ribonucleic Acid from Lymph Nodes and Spleen of Immunized Rabbits (Histochemical, Biochemical and Immunological Investigation).**

The main task of our investigation is to initiate synthesis of specific antibodies *in vivo* and *in vitro* by purified RNA from the tissues of immunized animals. In these experiments 130 rabbits were used as donors and recipients. *In vivo* experiments (administration of 10—20 mg of RNA or lymph node cells incubated *in vitro* with RNA for a period of 12 hours at $+ 2^0$ C) showed anti-BSA production. Serum and tissue fixed antibodies (titers 1/10—1/80) were demonstrated beginning from 12 hours up to 5 days after administration of RNA according to the method by Boyden-Stavitsky. A control group of rabbits (12 animals) was injected with RNA treated by RNA-ase ($5—10 \gamma$/ml). In 10 of them no antibodies were formed. In the RNA preparations, non-treated and treated by RNA-ase and then dialysed anti-BSA was not demonstrated. In *in vitro* experiments the cell suspension was prepared from lymph nodes and then incubated with RNA and leucin C^{14}. Anti-BSA was demonstrated in incubation media and in the cells by the Boyden-Stavitsky method (titers 1/4—1/64). Experiments with leucin C^{14} have not been completed. — Histochemical analysis showed accumulation of RNA in sinuses of regional lymph nodes and then in reticular cells (from 12 to 72 hours). The comparative histochemical and microspectrophotometric investigations of three groups of cells were made (small lymphocytes, blasts and plasma cells). The number of blasts and plasma cells with increased content of DNA were not demonstrated until anamnestic stimulation had taken place. There was no induction of plasma cell development after administration of RNA. But a decrease in the number of cells having a significant content of DNA could be observed in the group of small lymphocytes after anamnestic stimulation and also after RNA had been administered. — Our experiments permitted the demonstration of specific antibody formation induced by RNA. The possibility of trans_ferring information in the organism by means of RNA has been examined and discussed.

The Dept. of Experimental Biology, Institute of Cytology and Genetics of the Academy of Sciences of USSR, Novosibirsk 72 (USSR).

2.

M. I. Parfanovich and N. N. Sokolov: **Distribution of Nucleic Acids and Specific Antigens in Cells in Cases of Mixed Virus Infections by Means of Acridine Orange Staining and Immunofluorescent Technique.**

The investigations were carried out in cells from tissue cultures and in brain neurons of experimental animals with different forms of mixed infections, provoked by two DNA-viruses (herpes simplex and vaccinia), two RNA-viruses (arborvirus and poliovirus), two different — DNA and RNA — viruses (arborvirus and vaccinia; arborvirus and herpes; poliovirus and herpes). — In the case of mixed infections there takes place a different distribution and accumulation of nucleic acids and specific antigens — than in the case of the same infections developing separately. For example RNA and antigen of tick-borne encephalitis virus were usually accumulated in the cytoplasm, but in the case of mixed infection (tick-borne encephalitis virus and vaccinia) — in the nuclei as a rule. DNA of herpes virus usually takes place in the nuclei and in the cytoplasm, but in the case of mixed infection (herpetic and tick-borne encephalitis) — only in the nuclei. — One could think that in the case of mixed infection the interaction of two viruses in cells is a struggle for the synthesis of nucleic acid and protein.

The D.I.Ivanovsky Institute of Virology, Academy of Medical Sciences, Moscow (USSR).

14*

3.

J. Sri Ram and G. B. Pierce, Jr.: Use of Antibody-Ferritin Conjugates in Immunoelectron Microscopy*.

Specific antibodies labeled with fluorochromes have proved invaluable tools for histochemical localization of a variety of hormones and proteins. Analogous techniques for labeling antibodies with electron dense tags are in demand for use in immunoelectron microscopic identification and sub-cellular localization of antigens. The bifunctional reagent, p,p'-difluoro-m,m'-dinitrodiphenylsulfone, was used for conjugation of antibodies to the electron dense protein, ferritin, in a simple and mild procedure. The pure conjugate was separated from the unconjugated antibody and ferritin by a procedure combining fractional precipitation and electrophoresis in agar gel. The extreme bulk of the antibody-ferritin conjugate (molecular weight in excess of 500,000) imposes an inherent limitation as the conjugate diffuses rather sluggishly in intracellular spaces and fails to penetrate intact cellular membranes. — A specific antibody-ferritin conjugate was employed in studies on the histogenesis of basement membranes. The sub-cellular sites of synthesis of the basement membrane antigens will be demonstrated and the use and limitations of antibody-ferritin conjugates in immunoelectron microscopy will be discussed.

* Aided by grants E-4401 and CY-6113 from the U.S. Public Health Service.

Dept. of Pathology, The University of Michigan, Ann Arbor, Michigan (USA).

4.

Klaus Becker: Quantitative immunochemische Bestimmungen von Serum-Eiweißkörpern in Blutgefäßwänden und Entzündungsgewebe.

Die Frage der Wechselwirkung zwischen Gefäßwand und Serum bei der Arteriosklerose, sowie der Penetration von Serumeiweißkörpern in das Gewebe, ihrem Abbau bei der Wundheilung und Entzündung sind von großem Interesse. Nur exakte Bestimmungen von höhermolekularen Serumbestandteilen, also Eiweißkörpern, gestatten eindeutige Aussagen über die Änderung der Permeabilität der Gefäßwände bei diesen pathologischen Vorgängen. Eine eindeutige Identifizierung von Eiweißkörpern im Gewebe als Serumproteine ist nur mit immunologischen Verfahren möglich. Nur mit Präcipitationsverfahren ist eine quantitative Bestimmung möglich. Der nephelometrische Nachweis von Antigen-Antikörper-Präcipitaten erlaubt eine Bestimmung von wenigen Mikrogramm, so daß ein Nachweis in wenigen Kryostatschnitten möglich ist und an Parallelschnitten gleichzeitig eine histologische Lokalisation der Befunde erfolgen kann. — Die bisherigen Ergebnisse zeigen, daß in den ersten Stadien der Entzündung und der Arteriosklerose eine erhöhte Permeation von allen Serum-Eiweißkörpern, vornehmlich dem Albumin, stattfindet. Bei Fortschreiten der Veränderungen kommt es zu einer Verschiebung der Albumin:γ-Globulin-Reaktion zugunsten der γ-Globuline. Es kann nicht gesagt werden, ob dies die Folge eines erhöhten Albuminkatabolismus oder einer elektiven γ-Globulinadsorption an das Gewebe ist. Bei Nekrosen findet man wieder normale Relationen. Artefakte müssen durch Verwendung frischen Materials ausgeschlossen werden. Von Bedeutung ist, daß die Ergebnisse bei der menschlichen Arteriosklerose und Entzündung gezeigt werden. Gerade bei der Arteriosklerose sind tierexperimentell gewonnene Befunde nur bedingt auf die menschliche Pathologie übertragbar.

Quantitative Immunochemical Determinations of Serum-Proteins in Blood Vessels and Inflammatory Connective Tissue. The interdependent relations of blood vessel - walls and serum in arteriosclerosis, and of the penetration of serum proteins into the tissue in wound healing and inflammation are of great interest, as is also the catabolism of these serum proteins in the tissue. Only exact determinations of higher molecular substances of the serum, that is to say of serum proteins, give information about the alteration of the permeation rates in these pathological states. An exact identification of serum proteins is only possible by immunological methods. Only by antigen-antibody-precipitation procedures is a quantitative determination possible. The nephelometric estimation of antigen-antibody complexes allows the determination of a few micrograms so that it can be performed in some cryostat sections. In parallel sections a histological localization of the results is possible. — Our results show that in the first stages of inflammation and arteriosclerosis an enlarged permeation of all serum protein fractions, mainly of albumin is seen. In later stages the relation albumin: γ-Globulin is shifted to the globulin. We cannot say if this is a result of an enlarged albumin catabolism or an elective adsorption of γ-Globulin. In necrosis the relations are normal. Artefacts must be excluded by only taking fresh material. Of great importance is that these results have been gained on human arteriosclerosis and inflammation. Especially in arteriosclerosis research we know that results of animal experiments are of restricted value for the human pathology.

I. Medizinische Klinik der Universität, 2 Hamburg (Deutschland).

5.

G. GILLISEN: Fluorescenzserologische Darstellung der Komplementbindung durch sessile Antikörper.

Am Beispiel der Tuberkulinallergie wird gezeigt, daß sich an transfer-kompetenten Zellen sessile Antikörper befinden. Diese Antikörper binden fluorescenzserologisch nachweisbares Tuberkulin. — Transfer-aktive Zellen von Tb-sensibilisierten Kaninchen, die *in vitro* mit Tuberkulin behandelt wurden, binden in spezifischer Weise Meerschweinchenkomplement. Das gebundene Komplement kann fluorescenzserologisch nachgewiesen werden.

Complement Fixation by Sessile Antibodies and its Histochemical Demonstration. Lymphoid cells of tb-sensitized rabbits are able to fix Guinea pig complement after a previous *in vitro* contact with tuberculin. This finding indicates the presence of sessile antibodies against tuberculin on those cells which are able to transfer tuberculin hypersensitivity. The fixation of complement to antigen-coated cells of sensitized animals is undistinguishable from the complement fixation to antigen-antibody aggregates.

Institut für Medizinische Mikrobiologie, Langenbeckstr. 1, 65 Mainz (Deutschland).

6.

V. SEROV, Moscow/USSR: Immuno-Histochemical Studies in the Mechanism of Amyloid Formation.

7.

HOLDE PUCHTLER, SANFORD I. ROSENTHAL and FAYE SWEAT: Amyloid and Cellulose: A Re-Evaluation of Virchow's Concepts*.

Histochemical investigations of the Congo red stain for amyloid indicated similarity of the staining properties of amyloid and cellulose. Modern textile dyes for cellulose stained amyloid selectively without differentiation. The similarity of amyloid and cellulose had been emphasized by VIRCHOW, but his concept was abandoned by pathologists. We reinvestigated VIRCHOW's work from the viewpoint of histochemistry. — VIRCHOW applied iodine to corpora amylacea because of their structural similarity to starch and was astonished to obtain, upon addition of sulfuric acid, the microchemical reaction of cellulose. A similar substance was found in spleen. VIRCHOW's discussion of the terminology shows that he was well acquainted with the term amyloid and equated it with "cellulose-like". Throughout his writings VIRCHOW stressed the microchemical similarity of cellulose and amyloid but also emphasized the presence of nitrogen-containing substances in deposits of amyloid. — Since VIRCHOW (1853) purposely applied the microchemical reactions for starch and cellulose to tissue sections, he is apparently the first pathologist who employed microscopic histochemistry. Our histochemical studies support VIRCHOW's findings that amyloid and cellulose share certain properties. It seems possible that the stereochemical configuration of polysaccharides in amyloid may resemble cellulose.

* Supported by USPHS Grant No GM 07303 and Medical College of Georgia Professional Research Fund Grant No 05-5216.

Dept. of Pathology, Medical College of Georgia, Augusta, Georgia (USA).

8.

YUTAKA KIKKAWA, KUNIHIKO SUZUKI, and BORIS GUEFT: The Composition of Amyloid Deposits*.

Splenic amyloid from a case of tuberculosis was extracted by 6 molar urea several times and dialyzed according to the method of Benditt. The extracts showed the typical 75 Å fibril with rough periodicity and double-filament or possibly tubular structure. No structural evidence was found for a pre-amyloid fibril analogous to tropocollagen. — Amino acid analyses following acid hydrolysis and 2 dimension paper chromatography yielded the following tentative results (gms/100 gms): Alanine 10.90, Arginine+Lysine 9.06, Aspartic acid 9.17, Glutamic acid 10.82, Glycine 5.57, Histidine 2.73, Leucine 13.02, Phenylalanine 4.63, Serine 8.19, Threonine 1.26, Tyrosine 4.56, Valine 4.12, Cystine 3.81. — Microspectrophotometric studies showed marked 2800 Å absorption in formalin fixed paraffin embedded sections, interpreted as indicating tyrosine and tryptophan although no tryptophan notch at 291μ was seen. Electron staining was marked with lead hydroxide, lead subacetate, uranyl acetate, uranyl nitrate and phosphotungstic acid. These findings are consistent with the arginine and aspartic and glutamic acid content. — Micro X-ray diffraction studies with the Chesley micro camera on cardiac amyloid frozen sections, and wide and low angle studies on "pure" extracted amyloid did not yield results clear enough to report, although efforts to increase selective orientation are in progress.

* Aided by a grant from the National Institutes of Health, AM-02967-C5, USA

Depts. of Pathology and Neurology, Albert Einstein College of Medicine, New York 61, N.Y. (USA).

9.

A. NIKULIN* und H. G. SCHIEMER: Interferenzmikroskopische Untersuchungen der Endothelzellen bei akuter Histaminfreisetzung.**

Interferenzmikroskopisch wurde der Einfluß des endogenen, durch den „Histamin-Liberator Polymyxin B" freigesetzten Histamins auf Volumen, Trockensubstanz und Wassergehalt von Endothelien — jeweils getrennt für Cytoplasma und Kern — untersucht. — Kaninchen beiderlei Geschlechts wurde Polymyxin B i.v. injiziert. Einschließlich der Kontrolltiere wurden insgesamt 2600 Endothelzellen der Aorta und A. pulmonalis mit dem Zeiss'schen Interferenzmikroskop analysiert. Die Ergebnisse wurden mit Hilfe der elektronischen Datenverarbeitungsanlage IBM 7090 des Deutschen Rechenzentrums errechnet (Programm SCHIEMER). — Die an den Endothelien der Aorta und der A. pulmonalis gemessenen Werte: Volumen, Trockengewicht, %-Trockengewicht und spezifisches Gewicht stimmten im wesentlichen miteinander überein. Nach der Histaminfreisetzung verlieren die Endothelien an Substanz, das Volumen ändert sich dagegen nur wenig, der %-Wassergehalt nimmt zu. Die Unterteilung der Versuche in 4 Gruppen: a) ohne Polymyxin B-Injektion, b) 5—14 min, c) 15—29 min und d) 30—50 min nach der Histaminliberation läßt unter Heranziehung der gemeinsam mit H. LAPP durchgeführten elektronenmikroskopischen Untersuchungen gewisse Rückschlüsse auf den Wasser- und Partikeltransport durch die Endothelschranke zu, und zwar sprechen die Ergebnisse für einen aktiven transcellulären und gegen einen passiven intercellulären Transport.

An Interference Microscopic Investigation of the Influence of Histamine Liberation on Endothelial Cells. The influence of endogenous (i.e. Polymyxin B-liberated) histamine on the volume, dry weight and water content of endothelial cells was examined by means of interference microscopy. The examination was carried out separately for nuclei and cytoplasm. Polymyxin B was injected (i.v.) into rabbits of both sexes. 2600 endothelial cells — including controls — originating from the aorta and pulmonary artery were analyzed with the Zeiss interference microscope. The results were calculated with the electronic computer IBM 7090 of the "Deutsches Rechenzentrum" (programme SCHIEMER). With regard to volume, dry weight, %-dry weight and specific gravity the results for endothelial cells from the aorta and pulmonary artery were almost identical. After histamine liberation the endothelial cells lost substance. The volume, however, changed only slightly and the %-water content was increased. The subdivision of the test into 4 groups: a) no Polymyxin injection, b) 5—14 minutes, c) 15—29 minutes, d) 30—50 minutes after histamine liberation and additional electron microscopic examinations (carried out in co-operation with H. LAPP) enabled us to draw certain conclusions concerning the transport of water and particles through the endothelial barrier. The results obtained were in favour of an active trans-cellular transport and against a passive intercellular transport.

* Stipendiat der A. v. Humboldt-Stiftung.
** Mit Hilfe der Deutschen Forschungsgemeinschaft.

Pathologisches Institut der Universität, Ludwig-Rehn-Str. 14, 6 Frankfurt a.M. (Deutschland).

Zelle und Zellbestandteile · The Cell and its Components
La cellule et ses éléments

1.

V. YA. BRODSKY: Protein Synthesis in the Nucleus of the Glandular Cell, and Some New Data on the Dynamics of the Process of Secretion.

The cytophotometric, interferometric and histochemical study of the parotid and exorbital glands of the rat allowed us to draw the following conclusions: 1) dry weight of the nucleus of the glandular cell undergoes changes during the secretion cycle, 2) proteins excreted from the nucleus resemble those of the basal cytoplasm (ergastoplasm) by their amino acid composition, 3) the true synthesis of proteins, differing by their properties from the proteins of the ergastoplasmic and nuclear origin, proceeds in the apical cytoplasm, 4) "the secretory protein" is a mixture of proteins synthesized in the ergastoplasm, nucleus and apical cytoplasm, 5) the synthesis of protein and RNA in the glands under study is stimulated not only by an unconditioned reflex but by the conditioned one as well, the character of changes being similar in both cases, 6) the cycle of secretion in the parotid exorbital glands proceed according to the scheme "excretion-restoration", but the cycle of digestion includes 2—3 cycles of the production of secretion, the maximal accumulation in the cell exceeding the amount required for the restoration.

Institute of Animal Morphology, Academy of Sciences of USSR, Vavilow street 12/2, Moscow B-113 (USSR).

2.

ANDERS ZETTERBERG: **Kinetics of Protein Synthesis during the Interphase of Mouse Fibroblasts in vitro.**

The rate of protein synthesis during the interphase of mouse fibroblasts *in vitro* (L-cells) has been studied. Asynchronously growing cells in the logarithmic phase of growth were fixed according to the freeze substitution method. About 200 randomly selected cells were analyzed individually for protein, RNA and DNA content as represented by their dry mass, total absorption at 265 mμ and at 546 mμ after Feulgen staining. Quantitative micro-interferometric and microspectrophotometric methods were used. Mathematical methods described by WALKER, SCHERBAUM and RASCH were used for calculation of the DNA, RNA and protein synthesis curves from the DNA, RNA and protein histograms. — The amount of DNA per cell doubled during one third of the interphase (the S-period), prior to which (the G_1-period) and after which (the G_2-period) the amount of DNA was constant. RNA and protein content per cell increased during the whole interphase. The rate of protein synthesis was constant during the G_1-period, but increased by a factor of approximately two during the S- and G_2-period. — The rate of protein synthesis was thus well correlated to the amount of DNA in the cell during the interphase. Possible factors regulating the rate of protein synthesis will be discussed.

Dept. for Cell Research, Karolinska Institutet, Stockholm (Sweden).

3.

TIBOR BARKA and HENDRIKA VAN DER NOEN: **Cell Proliferation Induced by Isopropylnoradrenaline*.**

The uptake of H^3-thymidine by the salivary gland and liver and the DNA turnover-rate in these organs were studied in rats treated with Isoproterenol-HCl (IPR), a drug known to cause enlargement of the salivary gland (SELYE et al., Science *133*, 44, 1961). In both the submaxillary gland and liver maximal labeling occurred after 2 days of IPR administration, when 16.3% of the acinar cells and 2.9% of the hepatocytes were labeled representing 18 and 3.7 fold increases respectively. In the salivary gland smaller peaks of labeling were found after 1 and 3 weeks of the administration of the drug. Maximal labeling of the interstitial and duct cells was seen at the 7th day. The increase in weight of the salivary gland was paralleled by a 2.5-fold increase in total DNA content after 3 weeks of IPR administration. The total uptake of thymidine per gland was the highest (12.5 times control) on the 4th day. It declined sharply after the 4th day but it was still 7-times greater than that of the control on the 21st day. The specific activity of DNA reached its maximum on the 2nd day. Withdrawal of the drug after 9 days of administration caused a rapid fall in the weight of the gland. Since the amount of total DNA decreased at a slower rate there was a temporary two-fold increase of the DNA-concentration. Weight, DNA-content and DNA-concentration returned to normal within 3 weeks. Calculated from the total activity of DNA the half-life of DNA of salivary gland cells of normal and IPR-treated rats were not different. When the IPR administration was discontinued the radioactivity per gland decreased to 66% within a week and to 48% within 4 weeks. These results indicate that: 1) there is no significant cell necrosis during IPR treatment; 2) some cells depend on the continuous administration of the drug; and 3) the increase in weight of the salivary gland is caused both by the proliferation of cells and their enlargement. The effect of IPR is not restricted to the salivary gland. The IPR induced cell-proliferation seems to be a favorable model in studying the control mechanisms of DNA synthesis.

* Supported by Grant (A-3346) from the National Institutes of Health, U.S. Public Health Service.

The Mount Sinai Hospital, 100th Street and Fifth Avenue, New York 10029, N.Y. (USA).

4.

Giuseppe Gerzeli and **Eugenio Mira**: **Microrefractometric Characteristics of the Various Cell Types in the Salivary Glands of Rat under Different Experimental Conditions.**

Dry mass concentration in single histological elements of the submaxillary and sublingual glands of adult male rats fasted for 5 hrs or 24 hrs and 90 min or 7 hrs after intraperitoneal injection of pilocarpine (50 mg/kg) has been studied. Cryostat sections of $7\,\mu$, either unfixed or fixed in formol or ethanol, using glycerol buffered at pH 7.4 or distilled water as immersion medium, were examined under the Leitz double beam microrefractometer. The results may be summarized as follows: 1) Prolonged fasting and pilocarpine treatment cause clear and characteristic alterations in dry mass per unit area. The modifications observed in unfixed sections are given as illustration: increase of nuclei (from 13% to 40%), of cytoplasma of ducts (from 10% to 32%), of mucous secretory units of the sublingual gland (from 11% to 43%), of sero-mucous acini of the submaxillary gland (from 7% to 31%) and of convoluted granular tubules of the submaxillary gland (up to 17%). 2) Histological fixation can cause reduction in dry mass. Under fasting conditions this reduction is zero or negligible after ethanol fixation but varies up to a maximum of 25% after formol fixation. Under conditions of pilocarpine stimulation, the reduction in dry mass is on average 20% after ethanol fixation and 35% after formol fixation. 3) Immersion of sections in distilled water cause a reduction in dry mass which in general amounts to 36% in the unfixed sections, 33% in the formol fixed sections and 24% in the ethanol fixed sections.

Institute of Comparative Anatomy and Center of Histochemistry of C.N.R., University of Pavia (Italy).

5.

P. O'B. Montgomery and **William A. Bonner**: **Ultraviolet Time Lapse Motion Pictures of Mitosis *.**

The ultraviolet flying spot television microscope is a technique which will permit prolonged observations of living cells as they appear when illuminated with light at 2600 Å wavelength. This communication will begin with a brief presentation of the apparatus necessary for ultraviolet flying spot television microscopy. The principal portion of the presentation will consist of the projection of a silent 16 mm black and white time lapse motion picture of the ultraviolet absorption images of living cells as they pass through the mitotic cycle. The film will demonstrate the intense and unchanging ultraviolet absorption of the chromosomes of the endothelial cells of the newt heart as they move through the various stages of mitosis. At the time of telophase, as the new nucleus is reconstructed, the chromosomes undergo a marked change in ultraviolet absorption. During the process of mitosis the cytoplasmic absorption does not change.

* Aided by a grant from the Damon Runyon Memorial Fund for Cancer Research.

Dept. of Pathology, The University of Texas, Southwestern Medical School and Woodlawn Hospital, Dallas, Texas (USA).

6.

U. Glas, **G. F. Bahr** and **E. Zeitler**: **Cytochemistry of Mitochondria.**

The dry mass of individual rat liver mitochondria was determined using quantitative electron microscopy. These values were compared to values of average protein content obtained by counting mitochondria in a Coulter Counter and/or in Petroff-Hausser bacteria counting chambers and determining the nitrogen content in aliquots with a micro Kjeldahl apparatus. — It was found that differences in the homogenization procedure strongly influenced the average protein N/mitochondria in that increasing the proportion of liver tissues (in 0.44 M sucrose) with 0.9% KCl proved to be suitable for counting mitochondria with the Coulter Counter. Lower electrolyte concentrations were found to be incompatible with proper counter function. Mitochondrial morphology was adequally preserved in this mixture as could be ascertained by electron microscopy. The Coulter Counter recorded typically skewed size distribution curves well in accord with results of quantitative dry weight determinations with the electron microscope. — The counting of mitochondria in bacteria counting chambers using phase microscopy, proved to be a rather unreliable technique. Although average protein values based on such counts and nitrogen determinations were in fair agreement with data from the Coulter Counter and electron microscopy errors as large as $\pm 30\%$ for one animal were not uncommon. — Graphs are demonstrating: a) The influence of the composition and concentration of the homogenization medium, on mitochondrial dry mass and counts. b) The influence of the centrifugation procedure on mitochondrial dry mass and the elimination of nonmitochondrial material from preparations. The average total dry weight of the normal rat liver mitochondrion was found to be 11×10^{-14} gm.

Armed Forces Institute of Pathology, Washington, D.C. (USA).

7.

N. A. TCHERNOGRJADSKAJA, E. M. BRUMBERG, V. M. BRESSLER, E. M. PILSHTCHIK, M. S. SHUDEL, M. V. KUDRJAVTSEVA and T. P. ASTASHINA: Some Data on the Inherent Ultra-Violet Fluorescence of Mitochondria of Living Cells.

The inherent ultra-violet fluorescence of living cells from different tissues was studied by the original method worked out by BRUMBERG (1956). Simultaneously with this method the technique of vital detection of mitochondria with the aid of tetracycline was used. — The examination of the microphotographs obtained has shown that in comparison with other organoids mitochondria possess the most intensive autofluorescence in the ultra-violet region of the spectrum. — The application of specific inhibitors of respiration showed that the reduction of respiratory enzymes (action of KCN) caused an increase in the intensity of the ultra-violet fluorescence of cells. A full or partial oxidation of respiratory enzymes (action of 2,4-dinitrophenol, ADP) caused a decrease in the intensity of the ultra-violet fluorescence of cells. — The respiratory activity of cells in the control and in the experiment was determined by the intensity of blue autofluorescence of cells induced by DPN-H (CHANCE, 1954). — Hence, the character of cell ultra-violet fluorescence, to a great extent, is determined by the degree of reduction and oxidation of respiratory enzymes and reflects the functional state of mitochondria. — A carcinogenic agent (o-aminoazotoluene) was found to change the character of cell ultra-violet and blue fluorescence. O-aminoazotoluene 15 minutes after feeding to the tadpole caused a sharp decrease in the intensity of the ultra-violet fluorescence of liver cells. In an hour the ultra-violet fluorescence considerably exceeded the control level. 3 hours later a new sharp decrease in the fluorescence intensity was noted. At the same time morphological changes in mitochondria were detected. — A noncarcinogenic agent (diethylaminobenzene) similar in character caused a short-time slight decrease in the intensity of ultra-violet fluorescence. — Hence, the changes in the ultra-violet fluorescence of cells caused by o-aminoazotoluene are specific enough and are connected probably, with the injuring action of this agent on the respiratory enzymes of mitochondria.

Laboratory of Microscopy, Institute of Cytology of the Academy of Sciences of the USSR, Prospekt Maxlina 32, Leningrad F-121 (USSR).

8.

SIDNEY GOLDFISCHER*, NINA CARASSO and PIERRE FAVARD: Electron Microscopy of the Acid Phosphatase-Rich Ergastoplasm of Campanella umbellaria.

The Ciliate, *Campanella umbellaria*, incubated for acid phosphatase activity in Gomori's medium with β-glycerophosphate as substrate, displays high levels of enzyme activity throughout the "rough" surfaced ergastoplasm. Acid phosphatase appears to be transported from the ergastoplasm to two other organelles; the digestive vacuoles (lysosomes) and the Golgi apparatus. — The presence of acid phosphatase activity in certain food vacuoles and in the cisternae of ergastoplasm which envelope these vacuoles suggests that this lysosomal enzyme is fabricated in the endoplasmic reticulum and transferred to the food vacuole without recourse to the hydrolase "storage granules" found in other active phagocytes, e.g. rabbit polymorph leucocytes (COHN and HIRSCH, J. Exp. Med. *112*, 983, 1960). — A dilated cistern of endoplasmic reticulum delimited by both "rough" and smooth membrane is found close to the Golgi apparatus. Small vesicles which appear to "bud" from the smooth surface of this ambivalent portion of the ergastoplasm lie between the ergastoplasm and the Golgi saccules. Fine deposits of acid phosphatase reaction product are present in these vesicles and in the Golgi saccules and vacuoles. The latter often contain dense secretory material. Acid phosphatase is apparently carried to the Golgi apparatus from the endoplasmic reticulum in vesicles bounded by a membrane of ergastoplasmic origin.

* Post-Doctoral Fellow, USPHS. National Inst. of Arthritis and Metabolic Diseases.

Lab. de Synthèse Atomique et d'Optique Protonique, CNRS, Ivry, Seine (France).

9.

A. V. ZELENIN and E. A. LIAPUNOVA: On the Nature of Cytoplasmic Granules Having Red Fluorescence and Produced in Living Cells Treated with Acridine Orange.

A study of cytoplasmic granules having red fluorescence (RCG) and originating in living animal cells stained with acridine orange (AO) was undertaken. The dynamics of granule formation in the cells of tissue cultures under the action of different concentrations of AO was investigated in detail. The concentrations of AO resulting in the origination of green granules and their bright red fluorescence were determined to be 0.1 γ/ml and 0.5—1.0 γ/ml respectively. The possibility to detect in the cells simultaneously RCG and ribosomic RNA having red fluorescence was demonstrated. The data obtained speak in favour of a single substrate and common mechanism of the origination of cytoplasmic granules under the action of AO and neutral red. — A comparative study of the formation of granules in cells of different types was undertaken. It was shown that granules originating in the cells of the liver of an animal (mouse) injected with AO are not affected by homogenisation. Experiments on biochemical isolation of these granules were carried out and a subcellular fraction enriched with RCG was obtained and investigated. The effect of treating cultured cells with AO on their viability was studied. It was thereby autoradiographically proved that with AO in concentration of 1.0 to 2.0 γ/ml and above the incorporation of labelled amino acids in proteins is inhibited. It was also shown that at the same concentrations, AO exhibits a strong anti-mitotic effect. A correlation between the mitosis-suppressing action of AO and RCG formation was ascertained. An attempt was made to elucidate at what level the protein synthesis is inhibited by AO. That was done by cytological and autoradiographic comparison of the effect of AO and specific inhibitors of cell metabolism, such as actinomicin D, on the cells.

Institute of Physico-Chemical and Radiation Biology, Academy of Sciences of the USSR, Vavilov Street 18, Moscow B-312 (USSR).

10.

M. N. Meissel, V. M. Manteufel, V. E. Barskiy and M. N. Poglasova: **Fluorescent Cyto-chemistry of Cell Damage, Necrobiosis and Intracellular Digestion.**

The autofluorescence in the region of 310—380 mμ of cells on excitation by waves of length 265—300 mμ, has been investigated in relation to the functional state of the cells, their damage, especially by ionizing radiation, and subsequent necrobiosis. An appreciable increase in the ultraviolet autofluorescence of the cells of blood forming organs can be noted quite soon after irradiation. The increase in intensity of ultraviolet absorption and autofluorescence is especially marked in necrobiotic cells and also in micronecrotic foci, we have earlier described, arising in the blood forming organs as the result of irradiation. Cytochemical and fluorescent-cytochemical studies led to the conclusion that these foci are macrophages, phago-cytizing the damaged cells. In view of this, the change in character of the autofluorescence was investigated on macrophages and amoebae in the process of intracellular digestion and it was shown that fluorescent microscopy in the visible and ultraviolet spectral regions could be utilized for detecting and investigating the processes of activation of hydrolytic enzymes and also for studying the consecutive steps of the chemical degradation of phago-cytized cells and cell components. Experiments on the spectral properties of cells denaturized by various agents and on lysis and autolysis made possible determination of the relative part played by protein and nucleic acid components in the ultraviolet autofluorescence of cells.

Institute of Physico-Chemical and Radiation Biology, Academy of Sciences of the USSR, Vavilov Street 18, Moscow B-312 (USSR).

11.

Arun Kumar Sharma and Archana Sharma: **The Influence of Nucleoprotein on the Manifestation of Drug Effects on Chromosomes of Plant Cells.**

In order to work out the effect of different narcotics as influenced by nucleic acid and protein on chromosomes, the present work was planned. In this investigation, somatic cells of plants (roots) were subjected to treatment with narcotics before and after nucleic acid or protein digestion. Deoxyribonuclease and Ribonuclease were applied for the digestion of two types of nucleic acids and Pepsin and Trypsin for removing the non-basic and basic proteins. — It has been observed that removal of RNA before the application of narcotics does not affect metaphase arrest and polyploidy. On the other hand, if it is removed later i.e. post narcotic application, immediately the division of giant polyploid cells are checked. The result of both DNA and RNA digestion shows that the latter component is of essential necessity in maintaining polyploidy at least at the initial phase. — It has been suggested that by the action of several narcotics, e.g. Colchicine etc. the formation of fibrous protein from corpuscular form — an essential necessity for spindle production — is checked, thus causing spindle inhibition.

Cytochemistry Laboratory, Dept. of Botany, Calcutta University, 35 Ballygunge Circular Road, Calcutta 19 (India).

Muskulatur · Musculature · Musculature
Verschiedenes · Divers · Divers

1.

ERNST ZEBE: The Fine Structural Localization of Adenosine Triphosphate Splitting Reactions in Insect Flight Muscle*.

The site of ATPase activity and the conditions under which it can be demonstrated in indirect flight muscles of the blowfly *Phormia regina* were studied after incubation in a modified Gomori medium. 3 preparations were used: tissue blocks treated shortly with fixing agents, frozen sections and fresh or pretreated isolated myofibrils. The deposition of lead phosphate indicating the hydrolysis of ATP is demonstrated in the contractile apparatus and in the mitochondria. — (1) Myofibrils. In all preparations ATPase activity is found in 2 different parts of the sarcomeres: in the A band and in the Z disc. The same distribution is shown in experiments where calcium is used as precipitating agent instead of lead. The particles precipitated in the A band are frequently arranged in longitudinal rows which correspond to the position of the thick filaments. Under no conditions is lead phosphate found in the H zone. The A band contains apparently 2 regions of different ATPase activity: the center—often divided by the H zone—is densely covered with precipitate while narrow zones bordering the A-I junction contain significantly less lead phosphate. The ATPase of the Z disc is distinctly different from that in the A band, since it can be demonstrated separately under certain conditions. (2) Mitochondria. The splitting of ATP occurs mainly on the cristae mitochondriales. Little or no activity is found associated with the outer membranes. There seem to be different patterns of mitochondrial precipitation. Some mitochondria contain only a few large cristallites while others next to them have the cristae dotted with a great number very small ones.

* Supported by the Deutsche Forschungsgemeinschaft.

Zoologisches Institut der Universität, 69 Heidelberg (Germany).

2.

BARBARA SMITH: Histochemical Changes in Experimental Corticosteroid Myopathy.

Triamcinolone (fluoro-16-hydroprednisolone) was given to rabbits and extensive mucsle necrosis was produced. This differed from that in animals given cortisone acetate in its greater severity, its irreversibility and histologically by the absence of regeneration and inflammatory response. The histochemical changes which occurred before histological changes could be seen were patchy loss of phosphorylase activity in single or small groups of fibres. There was a loss of the normal checkerboard appearance given by oxidative enzymes due to an increase in mitochondrial activity in "white" fibres which normally have very little. In the later stages the fibres were uniformly dark with mitochondrial substrates and phosphorylase activity was greatly reduced. As frank muscle necrosis developed at first single fibres and eventually large areas of the muscle became positive for acid phosphatase, the deposit being apparently in lysozomes. The relation of the enzyme changes to the corticosteroid administration will be discussed.

Dept. of Pathology, St. Bartholomew's Hospital, London, E.C.1. (England).

3.

IMRE ZS. NAGY: Histochemische Untersuchungen über die Cholinesterase-Aktivität der Schließmuskeln der Süßwassermuschel Anodonta cygnea.

Nach unseren früheren Untersuchungen ähneln die in isometrischer Kontraktion fixierten Schließmuskeln (Schm) der *Anodonta cygnea* glatten Muskelzellen. Im Zustand der isotonischen Kontraktion weisen die Schm abhängig vom Grad der Kontraktion „spiralförmige" oder atypische, quergestreifte Strukturen auf. Über die Cholinesterase-Aktivität der Schm von Lamellibranchiaten gibt es keine Angaben in der Literatur. Bekannt sind dagegen die cholinergischen Eigenschaften der ABRM (anterior byssus retractor muscle) von *Mytilus edulis*. — Mit den Methoden von KOELLE-FRIEDENWALD, KOELLE, GEREBTZOFF, BARRNETT und SELIGMAN bzw. CREVIER und BÉLANGER durchgeführte histochemische Untersuchungen zeigten, daß das Sarkolemma der Schm von *Anodonta cygnea* eine diffuse, granulierte Cholinesterase-Aktivität aufweist. Diese Cholinesterase hydrolysiert das Acetylthiocholin, sowie das Butyrylthiocholin. DFP einer Konzentration von 10^{-6}M übt hemmende Wirkung nur auf die Butyrylthiocholin-Hydrolyse aus. Das Sarkolemma der Schm enthält also spezifische, sowie unspezifische Cholinesterase in diffuser Form. Auch diese Eigenschaft der Schm unterstützt unsere Meinung, die wir von histologischen, elektronenmikroskopischen und physiologischen Untersuchungen gewonnen haben, daß die Schließmuskulatur der

Anodonta cygnea ein spezielles Muskelgewebe ist, und man sie weder mit glatten, noch mit quergestreiftem Muskelgewebe identifizieren kann.

A Histochemical Investigation of Cholinesterase Activity in the Constrictor Muscles of a fresh water mussel Anodonta cygnea. According to our previous findings the constrictor muscles (CM) of *Anodonta cygnea*, if fixed in isometric contraction, resemble smooth muscle. In isotonic contraction, however, depending on the degree of contraction, the muscles show a convolute or atypically striated structure. No data exist so far in the literature concerning the CM cholinesterase activity of Lamellibranchiates. The cholinergic properties of the anterior byssus retractor muscle of *Mytilus edulis* are, on the other hand, well known. Histochemical determinations (using the methods described by KOELLE-FRIEDENWALD, KOELLE, GEREBTZOFF, BARRNETT and SELIGMAN, and CREVIER and BÉLANGER) show a diffuse granulated cholinesterase activity in the sarcolemma of the CM of *Anodonta cygnea*. This cholinesterase hydrolyses acetylthiocholine and butyrylthiocholine. DFP (in a concentration of 10^{-6} M) inhibits the hydrolysis of butyrylthiocholine only. This proves that specific and non-specific cholinesterases are present in the sarcolemma of the CM. Together with histological, physiological and electron microscopic evidence, these histochemical findings support our view that the CM of *Anodonta cygnea* is a special muscle tissue, that cannot be identified with either smooth or striated muscle.

Biologisches Forschungsinstitut der Ungarischen Akademie der Wissenschaften, Tihany (Ungarn).

<h2 style="text-align:center">4.</h2>

GERALDINE F. GAUTHIER and HELEN A. PADYKULA: **Cytochemical Studies of Adenosine Triphosphatase Activity in the Sarcoplasmic Reticulum of the Gas Bladder Muscle of Opsanus tau.**

ATPase activity was previously demonstrated at a site which appeared to correspond to the sarcoplasmic reticulum in the rat diaphragm (PADYKULA and GAUTHIER, 1963). Localization was not conclusive because the transverse distribution of mitochondria and sarcoplasmic reticulum are similar. More precise localization of a similar ATPase was achieved by studying a fish muscle in which mitochondria are separated spatially from the sarcoplasmic reticulum. Sites of enzymic activity could therefore be distinguished with the light microscope, which facilitates analysis of cytochemical properties. — In the striated muscle fibers of the gas bladder of *Opsanus tau*, mitochondria are usually confined to a thin circumferential rim and small central core of cytoplasm. The remainder of the fiber consists of radially arranged sheets of myofibrils separated by sarcoplasmic reticulum (FAWCETT and REVEL, 1961). ATPase activity distinct from that of mitochondria or myofibrils is located at sites occupied by the sarcoplasmic reticulum. This activity occurs at pH 7.2. It is sulfhydryl dependent, and is enhanced by the presence of magnesium ion. Response to ADP, AMP, or glycerophosphate suggests that the enzyme demonstrated is a true ATPase. Preliminary ultrastructural localization indicates an especially high ATPase activity in the sarcoplasmic reticulum at the H bands of the myofibrils.

Dept. of Biology, Brown University, Providence, Rhode Island (USA) and Dept. of Anatomy, Harvard Medical School, Boston, Massachusetts (USA).

<h2 style="text-align:center">5.</h2>

VICTOR DUBOWITZ: **Histochemical Studies of Developing Human and Animal Muscle*.**

Adult human and mammalian skeletal muscle contains two fibre types, one rich in phosphorylase and poor in some oxidative enzymes, the other rich in the oxidative enzymes and poor in phosphorylase. — In fullterm human infants the muscle has already differentiated into these fibre types at birth and resembles adult muscle. In premature infants, of 28 to 38 weeks gestation, differentiation of muscle was present but was less marked in the earlier cases. — Various skeletal muscles from 10 human foetuses, of 11 to 20 weeks gestation, showed an overall uniformity of enzyme activity. Some muscles showed patchy differentiation into 2 fibre types. The fibres were rounded in section and not arranged in compact bundles. — In the Guinea pig, rabbit and hamster differentiation of muscle fibres was present at birth, particularly in the Guinea pig where the polygonal fibres resembled mature adult muscle. Rabbit and hamster fibres were rounded and less well differentiated. Rat muscle showed uniform enzyme activity at birth and only differentiated into 2 types at 7 to 10 days. By 14 days it resembled mature muscle. The degree of enzyme maturation in these animals correlates partly with their respective gestation period and with their general maturity, activity and fur development at birth.

* Aided by grants from the Muscular Dystrophy Group of Great Britain and the Sheffield University Research Fund.

Dept. of Child Health, University of Sheffield, Sheffield, 10 (Great Britain).

6.

NICHOLAS CHRISTOFF, SUN K. SONG and **PAUL J. ANDERSON: Electrophoretic and Histochemical Evaluation of Skeletal Muscle Esterase Inhibitors*.**

The effects of inhibitors on the esterase activities of skeletal muscle from various muscle groups were studied on tissue sections and electropherograms of tissue homogenates. Simultaneous coupling azo dye, indoxyl, thiocholine and thiolacetic acid methods were utilized for staining. Electrophoresis was carried out on polyacrylamide gel. Inhibitors, including eserine, quaternary ammonium compounds (Mestinon, Ubretid, RO 2-1250, etc.) and organophosphorus reagents (E 600, DFP, etc.) were used both *in vivo* and by pretreatment of the sections or electropherograms prior to staining. — Multiple zones of cholinesterase and aliesterase activity were separated electrophoretically. These showed differing sensitivities to inhibitors, but their reactions were in general consistent with the inhibitor's specificity. The numerous bands of nonspecific esterase activity in the electropherograms could not be distinguished or separately identified in the histochemical preparations where most of the activity was visualized at the motor end-plates. The effectiveness of the inhibitors could generally be estimated by titration against the esterase activity at these sites. Some specific inhibitors, however, had little or no effect on the histochemical activity at the end-plate, e.g. Tensilon, 62-C-47 and iso-OMPA.

* Aided by a grant from the National Institute of Neurological Diseases and Blindness.

Mount Sinai Hospital, New York 29, N.Y. (USA).

7.

GAËTAN JASMIN: Staining Technics that Find their Application in the Diagnosis of Degenerative Muscle Diseases.

The conventional trichrome methods (hemalum-phloxine-saffron, ponceau-fuchsin-aniline blue and phosphotungstic hematoxylin) have been used as routine stains in our laboratory for diagnosis of various types of degenerative muscle diseases. Hyaline, floccular and lacunar necrosis of muscle fibres and proliferation of interstitial tissue can be well characterized. To better define the exact nature and the progression of degenerative changes in skeletal muscle, we have also been using the staining combination of luxol-PAS-hemalum. Using Rossman fluid as a fixative, it is possible to demonstrate that glycogen tends to accumulate in metabolically deficient fibres and to disappear in fibres that undergo atrophy. Luxol fast blue, on the other hand, more accurately discloses the degree of hyalinization and stains the A bands electively. PAS stains the Z discs and also shows hyaline granules, which are demonstrable in fibres that slowly degenerate. It should be pointed out that the two stains quite clearly delineate the myofibrils from the sarcoplasm. — In addition to this trichrome technic, several enzymes were investigated in frozen sections, to further characterize lesions of different types of myopathies. Phosphorylase and a number of dehydrogenases served as a sensitive index, taking into consideration their distribution between red and white fibres and also their location within each individual fibre. Naphthylamidase, which is normally absent in skeletal muscle, proved to be particularly useful for the demonstration of phagocytic activity that culminates at certain stages of even primary muscle diseases.

Dépt. d'Anatomie pathologique, Université de Montréal (Canada).

8.

N. W. RUNHAM: Histochemistry of the Odontophore Cartilages of Patella vulgata.

The structure of the odontophore cartilages of *Patella* closely resembles vertebrate cartilage. The chemistry of the molluscan cartilages has not however, been extensively investigated. GABE and PRENNANT (1955), demonstrated histochemically the presence of an acid mucopolysaccharide and glycogen in the cartilages of several molluscs. Biochemically LASH and WHITEHOUSE (1960), detected polyglucose sulphate, and BARRY and MUNDAY (1959), glycogen in molluscan cartilage. The detailed histochemistry of the cartilage is reported here. The cartilage consists of cartilage cells, extracellular capsules and intercapsular material. It was difficult to differentiate the cytoplasm from the inner layers of the capsule. Within the cytoplasm large masses of glycogen were found. The cytoplasm nearest the capsule gave a uniform strong PAS reaction, but the Hale technique revealed numerous positive granules. The other parts were strongly PAS positive, but negative for acid mucopolysaccharides. Protein reactions were found in all regions — the cytoplasm for carboxyl, tryptophan, tyrosine, and amino groups, the capsule very strong tyrosine and carboxyl reactions, and the intercapsular material very strong tyrosyl and amino with weak carboxyl and tryptophan reactions. The interpretation of these results will be discussed.

Dept. of Zoology, University College of North Wales, Bangor, Caerns., Wales (U.K.).

9.

F. M. HELMY and M. H. HACK: Histochemistry of Bidder's Organ*.

This communication is a direct extension of our histochemical studies on the melanin pigment cell system of various vertebrate livers (Acta histochemica, in press) into other situations where melanin pigment cells are a normal characteristic of the tissue. — Male specimens of *Bufo marinus* and *Bufo valiceps* were castrated under viadril anesthesia and allowed to recover in the laboratory for several months during which time they were adequately fed. Cryostat sections of fresh frozen tissue from normal and castrated specimens were examined by standard histochemical techniques for carbohydrate, plasmalogen, succinic dehydrogenase and nucleic acids. Since removal of the testes can result in the development of Bidder's organ into a functional ovary the ovaries of female specimens of both species of toad were also examined by these same techniques. The pertinent histochemical observations will be discussed and related to the melanin pigment present in both ovary and Bidder's organ.

* Aided by a grant from the National Heart Institute, U.S.P.H.S.

Dept. of Medicine, Tulane University School of Medicine, New Orleans, Louisiana (USA).

10.

V. D. MARZA, H. ARATEI et C. DIMITRIU, Jasi/La Roumanie: Historéaction du cholestérol au cours de l'evolution des follicules ovariens de quelques Mammifères.

11.

BOJAN RODE and MILAN ARKO: The Re-Evaluation of Acid Phosphatase Staining with Unsubstituted Naphthol-AS-Phosphoric Acid.

A number of substituted naphthol-AS-phosphates has been particularly recommended for acid phosphatase staining in tissue cells (BURSTONE 1958, 1961), however, these substrates revealed certain disadvantages, e.g. slow rate of enzymatic hydrolysis and variability in staining results (BARKA and ANDERSON 1962). An attempt has therefore been made to reexamine the value of unsubstituted Naphthol-AS-phosphate in the acid phosphatase staining technique. In our staining experiments in the proximal kidney convolutions of the rat and in a great number of various tissue cells, we have found that staining results with unsubstituted Naphthol-AS-phosphoric acid prepared in our laboratory, largely depended on the coupling reagent selection. Formol-calcium fixed frozen sections were incubated in the medium containing substrate diluted in the appropriate buffer-solutions (0,2 mg/ml) and the following coupling agents respectively: Fast Dark Blue R salt (I), Fast Red Violet LB salt (II) and hexazonium pararosanilin (III). "Lysosomes" and their derivatives were optimally stained after 5 to 10 minutes incubation time without, or with insignificant amounts of diffusion artefacts when I and III were used, but considerable diffusion staining always occured with II salt. The staining results can be employed for critical evaluation of the acid phosphatase localization and additional staining for alkaline phosphatase can be performed with the same substrate.

University of Zagreb, Faculty of Veterinary Medicine, Zagreb (Yugoslavia).

Enzyme · Enzymes · Enzymes

1.

R. WEGMANN: Corrélations entre les cycles métaboliques et les niveaux histoenzymologiques.

Par quelques exemples choisis au niveau de différents organes (muscles striés et lisses, cervelet, foie, peau) on démontre les corrélations remarquables existant sur le plan enzymologique entre les données d'ordre biochimique et les constatations d'ordre histochimique. La nécessité du support biochimique est donc un facteur important pour le progrès des recherches histochimiques, tant par les déductions directes que par celles indirectes qu'elles permettent. L'histochimie s'oriente ainsi imperceptiblement mais inéluctablement vers des recherches métaboliques qui, plus que toutes autres, sont rentables tant sur le plan expérimental que sur le plan descriptif pur. — Cette concordance est une preuve de la fiabilité d'une grande partie des techniques actuelles. On observe cependant des cas ou ce parallélisme entre les deux matières n'est pas respecté et on peut s'interroger sur les causes des divergences. — L'histochimie devient plus cohérente mais de grands progrès restent a faire.

Correlations between Metabolic Pathways and Histoenzymological Levels. With some well choosen examples on the level of different organs (striated and smooth muscles, hind brain, liver and skin) the noteworthy correlations between biochemistry and histochemical findings in the field of enzymology are demonstrated. The necessity of a biochemical support is an important element on progress of histochemical research, as by the direct deductions as by the indirect ones they permit. The histochemistry is oriented imperceptibly and ineluctably in the sense of metabolic researches which reveal, better than anyone, as well in experimental as on descriptive level the true functionning of the cells. — This accordance is a proof of the fiability of a great part of the actual techniques. There exist, however, cases where the parallelism between both chemistry and histochemistry is not respected and we may ask us on the reasons of the discrepance. — The histochemistry is more and more coherent but many progresses have to be done.

Institut d'Histochimie Médicale, Faculté de Médecine, Paris (France).

2.

WILLIAM MEIER-RUGE: Die histotopochemischen Eigenschaften der Tetrazolsalze MTT, NBT und Tetra-NBT in der Retina.

An der Retina wurden die heute meist benutzten Tetrazolsalze wie MTT, NBT und Tetra-NBT an Hand von Fermentreaktionen (SDH, G-6-PDH etc.) bezüglich der histotopochemischen Qualitäten ihrer Formazane untersucht. Hierbei zeigte sich, daß MTT-Co-Formazan die schlechtesten lokalisatorischen Eigenschaften in der Retina besitzt. Selbst bei schonendster Gewebsbehandlung ist eine Beziehung weder zu den Mitochondrien der proximalen Zapfen- und Stäbchensegmente, noch zu letzteren selbst zu beobachten. Demgegenüber besitzt Tetra-NBT die besten histotopochemischen Qualitäten. Der Formazanniederschlag ist überaus feinkörnig und exakt auf Zapfen und Stäbchenkörper der Retina zu lokalisieren. Nitro-BT-Formazan hat im Vergleich dazu gering schlechtere Eigenschaften. Tetra-NBT-Formazan besitzt eine noch geringere Lipoidlöslichkeit als Nitro-BT-Formazan, wodurch keine Mitfärbung lipoidhaltiger Strukturen erfolgt. Damit hat Tetra-NBT besonders günstige Voraussetzungen für Fermentreaktionen in besonders fettreichen Geweben wie Nebennierenrinde oder braunem Fettgewebe. Es ist möglich, den Gewebsschnitt zu entwässern und in Balsam einzudecken, ohne daß TNBT-Formazan durch Lösung verloren geht oder die Gewebsbindung augenscheinlich verändert wird. TNBT gestattet eine optimale Überwindung der Schwierigkeit von Fermentaktivitätsbewertungen durch auftretende Monoformazanreduktionsstufen, die bei NBT vorkommen. TNBT-Formazan ist vor allem für eine cytophotometrische Auswertung von Tetrazolsalzreaktionen hervorragend geeignet. Die praktisch fehlende Lipoidlöslichkeit des TNBT-Formazans wirft erneut die Frage der Präzipitation bzw. Bindung von Formazan an lipoidhaltige Strukturen auf.

Histotopochemical Properties of the Tetrazolium Salts MTT, NBT and Tetra NBT in the Retina. By means of histochemical enzyme reactions (SDH, G-6-PDH etc.) the most frequently used tetrazolium salts — MTT, NBT and tetra-NBT — have been tested on the retina for the histotopochemical qualities of their formazans. It is found that MTT-Co-formazan has the least effective localizing properties in the retina. Even after the most careful treatment of the tissue it is not possible to observe either a relationship to the mitochondria of the proximal rod- and cone segments or to the latter themselves. Tetra-NBT, however, appears to have the best histotopochemical properties. The formazan precipitate is extremely delicate and can be demonstrated exactly on the rods and cones of the retina. In comparison with Tetra-NBT, Nitro-BT-formazan has slightly inferior qualities. As tetra-NBT-formazan has an even lower lipid solubility than Nitro-BT-formazan, lipid-containing structures remain unstained. For that reason tetra-NBT is very convenient for enzyme reactions in tissues which are particularly rich in fat, as for example adrenal cortex or brown adipose tissue. It is possible to dehydrate the tissue section and to mount it (balsam) without TNBT-formazan going back into solution or an apparent alteration in the tissue linkage taking place. TNBT overcomes the difficulties in accessing enzyme activities, which arise in the case of NBT through the formation of monoformazan reduction products. TNBT-formazan is especially suited for the cytophotometrical interpretation of tetrazolium salt reactions. The almost negligible lipid solubility of TNBT-formazan again raises the question of a precipitation and/or binding of formazan to lipid-containing structures.

Pathologisches Institut der Universität Basel und Laboratorium für experimentelle Pathologie der Sandoz A.G., Basel (Schweiz).

3.

P. LORBACHER und R. FISCHER: Über den sog. „Nothing Dehydrogenase"-Effekt an Gewebsschnitten und Blutausstrichen*.

Bei der histochemischen Darstellung Pyridinnukleotid-gebundener Dehydrogenasen mittels Tetrazoliummethoden ergeben sich eine Reihe von Problemen, zu denen auch der sog.

„Nothing Dehydrogenase“-Effekt gehört. ZIMMERMANN und PEARSE haben diese im alkalischen Inkubationsgemisch (pH 8—9) ohne spezifischen Substratzusatz auftretende Reaktion auf eine nicht-enzymatische Reduktion von DPN bzw. TPN durch proteingebundene SH-Gruppen im Gewebe zurückgeführt. — Im Rahmen eigener methodischer Untersuchungen zum Nachweis oxydativer Enzyme in Blutzellen konnten auch im Ausstrichpräparat unter Verwendung eines Inkubationsgemisches mit DPN und Nitro-BT, aber ohne Zusatz eines spezifischen Substrates, bei einem Reaktionsmaximum zwischen pH 8—9 im Cytoplasma der weißen Blutzellen feine Formazangranula mit charakteristischem Verteilungsmuster in den verschiedenen Blutzellreihen festgestellt werden. Durch Zusatz von SH-Inhibitoren zur Inkubationslösung wurde die Reaktion sowohl in Blutausstrichen als auch in Kryostatschnitten von Haut und Leber der Ratte gehemmt. Nach Hitzeeinwirkung, Spülung in Pufferlösung oder Fixierung fiel die Reaktion negativ aus. Die gleichen Vorbehandlungen blieben zum Teil ohne Einfluß auf die Nachweisbarkeit der DPNH-Diaphorase und der proteingebundenen SH-Gruppen. — Angesichts der großen Empfindlichkeit der „Nothing Dehydrogenase“-Reaktion gegenüber jeglicher Vorbehandlung sowie einer Lokalisation, die einerseits weitgehend der Verteilung der verschiedenen oxydativen Enzyme entspricht, die andererseits aber vom Verteilungsmuster der proteingebundenen SH-Gruppen, vor allem in den Blutzellen, häufig abweicht, wird das Vorliegen einer Pyridinnukleotid-abhängigen endogenen Dehydrogenaseaktivität mit einem Reaktionsoptimum zwischen pH 8 und 9 diskutiert.

The Effect of the So-Called "Nothing-Dehydrogenase" on Tissue Sections and Blood Smears. The histochemical determination of pyridine nucleotide-bound dehydrogenases by means of tetrazolium methods raised a number of problems which included the so-called "nothing-dehydrogenase" effect. ZIMMERMANN and PEARSE attributed this reaction, which takes place in an alkaline incubation mixture (pH 8—9) without the presence of a specific substrate, to a non-enzymatic reduction of DPN or TPN respectively by protein-bound SH-groups in tissues. During our own investigations concerning the demonstration of oxidizing enzymes in blood cells, we obtained the following results: If an incubation mixture (pH optimum between pH 8 and 9), containing DPN and Nitro-BT but lacking a specific substrate, was used on blood smears the cytoplasm of the white blood cells showed fine formazan granules which were arranged in the characteristic pattern. The addition of SH-inhibitors to the incubation mixture resulted in a suppression of the reaction not only in blood smears but also in cryostat sections of rat skin and liver. After heat treatment, washing in buffer solutions, or fixation the reaction became negative. To some extent the same preliminary treatment did not influence the determination of DPNH-diaphorase and protein bound SH-groups. — Considering the extreme sensitivity of the "nothing-dehydrogenase" reaction to any kind of pretreatment, and at the same time considering the localization (which on the one hand corresponded largely with the distribution of the various oxidizing enzymes but on the other hand deviated — especially in blood cells — from the distribution pattern of the protein-bound SH-groups) the existence of a pyridine nucleotide dependent endogenous dehydrogenase activity (reaction optimum pH 8—9) has to be discussed.

* Der Deutschen Forschungsgemeinschaft danken wir für die Unterstützung bei diesen Untersuchungen.
Pathologisches Institut der Universität, 53 Bonn (Deutschland).

4.

ALBERT W. SEDAR and RONALD M. BURDE: Localization of the Succinic Dehydrogenase System in Bacteria Using Combined Techniques of Cytochemistry and Electron Microscopy*.
The activity of the succinic dehydrogenase system was studied in *Escherichia coli* and *Bacillus subtilis* utilizing combined techniques of cytochemistry and electron microscopy. Organisms were incubated in a medium containing sodium succinate and tetranitro-blue tetrazolium (TNBT) which served as an electron acceptor. Enzymatic activity, as evidenced by deposition of aggregates of TNBT-formazan, was found associated with the site of the plasma membrane in *E. coli* and the mesosome in *B. subtilis*.

* Supported in part by a research grant (GM-04810-08) from the United States Public Health Service.
Jefferson Medical College, Daniel Baugh Institute of Anatomy, Philadelphia, Pennsylvania (USA).

5.

FRANK WOHLRAB: Histochemischer Nachweis der Betainaldehyd-Dehydrogenase in Rattenorganen.
Die Oxydation des Cholins zu Betain vollzieht sich in zwei Stufen. Während die Cholin-Dehydrogenase die erste Teilreaktion Cholin → Betainaldehyd katalysiert, wird der Betainaldehyd in einem zweiten Schritt durch die Betainaldehyd-Dehydrogenase, für die es bisher keinen histochemischen Nachweis gab, zu Betain oxydiert. — Das Inkubationsmedium

zum histochemischen Nachweis der Betainaldehyd-Dehydrogenase setzt sich aus folgenden Komponenten zusammen: Betainaldehyd (0,02 M); DPN (0,001 M); NaCN (0,01 M); Nitro-BT; Phosphatpuffer, pH 7,2. — Kontrollen mit Substrat in Abwesenheit von DPN und umgekehrt lassen erkennen, daß die Reduktion von Nitro-BT enzymatischer Natur ist. Der Reaktionsausfall ist anaerob stärker als aerob. Vorinkubation der Kryostatschnitte in p-Chlormercuribenzoat (0,001 M) und Atebrin (0,001 M) führt zur vollständigen Hemmung der Betainaldehyd-Dehydrogenase. — In Übereinstimmung mit den Ergebnissen biochemischer Untersuchungen findet sich in der Ratte die stärkste Betainaldehyd-Dehydrogenase-Aktivität in der Leber und in der Niere. Schwache Betainaldehyd-Dehydrogenase-Aktivität zeigt des weiteren die Schleimhaut von Magen und Duodenum sowie die Herz- und Skelettmuskulatur. Die topochemische Lokalisation in Leber und Niere stimmt mit der Lokalisation der Cholin-Dehydrogenase in diesen Organen überein. Auf Diskrepanzen hinsichtlich der intrazellulären Lokalisation der Betainaldehyd-Dehydrogenase wird im Vortrag eingegangen.

The Histochemical Demonstration of Betaine Aldehyde Dehydrogenase in the Organs of Rats. The oxidation of choline to betaine takes place in 2 stages: the first stage (choline → betaine aldehyde) is catalysed by choline-dehydrogenase; in the second stage of the reaction betaine aldehyde is oxidized to betaine by betaine aldehyde dehydrogenase. Up to now no histochemical method for the determination of betaine aldehyde dehydrogenase has been available. The incubation medium for the histochemical determination of betaine aldehyde dehydrogenase is made up from the following components: betaine aldehyde (0.02 M); DPN (0.001 M); NaCN (0.01 M); Nitro-BT; phosphate buffer, pH 7.2. Control media containing the substrate but not DPN (and vice versa) show that the reduction of Nitro-BT is due to an enzymatic reaction. The reaction is stronger under anaerobic than under aerobic conditions. If cryostat sections are incubated in p-chloromercuribenzoic acid (0.001 M) and atebrin (0.001 M) the activity of betaine aldehyde dehydrogenase is completely abolished. In agreement with biochemical investigations the strongest betaine aldehyde dehydrogenase activity is found in the liver and the kidney of rats. A weak betaine aldehyde dehydrogenase activity can be demonstrated in gastric mucosa, duodenum and cardiac and skeletal muscle. The topochemical localization in the liver and the kidney is in agreement with the localization of choline dehydrogenase in these two organs. Discrepancies in regard to the intracellular localization of betaine aldehyde dehydrogenase will be explained in the course of the lecture.

Karl-Marx-Universität Leipzig, Pathologisches Institut, Liebigstr. 26, Leipzig C 1 (Deutschland).

6.

J. V. DIENGDOH: Amine Oxidase and its in vivo Activation.

The localization of this enzyme, its response to various substrates and its activity at different pH values, have been examined in the skin of mice. Adrenaline has been found to be the most suitable substrate. In normal skin the enzyme was localised in discrete granules with little diffuse colour. However, injection of histamine produced a strikingly enhanced activity which was both granular and diffused. Such increased activity was observed even 15 min after the injection, and hence was unlikely to be due to synthesis of new molecules of amine oxidase. A similar response was found after croton oil was painted onto the skin. Thus it seems likely that the histochemical test is capable of demonstrating the extent to which this enzyme is being used in the tissues at the moment of biopsy.

Royal College of Surgeons of England, London W.C. 2 (Great Britain).

7.

R. DAOUST: Improved Resolution in the Localization of Ribonuclease Activity by the Substrate Film Method*.

Substrate film methods for the histochemical localization of enzyme activities consist essentially in placing a film of substrate in contact with tissue sections and, after exposure, staining the unaffected substrate remaining in the film. Comparison of the unstained areas in the film with corresponding tissue sections reveals the sites of enzyme activity in the tissue. — Original film methods and modifications proposed by different authors have been recently reviewed and discussed (R. DAOUST, Intern. Rev. Cytol., Acad. Press, New York, in press). Different procedures were assayed in this laboratory and improved results were obtained in the localization of RNAse activity by modifying various steps of the original procedure.

* Supported by The National Cancer Institute of Canada.

Laboratoires de Recherche, Institut du Cancer de Montréal, Hôpital Notre-Dame et Université de Montréal, Montréal (Canada).

8.

ST. LAGERSTEDT, Lund/Sweden: **Histochemical Studies on Ribonuclease in the Pancreas of Rat.**

9.

JOHAN AHLQVIST: **Tissue Lipids and Enzyme Histochemistry.**

Some observations indicate that tissue lipids might have an effect on the quantity of final reaction product formed or deposited in tissues in histo-enzymatic reactions. Results of studies on the mechanism of this phenomenon are presented. — The affinity of some free naphthols for tissues is abolished by preceding extraction of lipids from the sections. Such free naphthols and all hitherto studied substrates prefer lipid to water in mixtures of corn oil and water. By an UV absorption technique it is shown that α-naphthyl acetate distributes unevenly in tissues and that it has an affinity for structures containing lipids. — The influence of the findings on the interpretation of histochemical enzyme reactions is discussed.

Dept. of Pathology, University of Helsinki, Snellmansg. 10, Helsinki (Finland).

10.

MANFRED E. BAYER: **Electron Microscopy of Urinary Mucoprotein and Neuraminidase (RDE) *.**

An active mucoprotein has been prepared from human urine which exhibits the sites responsible for specific reactions between influenza viruses and red cells or host cells. The fibrous molecules of the mucoprotein have been seen to come in close contact with added virus particles and are subject to the action of the receptor destroying enzyme (RDE) of the virus. We report here the action of RDE preparations from *Vibrio cholerae*. By keeping the enzyme-receptor system at 1^0 C, elution of the enzyme is slowed and in negatively stained preparations the enzyme molecules can be seen attached to the mucoprotein molecule. The enzyme is of round or oval shape with a diameter of about 60 Å. Assuming a density of 1.3, one can estimate the molecular weight to be about 100,000. This is in agreement with the results of ultracentrifugal studies of PYE and CURTAIN. Moreover, the electron micrographs of the enzyme molecule show that it probably consists of four subunits.

* Aided by grant GB 982 from the National Science Foundation.

Dept. of Molecular Biology, The Institute for Cancer Research, Philadelphia 11, Pa. (USA).

Nervensystem · Nervous System · Système nerveux

1.

NIKOLAI N. DOEMIN: **Some Cytochemical Characteristics of Nervous Tissue.**

Microspectrophotometric study of neurons of various cellular layers of cat motor, visual and auditory cortex has shown (PEVZNER) that the nerve cells of inner layers contain more cytoplasmic RNA than upper ones. Higher concentration of RNA and DNA was revealed (BARANOV) in successive slices of the brain cortex by microchemical methods too. Acid ribonuclease activity in various cortical cellular layers has been determined (NECHAYEVA) to correlate positively with the corresponding values of cytoplasmic RNA content. Circulatory hypoxia of the brain decreased RNA content in the neurons of the very inner and the very upper layers of visual and motor cortex areas, this decrease being markedly less in auditory area (PEVZNER). The administration to animals both of stimulators (camphor, caffeine) and of narcotic (hexenal = evipan) increased the total content of nucleic acids also in the very inner and in the very upper layers of the cortex (BARANOV). A prolonged (3—4 hrs) electric stimulation of the cervical part of n. sympathici increased cytoplasmic RNA concentration in the neurons of the superior cervical sympathetic ganglion (PEVZNER) without influencing the total content of RNA, DNA and free nucleotides in the whole ganglion homogenates (BARANOV). Under these conditions the protein content in the cytoplasm of the single ganglionic neurons has been determined microinterferometrically to increase (PEVZNER, KOVAL and KOOCHIN).

Laboratory of Functional Biochemistry of the Nervous System, Pavlov Institute of Physiology, Academy of Sciences of the USSR, Nab. Makarova 6, Leningrad (USSR).

2.

K. D. ERISTAVI, L. K. SHARASHIDSE and R. V. BULUSASHVILI: **Cytochemical Evidence of Affection in Central Nervous System during Different Functional and Pathological Conditions of Organism.**

Histochemical changes of neurons in different parts of the central nervous system (cerebral cortex, subcortex, cerebellum, oblongata, spinal cord, spinal ganglion) in different animals (mouse, Guinea pig, rabbit, dog) have been studied during different functional and pathological conditions of the organism (narcosis, hibernation, hypothermia, anemia, extracorporal circulation, isolated perfusion, chronic infection, induced tumors). Histochemical peculiarity of neurons (alteration of IEP of RNP of mitochondria and tigroid, exposure of aminoacid and functional groups of proteins and establishment of activity of oxidative enzymes) were studied in the fresh frozen sections obtained in cryostat at -20°. — Earlier and minute histochemical changes included the dislocation of IEP of RNP of mitochondria and changes of oxidative enzymes. It is shown that the dislocation of IEP of RNP of the mitochondria to the acid side at $pH = 0.2—0.4$, corresponds to 160—250% of changes, accompanied by intensification of activity of oxidative enzymes, and it must be considered as a sign of increase of metabolic reaction of neurons; decreasing of IEP at $pH = 0.5—1.0$, corresponds to 300—1000% of changes accompanied by lowering of activity of oxidative enzymes and it must be considered as a sign of reversible depolymerisation of the ribonucleoprotein complex; more severe lowering IEP of RNP leads to the development of irreversible changes and finally to the necrosis of the neuron. — Out of these changes, which are not specific by themselves, for its peculiar combination (there is a certain relation between the depth of affection of neurons, extension of affection within the formation and position of studying formation within the central nervous system) are formed rather characteristic pictures for different functional and pathological conditions of organism.

Institute of Experimental and Clinical Surgery and Haematology, Academy of Medical Sciences of the USSR, Kamo 43, Tbilisi (USSR).

3.

VÁCLAV BARTONIČEK und ZDENĚK LOJDA: Enzyme im Plexus chorioideus und Ependym der weißen Ratte.

Wir untersuchten die Lokalisation der hydrolytischen Enzyme, Diaphorasen und Dehydrogenasen im Plexus chorioideus (P.) und Ependym (E.) aller Gehirnkammern der weißen Ratte. Der histochemische Nachweis wurde in fixierten sowie in nicht fixierten Schnitten mit allen Methoden durchgeführt, wodurch die gegenseitige Kontrolle ermöglicht wurde. Im Epithel des P. wurden hohe Aktivitäten von saurer Phosphatase, nicht spezifischer Esterase, beiden Diaphorasen und allen untersuchten Dehydrogenasen gefunden. Alkalische Phosphatase konnte nur in Endothelien von Kapillaren gezeigt werden. Die basalen Membranen des P. enthalten die adeninnukleotidespaltenden Enzyme. In den untersuchten Strukturen konnte keine Aktivität der Aminopeptidase gezeigt werden. Im E. außerhalb des P. sind alle Aktivitäten beträchtlich niedriger. In den entsprechenden Strukturen einzelner Gehirnkammern konnten keine Unterschiede in der Lokalisation und Aktivitäten der untersuchten Enzyme gezeigt werden. Interessant war die Schwankung von Aktivitäten im Bereich desselben P. und E., was als Ausdruck des aktuellen Funktionszustandes gedeutet wird und für eine gewisse Periodizität zeugt. Die Resultate werden vom Gesichtspunkte der aktiven Beteiligung des P. an der Bildung der zerebrospinalen Flüssigkeit kurz diskutiert.

Enzymes in the Chorioid Plexus and Ependyma of White Rats. We investigated the localization of hydrolytic enzymes, diaphorases and dehydrogenases in the chorioid plexus (P.) and the Ependyma (E.) in all cerebral ventricles of white rats. Employing standard methods histochemical tests were carried out on both fixed and unfixed sections in order to provide sufficient controls. The epithelium showed high activities of acid phosphatase, non-specific esterase, both diaphorases and several dehydrogenases. Alkaline phosphatase could be shown only in the endothelial cells of the capillaries. The basal membranes of the P. contained adenine nucleotide splitting enzymes. There was no aminopeptidase activity demonstrable in the tissues in question. In the P. outside the E. the enzyme activities were considerably lower. Corresponding regions of the various cerebral ventricles gave the same results in regard to localization and activity of the enzymes. Fluctuations of the activities in the area of the same P. and E. were of special interest; they were interpreted as an expression of the actual state of function, which proves a certain periodicity. The results are briefly discussed from the aspect of an active participation of the P. in the formation of the cerebrospinal fluid.

Psychiatrisches Forschungsinstitut, Prag, Angiologisches Laboratorium und I. Institut für Pathologie der Medizinischen Fakultät der Karls-Universität, Prag (ČSSR).

4.

WESLEY J. BIRGE: Histochemical Study of the Avian Choroid Plexus.

Comparative histochemical studies have been made on the epithelial cells of the choroid plexus of embryonic and adult stages of the domestic fowl (White Leghorn strain). During embryonic development a heavy condensation of cytoplasmic ribonucleic acid (RNA) develops in the choroidal epithelial cells, the bulk of it being rather sharply restricted to the basal (subnuclear) portion of the cell. This basal localization of cytoplasmic RNA is discernible by the 6th day of development, and it becomes dense by 10 to 12 days. This heavy condensation of RNA is sustained through the hatching period. During embryonic development a PAS-positive substance which is resistant to malt diastase accumulates in the outer (supranuclear) cytoplasm of the choroidal epithelial cells. This substance is first detectable on the 6th day, and a heavy condensation occurs by 12 to 14 days. Studies of adult choroid plexus tissue give generally similar though somewhat more variable results, except there is substantially less RNA present in the epithelial cells, and the PAS-positive substance is restricted to a much narrower (but often dense) band in the outer margin of the cytoplasm.

University of Minnesota, Division of Science and Mathematics, Morris, Minnesota (USA).

5.

J. DRUKKER and J. P. SCHADÉ: The Optic Ganglion of Cephalopods after Stagnant Anoxia.

In the optic ganglion two types of constituents are recognizable, islets of cells which are interspersed between bundles of unmyelinated fibres. — The perikarya in the islets contain relatively low amounts of several dehydrogenases (i.e. $NADH_2$- and $NADPH_2$-tetrazolium reductase, lactate, isocitrate, succinate, glucose-6-phosphate, glutamate, β-hydroxybutyrate, and α-glycerophosphate dehydrogenase). Formazan containing particles are found mainly in the periphery of the cells. The fibres have a much higher content of these enzymes. Astroglia-like cells found between the fibres with glial stain do not show up in histochemical preparations. Alkaline and acid phosphatase are found in the perikarya whereas the AMP-ase and ATP-ases occur in fibres as well as in islets. After a short period of stagnant anoxia ($^1/_2$ h) the region within each perikaryon in which dehydrogenase activity shows up increases, after longer periods (up to 3 hrs) the activity falls off completely, especially in the hilar region. After $^1/_2$ h of anoxia the fibres lose their dehydrogenase activity, however, the glia cells stand out distinctly after application of any of the dehydrogenase techniques. The content of acid phosphatase of perikarya and capillary walls and the content of Ca^{++}-activated ATP-ase of the glial fibres increases highly. — It is concluded that we have been studying a process of morphotrophic necrobiosis notwithstanding the fact that with normal histological procedures no alterations were demonstrable.

Netherlands Central Institute for Brain Research, Mauritskade 59b, Amsterdam (The Netherlands).

6.

V. V. PORTUGALOV, L. M. GERSHTEIN and M. S. GAEVSKAYA: Changes in the Nerve Cell Proteins in Dogs during Reanimation from the State of Clinical Death.

The content and distribution of proteins indicated by histochemical reactions for NH_2-, SH-, S—S-, COOH-groups, by the Danielli tetrazonium reaction and the reaction with Aamido black in the motor cortex of dogs during reanimation from the state of clinical death and in the first 24 hours after reanimation was studied. Histochemical distribution and the amount of proteins in the cortex of dogs in the state of clinical death do not differ from those in control animals. Changes are observed 30 minutes after the beginning of reanimation and are manifested in the increased intensity of histochemical reactions for proteins. Increased reactions for proteins and their reactive groupings do not develop in all layers of the cortex and in all cells simultaneously. The earliest changes are found in small and very small neurons of the upper layers of the cortex (layers II and III). Later (from 3 up to 24 hours) changes are found in neurons of layers V, VI, and VII. The intensity and extent of changes progress with the time elapsing from the moment of reanimation. The nature of protein changes (in connection with data of biochemical investigations) and their significance in the chain of events accompanying reanimation of the animals from the state of clinical death are discussed.

Institute of Brain, Acad. Med. Scien. of USSR. per Obucha 5, Moscow B-120 (USSR).

7.

D. A. CHETVERICOV and I. N. ULIBINA: The Influence of Hypoxia on the Histochemical Distribution of Lipids in Nerve Cells of Rats.

The content and the distribution of lipids in cytoplasm of neurons of the central nervous system and spinal ganglia of normal rats and of those having undergone acute hypoxia were studied by histochemical methods. The material for study was chosen immediately after 2 hours stay of rats in barochamber at the pressure 160—220 mm Hg and after 13 and

7 days. The lipids were demonstrated by BAKER's, ELFTMAN's and STÜLER's methods. — The lipid granules of 1—2 micron in size distributed evenly in cytoplasm were revealed in the majority of neurons in the ganglia of the cervical cord of the control rats. The positive reaction given by granules to the tests being used, indicates that the granules contain a rather great quantity of phospholipids. In the anterior horns of the spinal cord, medulla oblongata and nuclei of cerebellum only a few neurons contain these granules. They were not revealed in nervous cells of cerebral cortex and Purkinje cells of cerebellum. — The rats having undergone acute hypoxia, have an increased number of lipid granules in neurons of spinal ganglia; these granules can be seen more easily and in a greater quantity of neurons. These changes are seen best of all 24 hours after hypoxia; in 3 or 7 days the quantity and appearance of the granules is like that of the controls. The changes found in lipid structures of the neurons of the spinal ganglia are analogous to those which are seen after demasking of the intracellular lipids with phenol. — It is possible that acute hypoxia also causes the partial release of the lipids from lipoprotein complexes and thus reveals the lipid granules. In the neurons of the parts of the central nervous system examined in the posthypoxial period the changes in the number of lipid granules demonstrated were not seen. — The histochemical data show that hypoxia causes reversible changes of the properties of the lipid-containing structures, rich in phosphatides, in the neurons of the spinal ganglia. The presence of changes only in the neurons of the spinal ganglia suggests that these changes are not a direct consequence of hypoxia from which neurons suffer as well, and reflect the participation of the sensitive neurons of the spinal ganglia in the general reaction of the organism to hypoxia.

Pavlov's Institute of Physiology, the Academy of Sciences of the USSR, Nab. Makarova 6, Leningrad (USSR).

8.

C. SOTELO: Histochemical Studies on Carbohydrate Metabolism of the Neuroglia. Correlations between Neuroglial and Neuronal Metabolism.

The following enzymes of the neurons and their satellite cells in the cerebellum and in the spinal cord of rat and rabbit were studied: Phosphorylase, active and total, nicotinamide adenine dinucleotide phosphate tetrazolium reductase (NADPH reductase), glucose-6-phosphate- and 6-phosphogluconate dehydrogenases, nicotinamide adenine dinucleotide tetrazolium reductase (NADH reductase), lactic-, malic-, and succinic-dehydrogenases. These reactions give a general idea on the metabolic pathways and derivatives of the glycolysis. — The "neuronal atmosphere" of DE CASTRO is formed, i) for Purkinje cells by the perikarya of the Bergmann cells, and ii) for the neurons of the spinal cord by the prolongations of protoplasmic astrocytes and by the perikarya of the satellite oligodendrocytes. No important metabolic difference, as revealed by histochemical techniques, was found between the different types of cells which surround the neurons. — The hypothesis advanced by CAJAL on the existence of neurono-neuroglial symbiosis may have a biochemical basis. The neurons have always shown a strong activity of the enzymes of Krebs' cycle (exergonic way; liberation of energy) and a weak activity of the enzymes of pentose cycle (endergonic way; reduction of NADP). The satellite neuroglia, on the contrary, shows an opposite enzymatic pattern; feeble for the enzymes of Krebs' cycle and relatively intense for the enzymes of the pentose cycle. — The neuroglia may give NADPH to the neurons, coenzyme necessary for the principal reactions of cellular biosynthesis; while the neuron metabolism is directed toward the liberation of energy, necessary for the nervous function. The satellite neuroglia can be considered as an auxiliary metabolic part of the neurono-neuroglial axis.

Institut d'Histochimie Médicale (Directeurs: Prof. Verne et Dr. Wegmann) 45, rue des Saints-Pères, Paris VIe (France).

9.

BERTALAN CSILLIK and SOLOMON D. ERULKAR: Labile Stores of Catecholamines in the Central Nervous System*.

The CNS of rats and cats has been studied by means of the Falck fluorescence technique for tissue catecholamines. It has been shown that not only frozen-dried material but also cryostat sections are suitable for this purpose, provided the samples are mounted in isotonic saline. The microscopic patterns in the vegetative nervous system and in the hypothalamus, obtained by this simplified procedure, were identical in all details with those described by FALCK. It was found, furthermore, that there exists a large amount of fluorescent fibers in the hippocampus, geniculate body, sensory cortex and spinal cord, that contain the fluorescent substance in a highly labile form, demonstrable only in samples taken from living animals. 2—5 minutes after death, these fibers lose their fluorescent content. Depletion experiments, carried out with reserpine and alpha-methyl-*meta*-tyrosine prove that the fluorescent material is either norepinephrine or a compound structurally and pharmacologically closely related to norepinephrine. Microstructural relations and electrophysiological data suggest that labile catecholamines are located in dendrites, supporting an aspecific background level for CNS activity.

* Supported by a research grant (NB 00282-11) from the National Institute of Neurological Diseases and Blindness, NIH., US Public Health Service.

Dept. of Anatomy, University Medical School, Szeged (Hungary), and Dept. of Pharmacology, University of Pennsylvania, Philadelphia 4, Pa. (USA).

10.

Z. LODIN, J. FOLBERGROVÁ, V. NOVÁKOVÁ and J. PILNÝ: The Effect of Methionine-Sulphoximine on Protein and Nucleic Acid Metabolism of Brain in vivo.

The effect of methionine-sulphoximine (MSI), a substance the administration of which to experimental animals evokes epileptic seizures, has been studied on some aspects of protein and nucleic acid metabolism of the brain. — The incorporation of leucine-H³ into protein of some brain regions *in vivo* has been studied after the administration of MSI to experimental mice. Analyses have been performed during the periods corresponding to different functional states of the animals (so called preparoxysmal, paroxysmal and postparoxysmal period) and also after the simultaneous administration of MSI and methionine. — Protein metabolism of the brain has also been studied as far as the structural changes are concerned, namely by histochemical determinations of some functional groups of proteins. — Using histochemical methods the metabolism of nucleic acids has also been studied. — The mechanism of neurotoxic action of MSI is being discussed.

Laboratory of Cytophysiology and Cytochemistry of the Nervous Cell, Institute of Physiology, Czechoslovak Academy of Sciences, Prague 6 (ČSSR).

11.

J. J. GHOSH, R. K. DATTA, (Miss) SWATI GHOSH, (Miss) KRISHNA SIKDAR, (Miss) SUCHANDRA ROY and (Miss) DIPALI BHATTACHARYYA: Action of Phenothiazine Group of Tranquilizers on Cytoplasmic Ribonucleoprotein Constituents of Nerve Cell.

While studying the action of different neuropharmacologic drugs at the level of the structural constituents of nerve cell bodies, it has been found that treatment of rats with different doses of chlorpromazine and trifluoperazine produces nuclear swelling and distinct chromatolytic changes in the Nissl granules of nerve cell bodies in different areas of rat brain. Among the different areas examined, thalamus-hypothalamus regions showed earliest onset of chromatolytic changes with subsequent regeneration. Next in the order of chromatolytic changes were the nerve cell bodies in the brain stem regions which were followed by the motoneurons of spinal cord. — Further studies made to interpret chromatolytic changes in terms of the activity of different degrading enzymes viz., ribonuclease, 5′-nucleotidase and phosphatases, which are present in brain ribonucleoproteins, indicate that the activity of most of these enzymes are involved in the chromatolytic changes in the Nissl granules. Results of the stability study of cytoplasmic ribonucleoproteins from brain tissue (R. K. DATTA and J. J. GHOSH, J. Neurochem. *10*, 485, 1963) suggest that some non-enzymic factors such as changes of pH, ionic concentration, weakening of intramolecular hydrogen bonds etc. may also play important role in the decreased stability of Nissl granules. Further elucidation of the mechanism by which both enzymic and non-enzymic factors are involved in drug-induced chromatolysis may bring to light the fundamental mode of action of these neuropharmacologic drugs at the level of nerve cell constituents.

Dept. of Biochemistry, Calcutta University, Calcutta-9 (India).

Nucleinsäuren · Nucleic Acids · Acides nucléiques

1.

JAURÈS JORDANOV: Critical Remarks on the Feulgen Reaction Hydrolysis.

The effect of the acid hydrolysis on nuclear chromatin in histological sections of mouse spleen and liver and in smears of rat thymus was investigated by means of Feulgen reaction (FR) and some staining methods. The data of a number of authors for the loss of Feulgen-positive material (apurinic acid) during a prolonged hydrolysis with n HCl at 60⁰ C were confirmed. It was established that the high temperature of the hydrolyzing solution and not the acid itself was responsible for the removal of the apurinic acid from sections. In this respect warm water at 60⁰ C exhibits the same effect on apurinic acid as n HCl at 60⁰ C. The standard 8 to 12 minutes' hydrolysis with n HCl at 60⁰ C causes a transition from only a part of DNA to apurinic acid capable to give FR. This fact besides the possibility of a partial removal of apurinic acid from the sections for the same time makes the standard hydrolysis inconvenient for quantitative cytophotometric studies. Hydrolysis with 5nHCl at room temperature carried out on sections fixed in formol-containing fluids (SERRA, 10 per cent neutral formol, formol-acetic acid) achieves for about 35 minutes a complete converting of DNA into apurinic acid and fails to remove the latter for hours from the sections. Therefore such a hydrolysis procedure seems to be much more convenient for cytophotometric application of FR.

Inst. of Morphology at the Bulgarian Academy of Sciences, Sofia (Bulgaria).

2.

LUIGI ZANOTTI and GIOVANNI PRENNA: Observations on Absorption and Fluorescence Cytophotometry by the Feulgen Reaction Using Various Schiff-Type Reagents.

Cytophotometric measurements are valid, aside from other errors, only if taken at extinctions which are neither too high nor too low; it is known, for example, that by use of the conventional Feulgen reaction underestimations arise in very dense nuclei. To avoid this, some authors take measurements at a wave-length which does not correspond to the maximum absorption peak of the chromophore. Crushing methods are also used but these are limited to isolated cells. Since numerous Schiff-type reagents with various chromatic properties are now available, investigation of their use in quantitative cytophotometry has been made, with the hope of finding reagents, to replace the normal Schiff's reagent, which would give extinctions of the required intensity. Firstly, in standard solution the extinction coefficient of the reaction product from some of these reagents and formaldehyde was studied. Cytophotometric measurements were then made of the absorption of nuclei of supposed constant DNA content but at variable concentration. — Analogous considerations may be applied to cytofluorometric measurements obtained using fluorescent Feulgen reactions since at elevated densities autoabsorption and filter effects may arise with consequent underestimation.

Institute of Comparative Anatomy and Center of Histochemistry of C.N.R., University of Pavia (Italy).

3.

GIOVANNI PRENNA and LUIGI ZANOTTI: A New Cytofluorometric Method for Quantitative Determinations of DNA.

A new cytofluorometric method for quantitative determinations of DNA stained with Acriflavine-SO₂ in the Feulgen reaction is given. The fluorescence of Feulgen reaction is subject to photodecomposition of the asymptotic type, allowing reproducible measurements. The total fluorescent emission of the whole nucleus was measured with a recording-cytofluorometer of our own construction. We used smears of blood-cells (*Triturus cristatus*) of wide morphological variations. For objective verification of the accuracy of the new method, the cytofluorometric measurements were compared with absorption measurements at 465 mμ, executed with the Deeley integrating microdensitometer. — In resting cells of supposed constant DNA content (polymorphonuclear leucocytes) we found an analogous spread of cytofluorometric and scanning values (respectively SD% = ±4.51 and 5.15). A group of resting and dividing cells (polymorphs and erythroblasts) were photographed for identification and the same nuclei were measured successively with both methods; a high degree of correlation was found between the two sets of results (SD% of the ratios of cytofluorometric-scanning measurements = ±5.9). These results indicate that: 1) the cytofluorometric method is as accurate as the absorption scanning method and equally able to overcome the distributional error, confirming ORNSTEIN's theories; 2) difference of content between two nuclei of about 12% can be considered significant with either method. — The cytofluorometric measures at high extinctions are inaccurate because of the undervaluation, due to an inner-filter-effect; to avoid this, at least partially, we prepared a Schiff-type reagent (dehydrothio-p-toluidine-SO₂) which provides highly fluorescent reactions in U.V. and colorless in visible light: the cytofluorometric possibilities are under investigation.

Institute of Comparative Anatomy and Center of Histochemistry of C.N.R., University of Pavia (Italy).

4.

P. van Duijn and C. G. Struyk: The Application of a Model System in Combination with Two-Wavelength Cytophotometry of Feulgen-stained Nuclei for Cytochemical Determination of DNA in Absolute Units.

Cellulose films containing DNA, in combination with a special colorimetric arrangement (van Duijn et al., J. Histochem. Cytochem. *10*, 473, 1962) have recently been used to study the stoichiometry of the Feulgen reaction (Persijn and van Duijn, Histochemie *2*, 283, 1961). In the present investigation, DNA-cellulose films and microscopic preparations were stained according to Feulgen. After colorimetry of the films and cytophotometry of the nuclei by the two-wavelength method, the quantity of DNA in the nuclei can be computed relative to that in the membranes, provided the reaction in both objects has the same stoichiometry. — Since the membranes can be analysed for DNA phosphorus, the DNA content of the nuclei can be computed in absolute units. In this way, the DNA content of nuclei of frozen dried smears of chicken erythrocytes, and rat liver and kidney cells and buffy coat leucocytes was determined. It was approximately 25—40% lower than after fixation in methanol 35% formaldehyde (9:1, v/v). In the latter case, the values for the DNA content of rat nuclei were highly constant and in good accord with biochemical data. The DNA content of leucocyte nuclei did not differ significantly from that of liver and kidney nuclei, contrary to earlier reports in the literature.

Laboratory for Pathology, Biochemical Section, University of Leiden (The Netherlands).

5.

Z. Lodin, J. Pilný and J. Gregor: Contribution to the Analysis of the DNA Histogram in Cell Nuclei.

The aim of cytophotometric measurement of DNA in the population of cell nuclei is the quantitative determination of DNA in individual nuclei and to find out the fluctuation in the content of the given population. The knowledge of the distribution of measured values and of correct methods used for their evaluation excludes possible errors and enables to make further conclusions. — In earlier work we demonstrated that objective conclusions on the exact distribution of values of the DNA content in cell nuclei cannot be made with a small number of measurements. The conclusions of the present work are based on the evaluation of 2.000 measurements of smears of rat nuclei. For the performance of the Feulgen reaction we determined ourselves the optimum of hydrolysis (8′), the reaction was performed by means of basic Fuchsin Harleco. On twelve nuclei the spectral characteristics were determined and thus the maximum of the absorption curve (5.600 Å) achieved. In 80 measurements the distribution error was determined and was less than 2%. The measurement was performed on a cytophotometer of original construction. — The obtained data were evaluated by statistical methods and the results answer the following main questions: 1. What probability exists that the nucleus of the measured value of the DNA content is localized in the corresponding column of the histogram ? — 2. What probability exists on the distribution of fluctuations of nuclei with the given DNA content and how the real distribution of DNA is determined in the studied tissue ?

Laboratory of Cytophysiology and Cytochemistry of the Nervous Cell, Institute of Physiology, Czechoslovak Academy of Sciences, and High Technical School, Faculty of Mathematics, Prague 6 (ČSSR).

6.

Günter Kiefer und Walter Sandritter : Veränderungen der Nukleinsäure- und Proteinmenge während der Mitose.

Frühere mikrointerferometrische Messungen an lebenden HeLa-Zellen hatten gezeigt, daß das Trockengewicht der Gesamtzelle in der Interphase verdoppelt und dann während der Mitose ungefähr halbiert und auf die beiden Tochterzellen übertragen wird. Eine genauere Analyse ergibt nun, daß die Trockenmasse beider Tochterzellen zusammen in der Posttelophase etwas geringer ist als die der Ausgangszelle in der Prophase, d.h. das Trockengewicht nimmt während der Mitose ab. Von der Reduktion der Substanzmenge ist in erster Linie die Ribonukleinsäure (RNS) des Cytoplasmas betroffen, wie aus cytophotometrischen Messungen Gallocyaninchromalaun- gefärbter Zellen hervorgeht. Die Abnahme ist die Folge einer stark verminderten RNS-Synthese im Kern. Die geringere RNS-Menge führt zu einer verminderten Proteinsynthese, so daß auch die Eiweißmenge der Gesamtzelle am Ende der Mitose etwas geringer ist. Desoxyribonukleinsäure- (Feulgen-) und Histon- (Fast-green-) Menge bleiben von der Prophase bis zur Telophase konstant.

Changes of Nucleic Acid Content and Protein Content During Mitosis. Previous micro-interferometric measurements on HeLa cells had shown that the dry weight of the entire cell is doubled during interphase and then during mitosis it is nearly halved and distributed among both daughter cells. A precise analysis shows that the dry mass of both daughter cells is less in the posttelophase than that of initial cells in prophase, i.e. that the dry weight decreases during mitosis. Of all determined components (total protein, histones, RNA, DNA) the RNA shows the biggest decrease during mitosis as shown by cytophotometric measurements of Gallocyaninechromalum stained cells. The decrease is the result of a highly reduced RNA synthesis in the nucleus. The smaller RNA amount causes a reduction in the protein synthesis. This results in a slightly lower total protein amount at the end of mitosis. Des-oxyribonucleic acid amount (Feulgen) and histone amount (Fast green) remain constant throughout the prophase until the telophase.

Pathologisches Institut der Justus Liebig-Universität, 63 Giessen (Deutschland).

7.

DICK KILLANDER: A Quantitative Cytochemical Study on Age-Determined Mouse Fibroblasts in vitro.

The nucleic acid and protein synthesis during the interphase of mouse fibroblasts *in vitro* (L-cells) was studied. Age-determined fixed cells were analyzed individually for protein, RNA and DNA as represented by their dry mass, total extinction at 2650 Å and at 5460 Å after Feulgen staining. Age determination was made by time-lapse cinemicrography and quantitative measurements were made in the rapid scanning microinterferometer and the rapid scanning microspectrophotometer. The combination of age determination and quantitative cytochemistry for determining synthesis curves during the interphase is associated with errors factors such as variation in generation time and momentary size variation, i.e. size variation of cells in the same physiological state of the cell cycle. These factors were investigated and their significance and relationship will be discussed. — The cells which had just initiated their DNA synthesis were analyzed separately with the respect to age and dry mass content. The possible role of a critical mass for the initiation of DNA synthesis will also be discussed.

Dept. for Cell Research, Karolinska Institutet, Stockholm (Sweden).

8.

NILS R. RINGERTZ: Relationship between DNA-Duplication and Histone Synthesis.

Cytochemical methods have been used to study histone and DNA synthesis in three different experimental systems. The time relationship between synthesis of DNA and synthesis and DNA-binding of histones has been examined in the macronucleus of *Euplotes*. Macronuclei isolated by microdissection have been analyzed by microinterferometry and microspectrophotometry at various stages of reorganization, and the incorporation of labelled amino acids has been studied by autoradiography. — Ehrlich ascites tumor cells have been used to study changes in nuclear DNA, histone and dry mass during a period of irradiation induced mitotic inhibition. The results will be discussed with respect to the regulation of histone synthesis. — Cells grown in tissue culture have been pulse labelled with H^3-arginine and H^3-lysine. The distribution of label on chromatids in dividing cells at various time intervals after labelling will be discussed in relation to the duplication of the deoxynucleo-protein complexes.

Dept. for Cell Research, Karolinska Institutet, Stockholm (Sweden).

9.

RUDOLF RIGLER jr.: Mikrofluorometrische Untersuchungen an Akridinorange-Nukleinsäure-Komplexen in Kulturzellen und Mikrotropfenmodellen*.

Nach fluoreszenzmikrospektrographischen Untersuchungen an Zellen einer Mäusefibro-blastenkultur nach Fixierung durch Gefriersubstitution und an Mikrotropfenmodellen bindet sich das Fluorochrom Akridinorange (3,6-Bis-dimethyl-amino-acridinium-chlorid) an DNS in monomerer Molekülform mit einem Emissionsmaximum bei 530—535 nm und an RNS in assoziierter (dimerer) Molekülform mit einem Emissionsmaximum bei 580—590 nm. Die entsprechenden Absorptionsmaxima liegen bei 502 bzw. 465 nm. Die Bindung von Akridin-orange an Nukleinsäuren, die hauptsächlich durch die NH_3^+-Gruppen des Akridinorange an die PO_4^--Gruppen der Nukleinsäuren erfolgt und von deren sterischer Molekularstruktur abhängig ist, erweist sich bei pH 4,1 für Nukleinsäuren spezifisch. Andere in Zellen vor-kommende Polyanionen wie z. B. Polysaccharide (Heparin) unterscheiden sich infolge der Bildung höherer Akridinorange-Molekülassoziate sowohl durch das Emissions- ($\lambda_{max} = 600$

bis 605 nm) wie Absorptionsspektrum (λmax $= 450$ nm). Die Fluoreszenzintensität des mono-meren DNS-AO-Komplexes wie die des assoziierten RNS-AO-Komplexes ist der durch Röntgenabsorption und Interferenzmessung bestimmten Masse der betreffenden Nuclein-säuren proportional. Aus dem Quotienten der bei 590 nm und bei 530 nm bestimmten Fluoreszenzintensität, läßt sich, wie an Modellsystemen mit verschiedenen DNS-RNS-Mischungen gezeigt werden kann, das DNS/RNS-Verhältnis bestimmen. Das Verhältnis der bei 590 nm zur bei 530 nm gemessenen Fluoreszenzintensität steht dabei in reziprokem Verhältnis zur Ionenkonzentration des zur Färbung benützten Puffergemisches und zur Differenzierungs-(Dialyse-)zeit des Nukleinsäure-AO-Komplexes. Unter standardisierten Be-dingungen sind eine Bestimmung des DNS/RNS-Verhältnisses in animalen Zellen und Zell-strukturen sowie Aussagen über die sterische Struktur der vorliegenden Nukleinsäuren möglich.

Microfluorometric Studies on Acridine Orange-Nucleic Acid-Complexes in Culture Cells and Micro Drop Systems *. Acridine orange (3,6 bis-dimethyl-amino-acridinium-chloride) binds on DNA in monomer molecular form with an emission maximum at 530—535 mμ and on RNA in associated (dimer) molecular form with an emission maximum at 580—590 mμ as shown in cells of mouse fibroblast cultures after fixation by freeze substitution and in micro drop models by microspectrofluorometric studies. The respective absorption maxima lie at 502 and 465 mμ. The binding of acridine orange on nucleic acids, mainly due to the NH_3^+-groups of acridine orange and the PO_4^--groups of the nucleic acid chain and depending on its steric molecular structure, is shown to be specific for nucleic acids at pH 4.1. Other polyanions occurring in animal cells e.g. polysaccharides (heparin) forming higher acridine orange-molecular associates can be differentiated by the emission- (λmax $= 600$—605 mμ) and the absorption spectrum (λmax $= 450$ mμ). The fluorescence intensity of the monomer DNA-AO-complex and of the associated RNA-AO-complex is proportional to the mass of the respective nucleic acid determined by X-ray absorption and interference measurements. From the ratio of the fluorescence intensity measured at 590 mμ and 530 mμ the DNA/RNA ratio can be calculated as shown in model systems with different DNA-RNA mixtures. The quotient F_{590}/F_{530} is in reciprocal proportion to the ionic concentration of the buffer used and to the differentiation-(dialyzing)time of the nucleic acid-AO-complex. Under standard conditions the DNA/RNA ratio in animal cells and cell structures can be determined and statements on the steric structure of the nucleic acids present can be made.

* These studies were supported by grants from the Swedish Cancer Society and the Wallenberg Foundation.

Institute for Cell Research and Genetics, Karolinska Institutet, Stockholm 60 (Sweden).

10.

NIRMAL K. DAS, PETER LUYKX and MAX ALFERT: **RNA Synthesis in Urechis Eggs*.**

RNA synthesis in chromosomes, intact nucleoli, and disintegrating nucleoli was studied by using liquid emulsion autoradiography. Both chromosomes and especially the nucleolus are found to be labeled in unfertilized eggs exposed to H^3-uridine for 15 minutes. When the time of exposure to H^3-uridine is increased from 15 minutes to 1 hour, labeled RNA accumulates at a faster rate in the nucleolus than in chromosomes, perhaps due to a transfer of chromo-somal RNA to the nucleolus. Following fertilization, chromosomes as well as the disintegrating nucleolus at metaphase do not incorporate H^3-uridine. In experiments in which unfertilized eggs are labeled for 1 hour, washed, fertilized in the presence of excess unlabeled uridine, and sampled at first metaphase it is found that most of the label that was initially present in the intact nucleolus is still retained in the small part of the nucleolus which persists at metaphase. The larger portion of nucleolar material which disappears by metaphase seems to consists of ribonucleoproteins that have been synthesized at an earlier time.

* Aided by grants from the National Science Foundation (GB-1642) and the U.S. Public Health Service (RG-6025).

Dept. of Zoology, University of California, Berkeley, California (USA).

11.

A. A. ZOTIKOV and V. G. KONDRATENKO: **Cytofluorometric Study of X-Ray Irradiated Nucleo-proteins of Cell Nuclei.**

Radiation damage of the nuclear nucleoproteins and their subsequent changes can be determined by the fluorescence of the nuclei stained with acridine orange, which is known to be selectively linked with the nucleic acids and nucleoproteins. The degree of such nucleo-protein damage was estimated by the intensity of induced fluorescence of the nuclei. — It has been shown that the intensity of induced fluorescence of the nuclei of irradiated cells increases about 1.5—2 times more than that of the control. The intensity of induced fluorescence is dose dependent. The total fluorescence intensity of the nuclei is due to the complex of acridine orange with nuclear RNA as well as with DNA. By means of enzymatic treatment of normal and irradiated cells with crystalline RN'ase the nuclear RNA contribution to the total nuclei fluorescence was estimated. It was demonstrated that increased fluorescence of the nuclei of irradiated cells is partly caused by the affected ribonucleoproteins. Although after the removal of RNA from the nuclei of irradiated cells the intensity of their fluorescence remains increased. These data enable us to suggest that the remaining increase of the fluorescence of the nuclei is due to the radiation damage of the DNP especially its protein moiety. After the removal of RNA from the cells and following trypsin digestion we found, actually, a difference in the fluorescence intensity of the nuclei of irradiated and normal cells. It is considered that one of the reasons of the residual intensity of the nuclear fluorescence in irradiated cells is due to the change in the DNA-protein ratio as the result of partial deproteinization of the DNP-complex.

Institute of Radiation and Physico-Chemical Biology, Academy of Sciences of the USSR, Vavilov street 18, Moscow B-312 (USSR).

Read by Title

WALTER BAUDISCH: Einbau von ^{14}C-Prolin in die Speicheldrüsen von Acricotopus lucidus (Chironomidae).

Die Chironomiden stellen ein bevorzugtes cytologisches Untersuchungsobjekt dar. Besonders die Riesenchromosomen in den Speicheldrüsen sind hochpolytän und können in ihrer Struktur sehr gut sichtbar gemacht werden. Wie MECHELKE nachwies, bilden die Riesenchromosomen in den drei Lappen der Speicheldrüsen von *Acricotopus* ein unterschiedliches Strukturmuster aus, das durch Anschwellen einzelner Querscheiben oder Querscheibenkomplexe entsteht. In diesen aufgeblähten Querscheiben, den sog. Puffs und Balbiani-Ringen wird verstärkt RNS synthetisiert. In Analogie zu dem unterschiedlichen Strukturmuster der Riesenchromosomen in den einzelnen Speicheldrüsenlappen ließen sich chemische Unterschiede zwischen den Drüsenlappen feststellen. In Haupt- und Nebenlappen der Speicheldrüsen wurde Hydroxyprolin nachgewiesen; es wird hier wahrscheinlich synthetisiert. Dagegen konnte in dem Vorderlappen diese Aminosäure nicht gefunden werden. Hydroxyprolin kommt weder im restlichen Tierkörper noch in der Nahrung vor. Versuche mit ^{14}C-Prolin sollten nun zeigen, ob Puffs oder Balbiani-Ringen im Haupt- und Nebenlappen der Speicheldrüsen eine direkte Beteiligung bei der Hydroxyprolinsynthese zukommt. Die Autoradiographien der mit ^{14}C-Prolin inkubierten Speicheldrüsen ergaben aber keine spezifische Anreicherung der Aktivität an irgendeiner Stelle der Riesenchromosomen. Wie bei einer Verabreichung anderer aktiver Aminosäuren findet sich auch bei einer Verfütterung von ^{14}C-Prolin die Aktivität im Zellplasma und besonders in dem Sekret der Drüsenzellen.

The Incorporation of ^{14}C-Proline into the Salivary Glands of Acricotopus lucidus (Chironomidae). The Chironomids are a specially favoured object for cytological investigations. The giant chromosomes of the salivary glands are highly polytene and hence their structure may be made clearly visible. As established by MECHELKE, the giant chromosomes in the three lobes of the salivary glands of *Acricotopus* have a different pattern of organization, due to the swelling of some cross-bands, or complexes of several of these. In these swollen cross-bands, the so-called "puffs" or "Balbiani rings", RNA is synthesized to a greater extent. In analogy to the different pattern of organization of the giant chromosomes in the various lobes there are also chemical differences between the lobes of the gland. As hydroxyproline can be demonstrated in the principal and accessory lobe of the gland it can be assumed that these parts are the site of synthesis. An attempt to demonstrate this particular amino acid in the anterior lobe failed. Hydroxyproline is not found in any other part of the animal's body or in the fodder. Tests with ^{14}C-proline are meant to show whether the puffs or Balbiani rings participate directly in the synthesis of hydroxyproline. However, autoradiographs — taken after the incubation of the salivary glands with ^{14}C-proline — do not reveal a specific increase of radioactivity in any part of the giant chromosomes. As it is already known from the application of other isotopically labelled amino acids, radioactivity is found in the cytoplasm and especially in the secretion of the glandular cells if ^{14}C-proline is applied with the fodder.

Institut für Kulturpflanzenforschung, Gatersleben, Kr. Aschersleben, der Deutschen Akademie der Wissenschaften zu Berlin und Institut für Genetik der Martin-Luther-Universität, Ludwig-Wucherer-Straße 2, Halle (Deutschland).

MIROSŁAW BESKID: Histochemical Investigations on the Hyalinic Carcinoma of the Thyroid Gland.

Ten cases of thyroid hyalinic carcinoma, described in 1954 by LASKOWSKI, were examined. The main purpose of this work was to define some aspects of the cancer, particularly the hyalinic changes, which dominated in its microscopic appearance. It was found as a result of multiple examinations that among the four types of epithelial elements of cancer (the maternal cells, the spindle-shaped cells, the foam cells and the lumpy cells) only the lumpy cells have the ability to produce and to accumulate the hyalinic substance. In the initial period of production the hyalinic substance in the shape of grains or small drops has the great optical density. Its main compounds were neutral and acid mucopolysaccharides, glycogen and small amounts of proteins. At the late stage of its development the hyaline has the shape of spheroids, trabeculae or hyalin foci built up from the complexes of large molecules. It consists of the neutral mucopolysaccharides and great amount of proteins. The acid mucopolysaccharides and glycogen are absent. Also abundant amounts of amyloid may be present among the hyaline masses. — The process of production and transformation of hyaline was accompanied by the proliferation of the connective tissue. In the course of maturation of the lumpy cells of hyalinic thyroid carcinoma there are the processes, which lead to the production of connective tissue. As result of this the cancer cells are blocked by the proliferating connective tissue which leads to the condition suggesting dying out of the neoplasm. It determines the moderate grade of malignancy of the hyalinic thyroid carcinoma.

Institute of Oncology, Wawelska 15, Warsaw (Poland).

ERNST H. BEUTNER: Relation of Antigen of Rabbit Pituitary Autoantibodies and Growth Hormone*.

Previous studies in this laboratory (BEUTNER, DJANIAN and WITEBSKY, Immunol., in press) point to the fact that rabbits immunized with rabbit pituitary preparations develop autoantibodies specific for rabbit pituitaries which localize in acidophiles of the anterior lobe by indirect immunofluorescent (IIF) staining. Absorption studies had revealed that purified bovine growth hormone removes this IIF reaction while thyrotropin fails to do so. Now it is found that pituitary autoantibodies fix complement with bovine growth hormone but not with prolactin, luteinizing hormone, or thyrotropin. Rabbit antibodies to sheep pituitary preparations (i.e., heteroantibodies) yield strong IIF staining of acidophiles of rabbit pituitaries, but other glandular cells of the anterior pituitary are also stained. Stronger complement fixation with growth hormone occurred with these heteroantibodies than with autoantibodies, but reactions also occur with bovine prolactin, leutinizing hormone, and thyrotropin.

* Aided by a grant from the National Institutes of Health (CA 2357).

State University of New York at Buffalo, Buffalo, N.Y. (USA).

MATTI HÄRKÖNEN: Histochemical and Chemical Observations of Esterases and Oxidative Enzymes in Intact and Denervated Sympathetic Ganglia of the Rat.

The distribution of cholinesterases and non-specific esterases has been studied histochemically in the superior cervical ganglion of the rat, intact or at varying periods after pre- and postganglionic denervation. Esterases have also been quantitatively determined in extracts of the ganglion and their nature investigated by using starch gel electrophoresis. Distribution of different types of esterases have also been compared with the distributions of several dehydrogenases, amine oxidase and catecholamines. — Postganglionic denervation caused a loss of acetylcholinesterase, non-specific cholinesterase, non-specific esterase and amine oxidase activities, but an increase in activities of all dehydrogenases, as was demonstrated by histochemical demonstration of these enzymes in tissue sections. Chemical determinations of enzyme activities in homogenates confirm the loss of esterase activity, while a loss, no change, or an increase, was observed in the activities of different oxidative enzymes.

University of Helsinki, Dept. of Anatomy, Siltavuorenpenger, Helsinki (Finland).

A. JONECKO: Die unspezifischen Esterasen in der Lunge nach Vergiftungen mit Cholinesterase-Inhibitoren.

Zwanzig gleichartige Mäuse erhielten i.p. letale Dosen von Neostigmin 1,0, Physostigmin 2,0, NaF 140,0, E 600 1,3, DFP 6,0 mg/kg. Der Tod trat innerhalb von 15 min ein. Kontrolltiere erhielten die Lösungsmittel oder Azetylcholin. Fixierung bei 0^0 in Baker-Flüssigkeit. An Gefriermikrotomschnitten wurde die Esterasendarstellung mit der Thiolessigsäuremethode (CREVIER u. BÉLANGER 1955; WACHSTEIN, MEISEL u. FALCON 1961; WILSON 1951) ausgeführt. Die Ergebnisse waren folgende: 1. Bei Kontrolltieren besteht eine sehr starke Esterasenaktivität im Epithel der Lungenwege, eine weniger starke in den Phagen und eine schwache in den Alveolarwänden. 2. Neostigmin- oder Physostigminvergiftungen beeinflussen die unspezifischen Lungenesterasen nicht. Der Einfluß von NaF oder Azetylcholin ist schwach. 3. Nach E 600 ist eine deutliche Aktivitätsdepression festzustellen. Die im Epithel und in den Phagen erscheinende granuläre Restreaktion ist in den Lysosomen lokalisiert. 4. Das DFP bewirkt eine vollständige Depression der Esterasenaktivität. Die Präparate bleiben weiß. 5. Bei Vergiftungen von graviden Mäusen bewirken beim Fötus Neostigmin, Physostigmin, E 600 oder NaF keine Hemmung der Lungenesterasen; DFP dagegen inhibiert auch die Fötusesterasen. Die angewandte Methode eignet sich besonders zur Darstellung der fötalen Lungenwege. 6. Chronische Intoxikationen mit organischen Phosphorderivaten schädigen die Lysosomesterasen und die Lysosome selbst. Diese Beeinträchtigung der Lysosomesterasen und Hydrolasen durch Cholinesterase-Inhibitoren läßt sich mikroskopisch demonstrieren.

The Influence of Lethal Doses of Cholinesterase Inhibitors on the Nonspecific Esterases in the Lung. Twenty uniform mice were injected i.p. with lethal doses of Neostigmin (1.0), Physostigmin (2.0), NaF (140.0), E 600 (1.3), DFP (6.0) (mg/kg), which caused death within 15 minutes. A control group was injected with either the solvents or acetylcholine. Fixation was carried out in Baker's solution at 0^0. The thiolacetic acid method was used for the demonstration of the esterases in frozen sections (CREVIER and BÉLANGER, 1955; WACHSTEIN, MEISEL and FALCON, 1961; WILSON 1951). These were the results: 1) Control group: very strong esterase activity in the epithelium of the lung, less in the phagocytes and a weak reaction only in the alveolar walls. 2) Neostigmin and Physostigmin had no influence on the nonspecific esterases in the lung. The influence of NaF and acetylcholine was feeble. 3) After E 600 activity was markedly reduced. A residual granular reaction in epithelium and phagocytes was localized in the lysosomes. 4) DFP produced complete abolition of esterase activity. The sections remained white. 5) When Neostigmin, Physostigmin, E 600, or NaF were injected into pregnant mice, no inhibition of esterases could be demonstrated in the lungs of the foetus. DFP, however, inhibited the foetal esterases. The method employed was suitable for the demonstration of foetal bronchi. 6) Chronic intoxication with organic phosphorus derivatives damaged both lysosomes and lysosome esterases. This damage of the lysosome esterases and hydrolases by cholinesterase inhibitors was microscopically demonstrable.

Instytut Onkologii, Wybrzeże Armii Czerwonej 15, Gliwice (Poland).

JON J. KABARA and ROBERTA GLOCK-DEUKER: Specific and Quantitative Autoradiography.

While the introduction of autoradiography to biological problems, has provided a most helpful tool to the biochemist, it lacks two rather important qualities; the first is specificity, the second is quantification of the amount of radioactivity visualized on the developed film. These two limitations of the method have been overcome in our studies on cholesterol metabolism in tissue sections. By fixing cholesterol in the tissue section with a specific reacting reagent before lipid extraction, the radioactivity due to the sterol alone, could be visualized. Also, the distribution of radioactivity in a tissue slice, as reflected by labeled neutral or phospholipids, was determined by extraction, thin-layer chromatography, and liquid scintillation counting (LSC). The use of the latter technique enabled us to monitor histological manipulation of radioactive tissue sections and to quantitate the results. Results from LSC enable us to predict the time for exposures of radioautographs. — The concurrent use of histological sections for autoradiography, chemical assay, as well as radioactivity assay will be demonstrated. Data from biochemical experiments in which animals were starved for various time intervals and then given tritiated acetate will be presented. The effect of starvation on lipid biosynthesis will be discussed in terms of chemical and histochemical results.

Biochemistry Division, Dept. of Chemistry, University of Detroit, Detroit, Michigan (USA).

A. KOLÍN: Histochemischer Nachweis von dispersen hypoxydotischen Veränderungen und die Frage der morphologischen Darstellung der Herzmuskelfaserbeschädigung.

An 60 Herzen des laufenden menschlichen Obduktionsmaterials und an Herzen von Hunden mit experimenteller Noradrenalinhypoxydose wurden disperse Veränderungen der Succinodehydrogenaseaktivität (SDH) beobachtet und ihre Bedeutung für die Funktion und Vitalität der Herzmuskelfaser diskutiert. Die Veränderungen der SDH-Aktivität (Verminderung der Aktivität und Erscheinung von groben, unregelmäßigen Formazangranula) waren zum großen Teile reversibel und entsprachen verhältnismäßig gut den EKG-Veränderungen im Tierexperiment. Der Nachweis der SDH wird daher zur morphologischen Darstellung einer Schädigung der Herzmuskelfasern empfohlen.

Histochemical and Morphological Evidence of Disperse Hypoxic Transformations in Myocardial Lesions. Disperse changes in succinate dehydrogenase activity (SDH) have been observed in 60 human hearts (routine post mortem material) and in the hearts of dogs suffering from experimental noradrenalin hypoxemia. The significance of this alteration for the function and vitality of the cardiac muscle fibre is discussed. The changes in SDH activity (decrease of activity and appearance of coarse irregular formazan granules) are reversible and correspond relatively well with the changes in the E.C.G. as obtained in animal experiments. The determination of SDH is recommended for the morphological demonstration of cardiac muscle lesions.

Institut für Sera und Vaccinen, W. Piecka 108, Praha 10 (Tschechoslowakei).

ALEKSANDRA KRYGIER: Parallelanalyse von Chromosomen und DNA von Fibroblasten in Gewebekulturen.

Im Bindegewebe der Extremitäten und des Herzens von Hühnerembryonen überwiegen Zellen mit diploidem DNS-Gehalt; analog findet sich bei der Mehrzahl der Zellen ein diploider Chromosomensatz. Gleiches gilt für die Fibroblasten aus Extremitätenbindegewebe in primärer Gewebekultur und nach den ersten Passagen. Dagegen findet sich in primären Kulturen von Fibroblasten des Herzens eine größere Zahl von haploiden Formen, sowohl in bezug auf den DNS-Gehalt als auf die Chromosomenzahl. Die Ursache hierfür scheint in einer Schädigung der Fibroblasten des embryonalen Herzgewebes durch die Übertragung auf das künstliche Nährmedium zu bestehen. Das Fehlen dieser Vulnerabilität bei Fibroblasten aus Extremitäten kann durch eine bessere Adaptationsfähigkeit dieser Zellen erklärt werden. Höchstwahrscheinlich steht die Adaptationsfähigkeit in umgekehrtem Verhältnis zum Reifegrad der Fibroblasten. Die Veränderung in der Gesamtzahl der Chromosomen in den untersuchten Fibroblasten beruht auf Schwankungen in der Mikrochromosomenzahl. Die Zahl der Makrochromosomen wies keine erkennbaren Unterschiede auf.

Parallel Analysis of Chromosome Number and DNA of Fibroblasts in Tissue Culture. Predominance of cells with diploid amount of DNA was determined in embryonic connective tissues of chick thigh and heart as well as in primary cultures originated from them and in Ist passages. Similarly, a diploid number of chromosomes were displayed by the majority of cells in these tissues and cultures. Only the cells of primary cultures of heart fibroblasts disclosed a high number of haploid forms, which could be established both on the basis of DNA amount and the number of chromosomes. A reason for this phenomenon to appear may be found in a widespread damage affecting the fibroblasts of the heart embryonic tissue, which occurred in connection with their transfer onto an artificial medium. The absence of the damage in fibroblasts obtained from the embryonic thigh may be explained by the fact that these cells exhibit a better adapting ability. It is highly probable that the adapting faculty is inversely proportional to the maturity degree of the fibroblast. — The changes in the total number of chromosomes of fibroblasts studied referred to the number of microchromosomes. No essential deviations were found in the number of macrochromosomes.

Zaklad Anatomii Patologicznei, Pomorska Akademia Medyczna, Szczecin, ul. Unii Lubelskiej 1 (Polen).

SATIMARU SENO, MASANOBU MIYAHARA, EIICHI YOKOMURA, KENICHI MATSUOKA, YUKI TOYAMA and TAKASHI KATANO: Microspectrophotometric and Autoradiographic Studies on the Differentiation of Erythroid Cells.

The RNA and protein contents per cell have been measured by microspectrophotometry on the bone marrow erythroblasts of normal and anemic rabbits. The data showed that the RNA and protein contents in erythroblasts decrease with the advance of differentiation which is accomplished by the repeated cell division, the decrease in cell size, and the denucleation at some maturation stage. In the nucleated stage the RNA contents reach the minimum value at orthochromatic erythroblast. After denucleation RNA continues to decrease further and disappears completely with the maturation of reticulocyte. In normal animal the RNA contents of young reticulocytes are comparable to those of orthochromatic erythroblasts, but in a severe anemic state there appear abnormally basophilic big reticulocytes and some of them show very high RNA contents, comparable to those of polychromatic erythroblasts. The radioautographic studies on the cells incubated with tritiated uridine *in vitro* revealed that no new RNA is synthesized in the cells. The results show that in emergency erythropoiesis the denucleation of erythroblasts can occur at an earlier stage of differentiation. On these findings the role of the nucleus in the differentiation of erythroblasts will be discussed.

Dept. of Pathology, Okayama University Medical School, 164 Oka, Okayama (Japan).

HALINA SIERAKOWSKA and D. SHUGAR: Intracellular Localization of Phosphodiesterase I by a Cytochemical Method.

A cytochemical method for the localization of phosphodiesterase (PDase) I based on hydrolysis of α-naphthyl thymidine-5'-phosphate and coupling the enzymatically liberated naphthol with diazotate has been elaborated. α-naphthyl thymidine-5'-phosphate and α-naphthyl thymidine-3'-phosphate were prepared by phosphorylating 3'-acetyl thymidine and 5'-trityl thymidine, respectively, with α-naphthylphosphoryl dichloride. The specificity of α-naphthyl thymidine-5'-phosphate toward PDase I and resistance of α-naphthyl thymidine-

3'-phosphate to PDase II have been demonstrated and the results discussed. — Intracellular localization of PDase I in rat kidney, duodenum, pancreas, salivary glands and liver has been established by incubating formalin fixed frozen sections with α-naphthyl thymidine-5'-phosphate (2 mg/ml) and Fast Red TR (4 mg/ml) in 0.1 M tris HCl buffer pH 9. Incubation times ranged from 15 sec to 30 min at room temperature. Effects on PDase I of EDTA, AMP, NaF as well as alcohol and aceton fixation, followed by paraffin embedding, have been investigated. No PDase I diffusion was observed.

Institute of Biochemistry and Biophysics, Academy of Sciences, Warsaw (Poland).

JAN ZARZYCKI: Histochemische Untersuchungen über die Resorptionsprozesse in der Milchdrüse.

In den Endstücken der Milchdrüse finden nicht nur Sekretions-, sondern auch Resorptionsprozesse statt. Resorbiert wird das durch den Saugakt nicht entleerte Sekret. Die Resorptionsprozesse wurden histochemisch an Milchdrüsen laktierender Meerschweinchen untersucht, denen die Drüsenausführungsgänge unterbunden waren. Die Gewebsstückchen wurden in verschiedenen Zeitabschnitten post partum und vom Moment des Unterbindens an entnommen. Die durch Saugen entleerten Drüsen dienten als Kontrolle. Untersucht wurde das Vorkommen von Mucopolysacchariden, SH-Gruppen, Fetten und Fettsäuren sowie die Lokalisation und Aktivität von Succinodehydrase, Cytochromoxydase, alkalischer und saurer Phosphatase. Während der Resorption erscheinen in den Drüsenzellen PAS-positive Substanzen. Dies spricht dafür, daß mit Polysacchariden verbunden Eiweißfraktionen resorbiert werden können. Da keine Unterschiede zwischen dem Vorkommen von SH-Gruppen in resorbierenden und sezernierenden Zellen festgestellt werden konnten, ist anzunehmen, daß das Eiweiß in der sezernierten Form resorbiert wird. Die Fette werden hauptsächlich als Fettkügelchen sowie als Fettsäuren resorbiert. Das im Lumen der Drüsenendstücke angesammelte Sekret hemmt die Aktivität der Oxydationsfermente. Während der Resorption kann man keinen Unterschied im Gehalt von Phosphatasen feststellen. Die elektronenmikroskopischen Untersuchungen sprechen dafür, daß die Resorptionsprozesse als Pinozytose ablaufen.

Histochemical Investigations of Resorptive Processes in Mammary Glands. It has been found that both secretory *and* resorptive processes take place in the mammary gland ducts. The secretion which remains in the glands after lactation is reabsorbed. A ligature was applied to the glandular excretory duct of lactating Guinea pigs and the resorptive process was investigated by histochemical methods. Specimens were taken from the glands at various times post partum and from the moment the ligature was applied. Glands which were depleted of the secretion (by suckling) were used as controls. Tests were carried out for the determination of mucopolysaccharides, SH-groups, fat and fatty acids, as well as for the localization and activity of succinate dehydrogenase, cytochrome oxidase and alkaline and acid phosphatase. During resorption PAS-positive substances appeared in the glandular cells. This proved that protein-bound polysaccharide fractions can be reabsorbed. As there was no difference in the level of SH-groups in secreting and absorbing cells, it was assumed that proteins are reabsorbed in the excreted form. Fats were absorbed mainly as fat globules and fatty acids. The secretion which accumulated in the lumen of the ducts inhibited the activity of oxidative enzymes. No variation in the amount of phosphatases could be shown during the resorptive process. The results of electron microscopic examinations suggested that resorption is a pinocytotic process.

Zaklad Histologii i Embriologii Akademii Medycznej, Wroclaw (Polen).

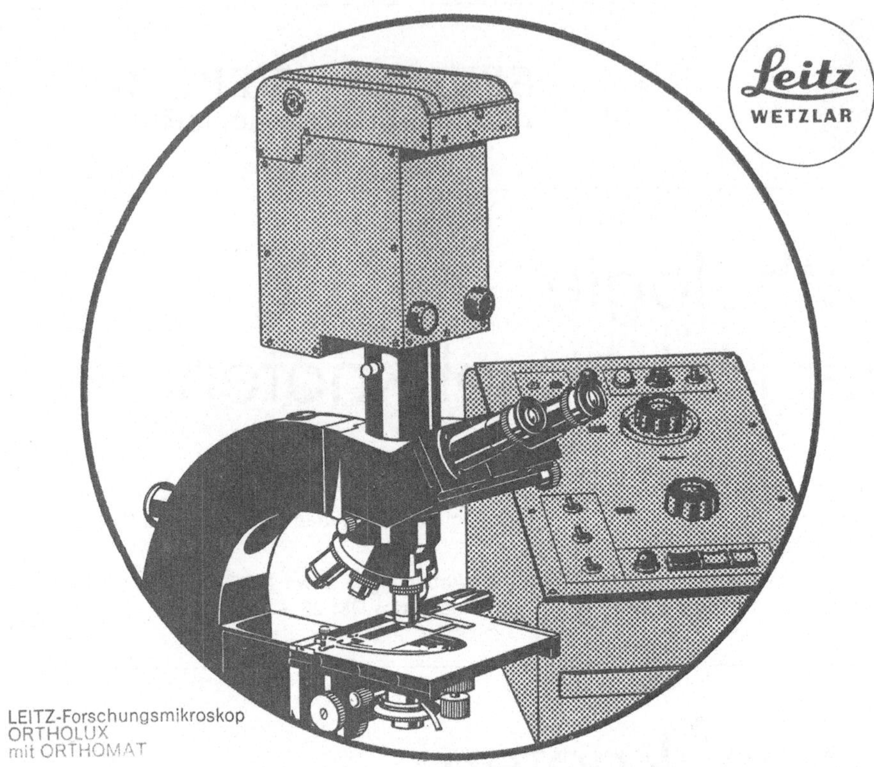

LEITZ-Forschungsmikroskop
ORTHOLUX
mit ORTHOMAT

LEITZ-ORTHOMAT

Die vollautomatische Mikroskopkamera für Forschung und Routinebetrieb

- Nur ein Tastendruck für jede Aufnahme.
- Kamera nach der Belichtung sofort wieder betriebsbereit.
- Ständige Aufnahmebereitschaft.
- Vollautomatische Belichtungsregelung- auch im Dunkelfeld, Phasenkontrast und Fluoreszenzlicht.
- Praktisch unbegrenzter Belichtungsspielraum von $1/100$ sec. bis zur Langzeitaufnahme schwach fluoreszierender Objekte.
- Belichtungsregelung individuell auf die Charakteristik jedes Präparates einstellbar.
- Helligkeitsänderungen des Objektes während der Aufnahme werden automatisch erfaßt.
- Prädestiniert für Bildserien veränderlicher Objekte.
- Keine Unterbrechung der mikroskopischen Beobachtung durch die Aufnahme, der Mikroskopierende kann sich ganz der eigentlichen Arbeit zuwenden.

ERNST LEITZ Geräte für die Histochemie und Cytochemie: Mikroskope ● Polarisationsmikroskope ● Stereomikroskope ● Interferenzmikroskop ● UV-Mikroskop
GMBH ● UV-Mikrospektrograph ● Mikromanipulator ● Mikrophotographische Apparate ● Mikrotome ● Kryomat für Mikrotome ● Photometer ● Kleinbild-
WETZLAR kamera L E I C A ● 8 mm Filmkamera L E I C I N A ● Projektoren.

Springer-Verlag, Berlin · Göttingen · Heidelberg.